Continuous Media with Microstructure

Bettina Albers

Continuous Media
with Microstructure

Dr.-Ing. Bettina Albers
Technische Universität Berlin
FG Grundbau und Bodenmechanik
Gustav-Meyer-Allee 25
13355 Berlin
Germany
E-mail: albers@grundbau.tu-berlin.de

ISBN 978-3-642-11444-1   e-ISBN 978-3-642-11445-8

DOI 10.1007/978-3-642-11445-8

Library of Congress Control Number: 2009943376

Cover Design: WMX Design, Heidelberg, Germany

Printed in acid-free paper

9 8 7 6 5 4 3 2 1

springer.com

This book is dedicated to
Prof. Dr. Krzysztof Wilmański
on the occasion of his 70th birthday.

Preface

This book is a collection of papers dedicated to Professor Dr. Krzysztof Wilmański on the occasion of his 70th birthday. The book contains 25 contributions of his friends and colleagues. He met the invited authors at different stages of his scientific career of almost 50 years so that the contributions cover a wide range of fields stemming from continuum mechanics. This happened at numerous universities and research institutes where he both taught and did his excellent research work, e.g.

- the University of Lódź, Poland, where he studied Civil Engineering and did his diploma work on *Elastic-plastic thermal stresses in a thin ring* and where he graduated with his PhD-work in the field of *Continuous Models of Discrete Systems,*
- the Institute of Fundamental Technological Research of the Polish Academy of Sciences in Warsaw, where he got his habilitation in the field *Nonlocal Continuum Mechanics* and where he was the head of the Research Group *Continuum Thermodynamics.* He collaborated with W. Fiszdon, Ł. Turski, Cz. Wozniak, H. Zorski and others on the topics *axiomatic and kinetic foundations of continuum thermodynamics, theory of mixtures, phase transformations in solids,*
- the Johns Hopkins University in Baltimore, US, where he worked together, e.g. with C. Truesdell, J. Ericksen and W. Williams, on *axiomatic and kinetic foundations of continuum thermodynamics,*
- the College of Engineering, University of Baghdad, Iraq, where he was a Visiting Professor and taught many courses,
- the University of Paderborn and the Technical University Berlin, Germany, where he had an Alexander von Humboldt Stipend and contracts as a Visiting Professor (works on *a model of crystallizing polymers, on a nonlocal thermodynamic model of plasmas and electrolytes and on martensitic phase transformations*),
- the Wissenschaftskolleg zu Berlin (Institute for Advanced Studies), Germany, where he worked together with e.g. I. Müller (TU Berlin, Germany),

R. Rivlin (Lehigh, USA) and J. Kestin (Brown, USA), on *martensitic phase transformations (SMA), non-newtonian fluids and acoustic waves in continua*,

- the Technical University of Hamburg-Harburg, Germany, where he did research on *crystal plasticity and the evolution of textures*,
- the University of Essen, Germany, where he worked on *thermodynamic models of porous materials*,
- the Weierstrass Institute for Applied Analysis and Stochastics, Berlin, Germany, where he was the head of the Research Group *Continuum Mechanics*, and where he continued his work on *thermodynamic models of porous materials* and
- the University of Zielona Góra (Poland) where he taught many courses in mechanics, physics of structures, thermodynamics and so on.

In all of these institutions and in many others all over the world, Professor Wilmański taught numerous courses. He is a brilliant teacher which he not only showed during the interesting courses but also in competent and patient answers to many questions of his students. From my own experience I can state that even attending more than once the same course was not boring as it has never been the same course. In every presentation there were new elements and ideas as well as improved approaches to the subject with a reference to newest developments. The same concerns his attendance of numerous scientific conferences and seminars where he was always a lively participant making comments on various subjects in which he used his extensive knowledge. For some years he was the secretary treasurer of the *International Society for the Interaction of Mathematics and Mechanics (ISIMM)*.

The contributions to the book concern various aspects of extension of classical continuum models. These extensions are related to the appearance of microstructures both natural as well as these created by processes. To the first class belong various thermodynamic models of multicomponent systems such as porous materials, composites, materials with microscopic heterogeneities (e.g. functionally graded materials). To the second class belong primarily microstructures created by phase transformations. Invited authors cover both fields of thermodynamic modeling and mathematical analysis of such continua with microstructure. In particular the following subjects are covered:

- thermodynamic modeling of saturated and unsaturated porous and granular media,
- linear and nonlinear waves in such materials,
- extensions of constitutive laws by internal variables, higher gradients and nonequilibrium fields,
- stochastic processes in porous and fractal materials,
- thermodynamic modeling of composite materials,
- mathematical analysis of multicomponent systems,
- phase transformations in solids.

I would like to thank all contributors for their willingness to write together this book. Thanks also to the editor in engineering of Springer Heidelberg, Dr. Christoph Baumann, who accepted with alacrity to publish this book. All of us wish Professor Wilmański the very best for his 70th birthday and many healthy, happy years to come! It has been a pleasure for all of us to work and to enjoy the leisure time with you.

Berlin, November 2009 Bettina Albers

Prof. Dr. Krzysztof Wilmański

Invited Contributors to the Present Book

Contents

List of Contributors

BETTINA ALBERS
Technische Universität Berlin, Institute for Geotechnical Engineering and
Soil Mechanics, Sekr. TIB1-B7, Gustav-Meyer-Allee 25, 13355 Berlin,
Germany,
e-mail: albers@grundbau.tu-berlin.de

ROMESH C. BATRA
Department of Engineering Science and Mechanics, Virginia Polytechnic
Institute and State University, Blacksburg, VA 24061 USA,
e-mail: rbatra@vt.edu

KAUSHIK DAS
Department of Engineering Science and Mechanics, Virginia Polytechnic Institute and State University, Blacksburg, VA 24061 USA, e-mail: kdas@vt.edu

ERICH BAUER
Graz University of Technology, 8010 Graz, Austria,
e-mail: erich.bauer@tugraz.at

JACOB BEAR
Department of Civil Engineering and Environmental Engineering,
Technion - Israel Institute of Technology, Haifa 32000, Israel,
e-mail: cvrbear@techunix.technion.ac.il

LEONID G. FEL
Department of Civil Engineering and Environmental Engineering,
Technion - Israel Institute of Technology, Haifa 32000, Israel,
e-mail: lfel@techunix.technion.ac.il

NICOLA BELLOMO
Dipartimento di Matematica, Politecnico di Torino, Corso Duca degli
Abruzzi 24, 10129 Torino, Italy, e-mail: nicola.bellomo@polito.it

ABDELGHANI BELLOUQUID
University Cadi Ayyad, Ecole Nationale des Sciences Appliquées, Safi,
Maroc e-mail: bellouquid@gmail.com

ELENA DE ANGELIS
Dipartimento di Matematica, Politecnico di Torino, Corso Duca degli
Abruzzi 24, 10129 Torino, Italy, e-mail: elena.deangelis@polito.it

GIANFRANCO CAPRIZ
Dipartimento di Matematica, Università di Pisa, via F. Buonarroti 2,
I-56127 Pisa, Italy
and Accademia dei Lincei, via della Lungara 10, I-00165 Roma, Italy,
e-mail: gianfrancocapriz@nettare.net

MAREK ELŻANOWSKI
Portland State University, Portland Oregon, USA e-mail:
elzanowskim@pdx.edu

SERGE PRESTON
Portland State University, Portland Oregon, USA, e-mail: serge@pdx.edu

PASQUALE GIOVINE
Dipartimento di Meccanica e Materiali, Facoltà di Ingegneria, Università
Mediterranea, Via Graziella n.1, Località Feo di Vito, I-89122, Reggio
Calabria, Italy, e-mail: giovine@unirc.it

JOE D. GODDARD
Department of Mechanical and Aerospace Engineering 0411, University of
California, San Diego, 9500 Gilman Dr., La Jolla, CA, 92093-0411, USA,
e-mail: jgoddard@ucsd.edu

DIMITRIOS KOLYMBAS
University of Innsbruck, Institute of Infrastructure, Division of Geotechnical
and Tunnel Engineering, Technikerstr. 13, A-6020 Innsbruck, Austria,
e-mail: dimitrios.kolymbas@uibk.ac.at

PAVEL KREJČÍ
Institute of Mathematics, Academy of Sciences of the Czech Republic, Žitná
25, 11567 Praha 1, Czech Republic, e-mail: krejci@math.cas.cz

JÜRGEN SPREKELS
Weierstrass Institute for Applied Analysis and Stochastics (WIAS),
Mohrenstr. 39, 10117 Berlin, Germany, e-mail: sprekels@wias-berlin.de

ELISABETTA ROCCA
Dipartimento di Matematica, Università di Milano, Via Saldini 50, 20133
Milano, Italy, e-mail: elisabetta.rocca@unimi.it

MIECZYSŁAW KUCZMA
University of Zielona Góra, Institute of Building Engineering, ul. Z. Szafrana
1, 65-516 Zielona Góra, Poland, e-mail: m.kuczma@ib.uz.zgora.pl

MATTHIAS KUNIK
Otto-von-Guericke-Universität, Institut für Analysis und Numerik, Universitätsplatz 2, 39106 Magdeburg, Germany, e-mail: matthias.kunik@ovgu.de

CARLO G. LAI
EUCENTRE, Via Ferrata 1, 27100 Pavia, Italy, e-mail: carlo.lai@eucentre.it

JORGE G. F. CREMPIEN DE LA CARRERA
NORSAR, Gunnar Randers vei 15, P.O. Box 53, NO-2007 Kjeller, Norway,
e-mail: jorge.crempien@norsar.no

I-SHIH LIU
Instituto de Matematica, Universidade Federal do Rio de Janeiro, Caixa
Postal 68530, 21945-970 Rio de Janeiro, Brasil, e-mail: liu@im.ufrj.br

GÉRARD A. MAUGIN
UPMC Univ Paris 6, UMR CNRS 7190, Institut Jean Le Rond d'Alembert,
Case 162, Tour 55, 4 place Jussieu, 75252 Paris Cedex 05, France, e-mail:
gerard.maugin@upmc.fr

MARTINE ROUSSEAU
UPMC Univ Paris 6, UMR CNRS 7190, Institut Jean Le Rond d'Alembert,
Case 162, Tour 55, 4 place Jussieu, 75252 Paris Cedex 05, France,
e-mail: martine.rousseau@upmc.fr

INGO MÜLLER
Technische Universität Berlin, e-mail: ingo.mueller@alumni.tu-berlin.de

MARTIN OSTOJA-STARZEWSKI
University of Illinois at Urbana-Champaign, Department of Mechanical
Science & Engineering, 1206 W. Green Street, Urbana, IL, 61801-2906,
USA, e-mail: martinos@illinois.edu

MARIO PITTERI
DMMMSA–University of Padova, via Trieste 63, 35121 Padova, Italy,
e-mail: pitteri@dmsa.unipd.it

EVGENIY V. RADKEVICH
Moscow State University, Department Mech.-Math., Vorobievy Gory,
119899, Moscow, Russia, e-mail: evrad07@gmail.com

MILES B. RUBIN
Faculty of Mechanical Engineering, Technion - Israel Institute of Technology,
32000 Haifa, Israel, e-mail: mbrubin@tx.technion.ac.il

TOMMASO RUGGERI
Department of Mathematics and Research Center of Applied Mathematics
(C.I.R.A.M.) University of Bologna, Via Saragozza 8, 40123 Bologna, Italy,
e-mail: ruggeri@ciram.unibo.it

GWIDON SZEFER
Cracow University of Technology, ul. Warszawska 24, 31-155 Kraków,
Poland, e-mail: szefer@limba.wil.pk.edu.pl

DOROTA JASIŃSKA
Cracow University of Technology, ul. Warszawska 24, 31-155 Kraków,
Poland, e-mail: jasinska@limba.wil.pk.edu.pl

ŁUKASZ A. TURSKI
Center for Theoretical Physics, Polish Academy of Sciences and Department
of Mathematics and Natural Sciences, Cardinal Wyszyński University.
Warsaw, Poland, e-mail: laturski@cft.edu.pl

CZESŁAW WOŹNIAK
Department of Structural Mechanics, Lodz University of Technology, al.
Politechniki 6, 93-590 Łódź, Poland, e-mail: czeslaw.wozniak@p.lodz.pl

JOWITA RYCHLEWSKA
Institute of Mathematics, Czestochowa University of Technology, ul.
Dąbrowskiego 73, 42-200 Częstochowa, Poland, e-mail: rjowita@imi.pcz.pl

Part I
SCIENTIFIC LIFE OF
PROF. DR. KRZYSZTOF WILMAŃSKI

Curriculum Vitae of Prof. Dr. K. Wilmański

1940	March 1; born in Lodz, Poland.
1957	Matura at T. Kosciuszko Gymnasium, Lodz, Poland.
1959	married to Anna Sosin; two sons: Pawel (1959), Jan (1962).
1962	MSc at Civil Engineering Department, Technical University of Lodz, Poland; diplom work on *Elastic-plastic thermal stresses in a thin ring*
1962-66	reader at the Technical University of Lodz (Poland).
1965	graduated at the Technical University of Lodz (Poland). PhD-work: *Some two dimensional problems of fibrous materials.* Field: Continuous Models of Discrete Systems.
1966-86	Associate Professor and Professor at the Institute of Fundamental Technological Research, Polish Academy of Sciences, Warsaw, Poland. Since 1970 the head of the Research Group: "Continuum Thermodynamics". Main fields of interest: axiomatic and kinetic foundations of continuum thermodynamics, theory of mixtures, phase transformations in solids. Works with Prof. Henryk Zorski.
1969-70	Postgraduate at the Johns Hopkins University, Baltimore (USA). Works on axiomatic and kinetic foundations of continuum thermodynamics with Prof. J. Ericksen and Prof. C. Truesdell (both Johns Hopkins University).
1970	Habilitation in 1970 in the Polish Academy of Sciences, Warsaw (Poland). Work: *Dynamics of Bodies with Microstructure.* Field: Nonlocal Continuum Mechanics.
1971	M. T. Huber Prize of the Polish Academy of Science for the work on axiomatic foundations of thermodynamics.
1972-74	Visiting Professor at the College of Engineering, University of Baghdad, Iraq.
1979	Nomination to the Professor (title) by the State Counsil of Poland.
1979-80	Alexander von Humboldt Stipend at the University of Paderborn and the Technical University Berlin.
1984	Fellow at "Wissenschaftskolleg zu Berlin", Institute for Advanced Studies. Works on martensitic phase transformations (SMA), non-newtonian fluids, acoustic waves in continua with Prof. Ingo Müller (TU Berlin, Germany), Prof. Ronald Rivlin (Lehigh, USA), Prof. Joseph Kestin (Brown, USA).

1985-86	Contract Professor at the Technical University of Berlin, Hermann-Föttinger-Institut. Work on a model of crystallizing polymers with Prof. Ingo Müller (TU Berlin).
1986-87	Visiting Professor at the University of Paderborn, Germany. Work on a nonlocal thermodynamic model of plasmas and electrolytes with Prof. J. Schröter.
1987-90	Contract Professor at the Technical University of Hamburg-Harburg, Germany. Work on crystal plasticity and the evolution of textures with Prof. O. Mahrenholtz.
1990-92	Contract Professor at the Technical University of Berlin, Germany. Work on the martensitic phase transformations with Prof. Ingo Müller.
1992-96	Lecturer at the University of Essen, Department of Civil Engieering. Work on thermodynamic models of porous materials.
1995	Habilitation (*Venia legendi*) at the University of Essen, Germany.
1996-2005	Professor at the Weierstrass Instutute for Applied Analysis and Stochastics, Berlin, Germany. Head of the Research Group "Continuum Mechanics" in WIAS (Berlin).
1998	Habilitation (*Venia legendi*) at the Technical University of Berlin, Germany.
1999	Conferment of the German citizenship.
2005	Professor of Mechanics at the Institute of Structural Engineering, University of Zielona Gora (Poland).

Teaching activities

1962-65	Undergraduate courses (lectures and tutorials) in Strength of Materials, Mechanics of Structures, Theory of Elasticity at the Technical University of Lodz (Poland) and its division in Plock (Poland).
1967-1984	Undergraduate courses (lectures, tutorials and some laboratory demonstrations) at the Technical University of Kielce (Poland) and College of Engineering, University of Baghdad (Iraq, 1972-1974). Subjects: Strength of Materials, Engineering Analysis.
1967-1992	Graduate courses at Polish Academy of Sciences (IPPT, Warsaw, Poland), Warsaw University (Poland), Technical University of Pozen (Poland), Technical University of Opeln (Poland), Technical University of Kielce (Poland), University of Paderborn (Germany), Technical University of Hamburg-Harburg (Germany). Subjects: linear and nonlinear mechanics of continua, phenomenological thermodynamics, theory of propagation of linear and nonlinear waves in continua, theory of mixtures and of multicomponent systems, magnetohydrodynamics and macroscopic theory of electrolytes, extended thermodynamics, kinetic theory of gases.

| 1992-2005 | Undergraduate courses at the University of Essen and the Technical University of Berlin. Subjects: as above. |
| 2005 | Undergraduateand graduate courses at the University of Zielona Gora (Poland). Subjects: Mechanics. Continuum thermodynamics, Physics of Structures, Dynamics of Structures, Theoretical Mechanics. |

Supervision of Ph.D. Theses

B. Uziemblo (Polish Academy of Sciences, Warsaw, Poland), M. Elzanowski (Polish Academy of Sciences, Warsaw, Poland), J. Frankowski (Technical University of Kielce, Poland), B. Albers (Technical University of Berlin, Germany).

Supervision of some 30 MSc Theses (Diplomworks) at various universities.

Some chosen lectures and courses

- 3 courses (two of them self-organized; 6 monographic lectures in each course with notes published by Springer Vienna) at CISM (International Centre for Mechanical Sciences) in Udine (Italy) in the years 1978, 1998, 2004.
- Graduate course of phenomenological thermodynamics (30 hours) at the Weierstrass Institute for Applied Analysis and Stochastics (Berlin, Germany), 1996-1997.
- General lecture at the 33rd Conference of Solid Mechanics in Zakopane (Poland), 2000 on: *Extended thermodynamics of porous and granular materials.*
- Graduate courses (6 hours) for architects within the European Program "*Structural Health Monitoring System*, Universita di Pisa (Italy), 2000 on: *Waves and nondestructive testing of porous materials.*
- Graduate course of thermodynamics at the XXVth Summerschool on Mathematical Physics, Ravello (Italy), 2000.
- General lecture at the conference "Modeling and Mechanics of Granular and Porous Materials", Reggio Calabria (Italy), 2000 on: *Mass, exchange, diffusion and large deformations of poroelastic materials.*
- Ph.D. Course (app. 15 hours) at the joint school of the Technical Universities of Turin, Milan and Rome, Turin (Italy), 2001 on: *Thermodynamics of multicomponent systems.*
- General lecture at the conference GEOMATH3 in Horton (Greece), 2002 on: *Microworld and macroworld – multiscaling problems of geophysics.*
- General lecture at the conference STAMM 2002, Maiori (Italy), 2002 on: *Acoustic waves and nondestructive testing of granular materials.*
- Lecture at ROSE School (European School for Advanced Studies in Reduction of Seismic Risk), Universita di Pavia (Italy), 2003 on: *Bulk and surface waves in saturated poroelastic materials – low frequency approximation.*
- General lecture (key-note lecture) and the conference CANCAM 2003 in Calgary (Canada), 2003 on: *Macroscopic modeling of porous and granular materials – microstructure, thermodynamics and some boundary-initial value problems.*

- Graduate Course (45 hours) at the Technion, Haifa (Israel), 2006 on: *Continuum Thermodynamics.*
- Graduate Course (15 hours) at the Technical University of Graz, 2007 on: *Continuum Thermodynamics.*

Some activities in corporations and organization of science

During the last 30 years organization and participation in the organization of numerous international scientific conferences. Some more important examples:

- Polish-American Symposium on *Physical Fields in Material Media.* Warsaw (Poland), August 25-30, 1975 (Scientific Committee: A. C. Eringen, W. Nowacki, B. A. Boley, S. Kaliski, S. L. Koh, H. Zorski, T. S. Chang).
- XVIIIth Semester in Banach Centre (Warsaw, Poland): *Mathematical models and methods in mechanics,* 1981. Lecturers: C. Atkinson, C. E. Beevers, G. Capriz, C. Cercignani, D. R. J. Chilingworth, M. Costabel, A. Damlamian, K. Hutter, E. Meister, Yu. Mitropolski, I. Müller, W. Nowacki, L. E. Payne, G. F. Smith, K. Stewartson, etc.
- Euromech Colloquium: *Mean field theory of shape memory alloys*, Jablonna (Poland), 1981.
- Euromech Colloquium: *Porous media – theories and experiments*, Essen (Germany), 1997.
- Conference of the Int. Soc. for Interaction of Mechanics and Mathematics: *Continuum Mechanics and Thermodynamics*, Potsdam, 2001.
- Chair of the Prenominated Session at 21st ICTAM – IUTAM Congress in Warsaw (Poland), 2004: *Impact and wave propagation* (with Prof. Andrew Norris, Rutgers University, USA).
- Organization of the CISM course in Udine (Italy): *Surface waves in geomechanics: direct and inverse modeling for soils and rocks* (with Prof. Carlo Lai, Universita di Pavia, Italy).
- Conference: *Continuous and Discrete Modelling in Mechanics* (CDMM2005), Warszawa (Poland), September 5-9, 2005.
- Co-chair of the Conference: *18th International Conference on Computer Methods in Mechanics, CMM2009*, in Zielona Gora (Polnad), May 18-21, 2009.

Publications of Prof. Dr. K. Wilmański

1. Elastic-Plastic Stresses in a Ring Produced by Action of a Temperature Field, *Rozp. Inz. (Engineering Trans.)*, **10**, 4, 715-729, 1962.
2. Dynamical Loading of Beams, Timoshenko Beam, *Mech. Teor. Stos.*, **2**, 2, 83-96, 1964.
3. The Plane Strain Problem in a Semi-Infinite Porous Elastic Medium, *Proc. Sem. Soil Mech. Found. Engn.*, Lodz, 483-493, 1964.
4. Ruled Coordinate Systems in the Geometry of the Middle Surface of Thin Shells, (with: Cz. Woźniak), *Mech. Teor. Stos.*, **4**, 1, 127-134, 1966.
5. On a Certain Fibrous Model of a Dense Grate, *Rozp. Inz. (Engineering Trans.)*, **14**, 3, 499-512, 1966.
6. Mechanics of Continuous Cosserat-Type Media, (with: W. Barański, Cz. Woźniak), *Mech. Teor. Stos.*, **5**, 2, 214-258, 1967.
7. Asymptotic Method in the Theory of a Body with Microstructure, Plane Problem, *Rozp. Inz. (Engineering Trans.)*, **15**, 2, 295-309, 1967.
8. On Geometry of Continuous Medium with Microstructure, (with: Cz. Woźniak), *Arch. Mech. Stos.*,**19**, 5, 715-557, 1968.
9. Microeffects of Size in Dynamics of a Continuum, *Rozp. Inz. (Engineering Trans.)*, **16**, 4, 545-557, 1968.
10. Dynamics of Bodies with Microstructure, *Arch. Mech. Stos.*, **20**, 6, 705-744, 1968.
11. Some Topological Properties of the Space of States of Nonisolated Systems, *Bull. Acad. Polon. Sci., Ser. Sci. Techn.*, **19**, 7-8, 305-312, 1971.
12. On Thermodynamics and Functions of States of Nonisolated Systems, *Arch. Rat. Mech. Anal.*, **45**, 4, 251-281, 1972.
13. The Local Form of the Entropy Inequality in Neoclassical Thermodynamics, *Bull. Acad. Polon. Sci., Ser. Sci. Techn.*, **20**, 9, 373-383, 1972.
14. Foundations of Phenomenological Thermodynamics, (in Polish) PSP, Warsaw, 1974.
15. Note on Clausius-Duhem Inequality for a Singular Surface, *Bull. Acad. Polon. Sci., Ser. Sci. Techn.*, **22**, 10, 493-500, 1974.
16. Thermodynamic Properties of Singular Surfaces in Continuous Media, *Arch. of Mech.*, **27**, 3, 517-529, 1975.
17. Foundations of Neoclassical Thermodynamics: Metrization of Direct Thermodynamic Processes, in: *Trends in Applications of Pure Mathematics to Mechanics*, G. Fichera (ed.), Pitman Publ., 425-445, 1976.

18. On the Galilean Invariance of Balance Equations for a Singular Surface in Continuum, *Arch. of Mech.*, **29**, 3, 459-476, 1977.

19. Continuity of Fluxes in Thermodynamics, *Lett. App. Engn. Sci.*, **16**, 321-333, 1978.

20. Localization Problem of Nonlocal Continuum Theories, *Arch. of Mech.*, **31**, 1, 77-89, 1979.

21. State Functions for a Pseudoelastic Body, (with: I. Müller), CNRS, #295, *Comp. Mech. Sol. Anis.*, 133-147, 1979.

22. Thermodynamic Foundations of Thermoelasticity, in: *Recent Developments in Thermomechanics of Solids*, G. Lebon, P. Perzyna (Eds.), 1-94, Springer Verlag, Wien-New York, 1980.

23. A Model for Phase Transition in Pseudoelastic Bodies, (with. I. Müller), *Il Nuovo Cimento*, **578**, 2, 283-318, 1980.

24. A Model for Creep and Strain Hardening in Martensitic Transformation, (with: M. Achenbach, I. Müller), *J. Thermal Stresses*, **4**, 3-4, 523-534, 1981.

25. Memory Alloys - Phenomenology and Ersatzmodel, (with: I. Müller), in: *Continuum Models of Discrete Systems 4*, O. Brulin, R. K. T. Hsieh (eds.), North-Holland Publ. Co., 495-509, 1981.

26. Dynamics of Spatially Curved Rods; Geometrically Non-Linear Theory, *IFTR Report*, #11, Polish Academy of Sciences, 8-34, Warsaw, 1981.

27. Propagation of the Interface in the Stress-Induced Austenite - Martensite Transformation, *Ing. Arch.*, **53**, 291-301, 1983.

28. Phenomenological Thermodynamics - Development and Trends, *Mech. Teor. Stos.*, **21**, 4, 655-678, 1983.

29. Phenomenological Thermodynamics, (in Polish) in: *Technical Mechanics*, H. Zorski (ed.), vol.1, 438-531, PSP, Warsaw, 1984; also: Englisch Translation, Elsevier Publ., 485-590, 1991.

30. Lecture Notes on Thermodynamic Models of Continuous Media, (in Polish) *Ossolineum*, Warsaw-Wroclaw, 1985.

31. Mathematical Models and Methods in Mechanics, *Banach Center Publications* #15, PSP, Warsaw, 1985 (edited with W. Fiszdon).

32. Extended Thermodynamics of a Non-Newtonian Fluid, (with: I. Müller), *Rheologica Acta*, **25**: 335-349, 1986.

33. Thermodynamics of Non-Newtonian Fluids, in: *Wissenschaftskolleg zu Berlin, Jahrbuch* 1984/85, 179-199, Siedler Verlag, Berlin, 1986.

34. Non-Newtonian Fluids of Second Grade - Rheology, Thermodynamics and Extended Thermodynamics, in: *Trends in Applications of Pure Mathematics to Mechanics*, Lecture Notes in Physics, 249, E. Kröner, K. Kirchgässner (eds.), Springer Verlag, Berlin-New York, 376-383, 1986.

35. Residual Deformation and Strain Recovery in Polymers with Folded and Oriented Crystals, *DFG-Bericht*, TU Berlin, 1-48, 1986.

36. The Passage from Memory Functional to Rivlin-Ericksen Constitutive Equations, (with: R. S. Rivlin), *ZAMP*, **38**, 624-629, 1987.

37. Toward Correlational Thermodynamics; Kinetic Foundations, (with: J. Schröter), *DFG-Bericht*, 1-109, Universität-GH-Paderborn, 1987.

38. Thermodynamics of a Heat Conducting Maxwellian Fluid, *Arch. of Mech.*, **40**,2-3,217-232, 1988.

39. A Microphysical Model of Crystallizing Polymers, in: *Constitutive Laws and Microstructure*, D. R. Axelrad, W. Muschik (eds.), Springer Verlag, Berlin-New York, 163-173, 1988.

40. Note on Simple Shear of Plastic Monocrystals, (with: O.Mahrenholtz), *Mechanics Research Comm.*, **17**(6), 393-402, 1990.

41. Statistical Model of Damage of Crystallising Polymers, in: *Free Boundary Problems: Theory and Applications*, vol.II, K. H. Hoffmann, J. Sprekels (eds.), Longman Scientific & Technical, New York, 822-829, 1990.

42. Large Elasto-Plastic Deformation of Two-Phase Alloys - Structural Continuous Model, *Proceedings of CSME*, Toronto, 1990; reprinted in: *Recent Developments in Micromechanics*, R. D. Axelrad, W. Muschik (eds.), 84-96, Springer Verlag, Berlin, 1990.

43. Physical and Continuumsmechanical Modeling of Mechanical Properties of Two-Phase - Systems, (with: O. Mahrenholtz), *DFG-Vorhaben* Ma 359/43, Abschlußbericht, Hamburg, 1990.

44. On Plasticity of Multicomponent Metallic Alloys - Evolution of Texture, in: *Trends in Applications of Mathematics to Mechanics*, W. Schneider, H. Troger, F. Ziegler (eds.),191-197, Longman Scientific & Technical, Essex, 1991.

45. Macroscopic Theory of Evolution of Deformation Textures, *Int. J. Plasticity*, **8**, 959-975, 1992.

46. On Pattern Formation in Stress-Induced Martensitic Transformation, in: *Non-Linear Thermodynamical Processes in Continua*, G. Maugin, W. Muschik (eds.), 88-105, TUB-Dokumentation, Berlin, 1992.

47. Symmetric Model of Stress-Strain Hysteresis Loops in Shape Memory Alloys, *Int. J. Engn. Sci.*, **31**, 8, 1121-1138, 1993.

48. Textures in Polycrystalline Metal Alloys - Structural Plasticity, in: *Continuum Models of Discrete Systems* 7, K.-H. Anthony, H.-J. Wagner (eds.), 361-370, Trans Tech Publications, Switzerland, 1993.

49. Note on the Model of Pseudoelastic Hysteresis, in: *Models of Hysteresis*, A. Visintin (ed.), 207-221, Pitman Research Notes in Mathematics Series, #286, 1993.

50. Papers on Ersatz-Model of Stress-Induced Pattern Formation in Shape Memory Alloys, *Schwerpunktprogramm der DFG: Anwendungsbezogene Optimierung und Steuerung*, Bericht #425, 1-113, Essen, 1993.

51. Skeleton as Reference for Two-Phase Porous Materials, *MECH-Bericht* 93/5, 1-23, Universität-GH Essen, Essen, 1993.

52. Acceleration Waves in Two-Component Porous Media Part I: The Model and Speeds of Propagation, *MECH-Bericht* 94/1, 1-23, Universität-GH Essen, Essen, 1994.

53. On Weak Discontinuity Waves in Porous Materials, in: *Trends in Applications of Mathematics to Mechanics*, M. Marques, J. Rodrigues (eds.), 71-83, Longman Scientific & Technical, Essex, 1995; also *MECH-Bericht* 94/3, 1-12 Universität-GH Essen, Essen, 1994.

54. Lagrangean Model of Two-Phase Porous Material, *J.Non-Equilibrium Thermodyn.*, **20**, 50-77, 1995.

55. The Thermodynamic Structure of the Two-Component Model of Porous Incompressible Materials with True Mass Densities, (with: J. Bluhm, R. de Boer), *Mechanics Research Comm.*, **22**, 2, 171-180, 1995.

56. Porous Media at Finite Strains. The New Model with the Balance Equation of Porosity, *Arch. of Mech.*, **48**, 4, 591-628, 1996, also: *MECH-Bericht* 95/12, 1-35. Universität-GH Essen, Essen, 1995.

57. On Incompressibility of True Components in Porous Materials, in: *Festschrift zum 60. Geburtstag von Prof. Reint de Boer: Beiträge zur Mechanik*, Forschungsbericht aus dem Fachbereich Bauwesen, #66, 423-434, Universität-GH Essen, 1995.

58. Acceleration Waves in Two-Component Porous Media Part II: Supplementary Remarks on the Model and the Evolution Equation of Amplitudes, *MECH-Bericht* 95/2, 1-20 Universität-GH Essen, Essen, 1995.

59. Two-Component Compressible Porous Materials - The Construction of the Thermodynamical Model, *MECH-Bericht* 95/1, 1-24. Universität-GH Essen, Essen, 1995.

60. Quasi-Static Plane-Strain Problems of the Linear Porous Material, *MECH-Bericht* 95/7, 1-19, Universität-GH Essen, Essen, 1995.

61. Was ist die erweiterte Thermodynamik? Beispiel: Starrer Wärmeleiter, *MECH-Bericht* 95/3, 1-14, Universität-GH Essen, Essen, 1995.

62. Porous Media at Finite Strains, in: *Continuum Models and Discrete Systems* CMDS8, K. Markov (ed.), World Scientific, 317-324, 1996.

63. Dynamics of Porous Materials under Large Deformations and Changing Porosity,in: *Contemporary Research in the Mechanics and Mathematics of Materials*, R. C. Batra, M. F. Beatty (eds.), Jerald L. Ericksen's Anniversary Volume, CIMNE, Barcelona, 343-356, 1996; also: *MECH-Bericht* 95/15, 1-25. Universität-GH Essen, Essen, 1995.

64. A Thermodynamic Model of Compressible Porous Materials with the Balance Equation of Porosity, *Transport in Porous Media,* **32**: 21-47, 1998; also: *WIAS–Preprint* No. 310, Weierstraß-Institut für Angewandte Analysis and Stochastik, 1997.

65. On the Time of Existence of Weak Discontinuity Waves in Poroelastic Materials, *Arch. of Mech.* **50**, 3, 657-669, 1998; also: *WIAS-Preprint* No.366, Weierstraß-Institut für Angewandte Analysis and Stochastik, 1997.

66. Sound Waves in Porous Materials - New Possibilities for Ultrasonic Tomography, in: *13. Symposium: Medizin und Mathematik*, Bad Honnef, 1-6, 23.- 24. Januar, 1997.

67. On the Acoustic Waves in Two-Component Linear Poroelastic Materials, in: *Ingo Müller's Anniversary Volume*, H. Struchtrup (ed.), TU Berlin, 1997; also: *WIAS-Preprint* No.312, Weierstraß-Institut für Angewandte Analysis and Stochastik, 1997.

68. Thermomechanics of Continua, Springer, Heidelberg, Berlin, N.Y., 1998.

69. An Axisymmetric Steady-State Flow through a Poroelastic Medium under Large Deformations, (with: B. Albers) *Archive of Applied Mechanics*, **69**, 121-132, 1998; also: *WIAS-Preprint* No. 406, Weierstraß-Institut für Angewandte Analysis and Stochastik, 1998.

70. Iterative Procedure for Multidimensional Euler Equations, (with: W: Dreyer, M.Kunik, W. Sabelfeld, N. Simonov), *Monte Carlo Methods and Applications*, Vol. 4, No. 3, 253-271 (1998); also: *WIAS-Preprint* No.445, Weierstraß-Institut für Angewandte Analysis and Stochastik, 1998.

71. Kinetic and Continuum Theories of Granular and Porous Media, CISM, Courses and Lectures #400, (edited with K. Hutter) Springer, Wien, N.Y., 1999.

72. Waves in Porous and Granular Materials, in: *Kinetic and Continuum Theories of Granular and Porous Media*, CISM, Courses and Lectures #400, K. Hutter, K. Wilmanski (eds.), 131 - 185, Springer, Wien, N.Y., 1999.

73. A Transient Model for the Sublimation Growth of Silicon Carbide Single Crystals, (with: N. Bubner, O. Klein, P. Philip, J. Sprekels) *J. Crystal Growth*, **205**, 294-304, 1999; also: *WIAS-Preprint* No.443, Weierstraß-Institut für Angewandte Analysis and Stochastik, 1998.

74. Surface Waves at a Free Interface of a Saturated Porous Medium, (with: I. Edelman, E. Radkevich) *WIAS-Preprint* No. 513, Weierstraß-Institut für Angewandte Analysis and Stochastik, 1999.

75. Surface Waves at an Interface Separating a Saturated Porous Medium an a Liquid, (with: I. Edelman) *WIAS-Preprint* No. 531, Weierstraß-Institut für Angewandte Analysis and Stochastik, 1999.

76. Note on the Notion of Incompressibility in Theories of Porous and Granular Materials, ZAMM, **81**, 37-42, 2001; also: *WIAS-Preprint* No.465, Weierstraß-Institut für Angewandte Analysis and Stochastik, 1998.

77. On a Homogeneous Adsorption in Porous Materials, ZAMM, **81**, 119-124, 2001; also: *WIAS-Preprint* No. 475, Weierstraß-Institut für Angewandte Analysis and Stochastik, 1999.

78. Elementary Thermodynamic Arguments on Non-Newtonian Fluids, (with: Ingo Müller) *WIAS-Preprint* No. 485, Weierstraß-Institut für Angewandte Analysis and Stochastik, 1999.

79. On a Stationary Axisymmetric Filtration Problem Under Large Deformations of the Skeleton, (with: B. Albers) in: *Trends in Applications of Mathematics to Mechanics*, G. Iooss, O. Guès, A. Nouri (eds.), Chapman & Hall/CRC, 47-57, 2000.

80. Toward Extended Thermodynamics of Porous and Granular Materials, in: *Trends in Applications of Mathematics to Mechanics*, G. Iooss, O. Guès, A. Nouri (eds.), Chapman & Hall/CRC, 147-160, 2000.

81. Surface Waves at an Interface Separating Two Porous Media (with: I. Edelman), *WIAS-Preprint* No. 568, Weierstraß-Institut für Angewandte Analysis and Stochastik, 2000.

82. Two Notes on Continuous Modelling of Porous Media: A Note on Objectivity of Momentum Sources in Porous Materials, (with: B. Albers), A Note on Two-component Model for Terzaghi Gedankenexperiment, *WIAS-Preprint* No. 579, Weierstraß-Institut für Angewandte Analysis and Stochastik, 2000.

83. A Riemann Problem for Poroelastic Materials with the Balance Equation for Porosity, (with. E. Radkevich), Part I: *WIAS-Preprint* No. 593, Weierstraß-Institut für Angewandte Analysis and Stochastik, 2000; Part II: *WIAS-Preprint* No. 594, Weierstraß-Institut für Angewandte Analysis and Stochastik, 2000.

84. Mathematical Theory of Porous Media - Lecture Notes. XXV Summer School on Mathematical Physics, Ravello, 2000. *WIAS-Preprint* No. 602, Weierstraß-Institut für Angewandte Analysis und Stochastik, 2000.

85. Radiation- and Convection-driven Transient Heat Transfer During Sublimation Growth of Silicon Carbide Single Crystals, (with: O. Klein, P. Philip, J. Sprekels), *Jour. of Crystal Growth*, **222**, 832-851, 2001; also: *WIAS-Preprint* No. 552, Weierstraß-Institut für Angewandte Analysis and Stochastik, 2000.

86. Sound and Shock Waves in Porous and Granular Materials, in: Proceedings "WASCOM 99", *10th Conference on Waves and Stability in Continuous Media*, V. Ciancio, A. Donato, F. Olivieri, S. Rionero (eds.), World Scientific, 489-503, 2001; also: *WIAS-Preprint* No. 563, Weierstraß-Institut für Angewandte Analysis and Stochastik, 2000.

87. Thermodynamics of Multicomponent Continua, in: *Earthquake Thermodynamics and Phase Transformations in the Earth's Interior*, J. Majewski, R. Teisseyre (eds.), 567-655, Academic Press, San Diego, 2001.

88. Some Questions on Material Objectivity Arising in Models of Porous Materials, in: *Rational Continua, classical and new*, M. Brocato (ed.), 149-161, Springer-Verlag, Italia Srl, Milano, 2001.

89. Propagation of Sound and Surface Waves in Porous Materials, *WIAS-Preprint* No. 684, Weierstraß-Institut für Angewandte Analysis und Stochastik, 2001.

90. Relaxation Properties of a 1D Flow through a Porous Material without and with Adsorption, (with: B. Albers), *WIAS-Preprint* No. 707, Weierstraß-Institut für Angewandte Analysis und Stochastik, 2001.

91. On the Onset of Flow Instabilities in Granular Media due to Porosity Inhomogeneities (with: Theo Wilhelm) *Int. Jour. Multiphase Flows*, **28**, 1929-1944, 2002; also: *WIAS-Preprint* No. 632, Flow Instabilities in Granular Media due to Porosity Inhomogeneities, Weierstraß-Institut für Angewandte Analysis und Stochastik, 2001.

92. Asymptotic Analysis of Surface Waves at Vacuum/Porous Medium and Liquid/Porous Medium Interfaces, (with: I. Edelman), *Continuum Mechanics and Thermodynamics*, **14**, 1, 25-44, 2002; also: *WIAS-Preprint* No. 695, Weierstraß-Institut für Angewandte Analysis und Stochastik, 2001.

93. Mass Exachange, Diffusion and Large Deformations of Poroelastic Materials, in: *Modeling and Mechanics of Granular and Porous Materials,* G. Capriz, V. N. Ghionna, P. Giovine (eds.), 211-242, Birkhäuser, 2002; also: *WIAS-Preprint* No. 628, Weierstraß-Institut für Angewandte Analysis und Stochastik, 2001.

94. Propagation of sound and surface waves in porous materials, in: *Structured Media*, B. Maruszewski (ed.), 312-326, Poznan University of Technology, Poznan, 2002; also: WIAS-Preprint No. 684, Weierstraß-Institut für Angewandte Analysis und Stochastik, 2001.

95. Acoustic waves in porous solid-fluid mixtures, (with: B. Albers), in: *Dynamic Response of Granular and Porous Materials under Large and Catastrophic Deformations*, K. Hutter, N. Kirchner (eds.), Lecture Notes in Applied and Computational Mechanics, Vol. 11, Springer, Berlin, 285-314, 2002.

96. O dispersii w modeli dla poristo-uprugich sred s urawnieniem balansa poristosti (in Russian), (with: E. Radkevich), in: *Trudy Seminara im. I. G. Petrowskogo*, wyp. 22, Moskwa, 74-104, 2002.

97. O globalnoj ustoicziwosti w zadacze dla poristo-uprugich sred s urawnieniem balansa poristosti (in Russian), (with: E. Radkevich) in: *Trudy Moskowskogo Matematiczeskogo Obszczestwa*, tom 63, Moskwa, 7-44, 2002.

98. Thermodynamical admissibility of Biot's model of poroelastic saturated materials, *Arch. of Mech.*, **54**, 5-6, 709-736, 2002.

99. Note on weak discontinuity waves in linear poroelastic materials. Part I: Acoustic waves in saturated porous media, *WIAS-Preprint* No. 730, Weierstraß-Institut für Angewandte Analysis und Stochastik, 2002.

100. Sound and surface waves in poroelastic media, (with: B. Albers), *WIAS-Preprint* No. 757, Weierstraß-Institut für Angewandte Analysis und Stochastik, 2002.

101. On thermodynamics of nonlinear poroelastic materials, *WIAS-Preprint* No. 792, Weierstraß-Institut für Angewandte Analysis und Stochastik, 2002.

102. On Dispersion in the Mathematical Model of Poroelastic Materials with the Balance Equation for Porosity, (with: E. Radkevich) *Journal of Mathematical Sciences*, 4, **114**, 1431-1449, 2003.

103. On thermodynamic modeling and the role of the second law of thermodynamics in geophysics, in: *Advanced Mathematical and Computational Geomechanics*, D. Kolymbas (ed.), Springer, Berlin, 3-33, 2003; also: WIAS-Preprint No. 813, Weierstraß-Institut für Angewandte Analysis und Stochastik, 2003.

104. On Thermodynamics of Nonlinear Poroelastic Materials, *Journal of Elasticity*, **71**, 247-261, 2003.

105. On Microstructural Tests for Poroelastic Materials and Corresponding Gassmann-type Relations, *Geotechnique*, **54**, 9, 593-603, 2004; also: On a micro-macro transition for poroelastic Biot's model and corresponding Gassmann-type relations, WIAS-Preprint No. 868, Weierstraß-Institut für Angewandte Analysis und Stochastik, 2003.

106. On Biot-like models and micro-macrotransitions for poroelastic materials, *WIAS-Preprint* No. 830, Weierstraß-Institut für Angewandte Analysis und Stochastik, 2003.

107. Macroscopic Modeling of Porous and Granular Materials - Microstructure, Thermodynamics and Some Boundary-Initial Value Problems, *WIAS-Preprint* No. 858, Weierstraß-Institut für Angewandte Analysis und Stochastik, 2003.

108. Objective relative accelerations in theories of porous materials, in: *Thermody-namische Materialtheorien, Oberwolfach Report No.* **55**, 36-38, 2004.
109. Surface Waves in Geomechanics: Direct and Inverse Modelling for Soils and Rocks, (edited with C. Lai), CISM: Courses and Lectures – No. 481, Springer-WienNewYork, 2005.
110. Elastic Modelling of Surface Waves in Single and Multicomponent Systems – Lecture notes (CISM). In: *Surface Waves in Geomechanics: Direct and Inverse Modelling for Soils and Rocks*, C. Lai, K. Wilmanski (Eds.), 203-276, SpringerWienNewYork, 2005; also: WIAS-Preprint No. 945, Weierstraß-Institut für Angewandte Analysis und Stochastik, 2004.
111. Modeling Acoustic Waves in Saturated Poroelastic Media, (with: B. Albers), *Jour. of Engn. Mechanics*, ASCE, **131**, 9, September 1, 974-985, 2005.
112. Monochromatic Surface Waves on Impermeable Boundaries in Two-Component Poroelastic Media, (with: B. Albers), *Cont. Mech. Thermodyn.*, **17**, 3, 269-285, 2005.
113. Tortuosity and Objective Relative Accelerations in the Theory of Porous Materials, *Proc. R. Soc. A,* **461**, 1533-1561, 2005; also: WIAS-Preprint No. 922, Weierstraß-Institut für Angewandte Analysis und Stochastik, 2004.
114. Thermodynamics of Simple Two-Component Thermo-Poroelastic Media, in: *Trends and Applications of Mathematics to Mechanics, STAMM 2002,* S. Rionero, G. Romano (Eds.), Springer, Italia, 293-306, 2005; also: WIAS-Preprint No. 901, Weierstraß-Institut für Angewandte Analysis und Stochastik, 2004.
115. Linear Sound Waves in Poroelastic Materials: Simple Mixtures vs. Biot's Model, in: *Trends in Applications of Mathematics to Mechanics, STAMM 2004,* Y. Wang, K. Hutter, (Eds.), Shaker Verlag, 569-578, 2005; also: WIAS-Preprint No. 950, Weierstraß-Institut für Angewandte Analysis und Stochastik, 2004.
116. Critical Time for Acoustic Waves in Weakly Nonlinear Poroelastic Materials, *Cont. Mech. Thermodyn.*, **17**: *171-181,* 2005.
117. Threshold to Liquefaction in Granular Materials as a Formation of Strong Wave Discontinuity in Poroelastic Media, in: *Poromechanics, III, Biot Cente-nial,* Y. Abousleiman, A. Cheng, F.-J. Ulm (Eds.), A. A. Balkem Publ., Leiden, *297-302,* 2005.
118. Thermodynamic Modelling of Saturated Poroelastic Materials - linear and non-linear effects, in: *Grenzschicht Wasser und Boden, Phänomene und Ansätze,* J. Grabe (Ed.), 87-106, TUHH, Hamburg, 2005.
119. Threshold to liquefaction in granular materials as a formation of strong wave discontinuity in poroelastic media, *WIAS-Preprint* No. 1003, Weierstraß-Institut für Angewandte Analysis und Stochastik, 2005.
120. A few remarks on Biot's model and linear acoustics of poroelastic saturated materials, *Soil Dynamics & Earthquake Engineering,* **26**, 6-7, 509-536, June-July, 2006.
121. Influence of coupling through porosity changes on the propagation of acoustic waves in linear poroelastic materials, (with: B. Albers), *Arch. Mech.* **58**, 4-5, 313-325, 2006.

122. Adaptive Linearly Implicit Methods for Linear Poroelastic Equations (with: B. Erdmann, J. Lang, S. Matera), Konrad-Zuse-Zentrum für Informationstechnik Berlin, ZIB-Report 06-37, 2006.

123. A few remarks on micro/macro transitions and Gassmann relations for poroelastic materials, *Mechanics of Solids and Structures*, **6**, 191-204, 2007.

124. Continuum Thermodynamics. Part I: Foundations, WorldScientific, Singapore, 2008.

125. On Waves in Weakly Nonlinear Poroelastic Materials Modeling Impacts of Meteorites, in: Proceedings "WASCOM 07", *14th Conference on Waves and Stability in Continuous Media*, N. Manganaro, R. Monaco, S. Rionero, (eds.), World Scientific, 589-597 2008.

126. Modeling of Thermomechanical Behaviour of Embankment Dams; One-component vs. Multicomponent Description, in: *Long Term Behaviour of Dams*, Proceedings of the 2nd Int. Conference LTDB09, E. Bauer, S. Semprich, G. Zenz, (eds.), Verlag der Technischen Universität Graz, 85-90, 2009.

127. Thermodynamic Modeling of Soil Morphology (with: Bettina Albers), in: *Long Term Behaviour of Dams*, Proceedings of the 2nd Int. Conference LTDB09, E. Bauer, S. Semprich, G. Zenz, (eds.), Verlag der Technischen Universität Graz, 437-442, 2009.

128. Diffusion and Heat Conduction in Nonlinear Thermoporoelastic Media: in: *Selected Topics in Mechanics of the Inhomogeneous Media,* M. Kuczma, R. Switka, Cz. Wozniak (eds.), University of Zielona Gora, 1-29, 2009.

129. Macroscopic Modeling of Porous Materials, in: *New Developments in Mathematical Modelling and Analysis of Microstructured Media. Professor Margaret Wozniak pro memoria*, K. Blazejowski, M. Wagrowska, Cz. Wozniak (eds.), 1-25, Warsaw, 2009.

130. Continuous Modeling of Soil Morphology - Thermomechanical Behaviour of Embankment Dams (with: B. Albers), *Frontiers of Architecture and Civil Engineering*, IFACE, (2009, in print).

131. *Computer Methods in Mechanics,* Lectures of the CMM 2009 (edited with M. Kuczma), Springer, Berlin, Heidelberg, 2009.

Part II
THERMODYNAMIC MODELING

On Pore Fluid Pressure and Effective Solid Stress in the Mixture Theory of Porous Media

I-Shih Liu

Dedicated to Krzysztof Wilmański on the occasion of his 70th birthday.

Abstract. In this paper we briefly review a typical example of a mixture of elastic materials, in particular, an elastic solid-fluid mixture as a model for porous media. Application of mixture theories to porous media rests upon certain physical assumptions and appropriate interpretations in order to be consistent with some better-known notions in engineering applications. We shall discuss, for instance, the porosity, pore fluid pressure, effective solid stress and Darcy's law in this paper.

1 Introduction

The essential features of theories of mixtures in the framework of continuum mechanics have been developed throughout the sixties and seventies. Here we briefly review a typical example of a mixture of elastic materials, in particular, an elastic solid-fluid mixture as a model for porous media. We shall outline a simple model of porous media by introducing the volume fraction, pore fluid pressure, effective solid stress, etc., as reinterpretations of the results of mixture theory to porous media. Some elementary results in soil mechanics are obtained, in particular, a generalization of the Darcy's law from the equation of motion of the fluid constituent, and the effective stress principle. As an example of an equilibrium solution in a linear theory for incompressible porous media, porosity distribution in saturated soil is considered.

I-Shih Liu

Instituto de Matemática, Universidade Federal do Rio de Janeiro, Rio de Janeiro, Brasil
e-mail: liu@im.ufrj.br

2 Mixture of Elastic Materials

For a mixture of N constituents, we introduce the following quantities for the constituent $\alpha \in N$ and the mixture:

ρ_α partial mass density of constituent α.
$\mathbf{v}_\alpha$ partial velocity of constituent α.
T_α partial stress tensor of constituent α.
$\mathbf{b}_\alpha$ external body force on constituent α.
$\mathbf{m}_\alpha$ interaction force on constituent α.
ρ mass density of mixture.
$\mathbf{v}$ velocity of mixture.
T stress tensor of mixture.
ε internal energy density of mixture.
$\mathbf{q}$ energy flux of mixture.
r energy supply of mixture.

Following the pioneering work of Truesdell ([5], see also [6]), the basic laws of a non-reacting mixture are given by the following balance equations for mass and linear momentum for each constituent:

$$\frac{\partial \rho_\alpha}{\partial t} + \operatorname{div}(\rho_\alpha \mathbf{v}_\alpha) = 0,$$
$$\frac{\partial \rho_\alpha \mathbf{v}_\alpha}{\partial t} + \operatorname{div}(\rho_\alpha \mathbf{v}_\alpha \otimes \mathbf{v}_\alpha - T_\alpha) = \rho_\alpha \mathbf{b}_\alpha + \mathbf{m}_\alpha. \tag{1}$$

and for the mixture as a whole, by defining

$$\rho = \sum_\alpha \rho_\alpha, \quad \mathbf{v} = \sum_\alpha \frac{\rho_\alpha}{\rho} \mathbf{v}_\alpha, \quad \mathbf{b} = \sum_\alpha \frac{\rho_\alpha}{\rho} \mathbf{b}_\alpha,$$

and

$$T = \sum_\alpha (T_\alpha - \rho_\alpha \mathbf{u}_\alpha \otimes \mathbf{u}_\alpha),$$

where

$$\mathbf{u}_\alpha = \mathbf{v}_\alpha - \mathbf{v}$$

is the diffusion velocity of constituent α, and by requiring

$$\sum_\alpha \mathbf{m}_\alpha = 0,$$

we obtain

$$\frac{\partial \rho}{\partial t} + \operatorname{div}(\rho \mathbf{v}) = 0,$$
$$\frac{\partial \rho \mathbf{v}}{\partial t} + \operatorname{div}(\rho \mathbf{v} \otimes \mathbf{v} - T) = \rho \mathbf{b}. \tag{2}$$

which together with the energy equation,

$$\frac{\partial \rho \varepsilon}{\partial t} + \mathrm{div}(\rho \varepsilon \mathbf{v} + \mathbf{q}) - \mathrm{tr}(T \,\mathrm{grad}\, \mathbf{v}) = \rho r. \tag{3}$$

constitute the basic balance laws of the mixture as a single body.

We consider a mixture of elastic materials characterized by the constitutive equation of the form:

$$f = \mathscr{F}(\theta, \mathrm{grad}\,\theta, F_\alpha, \mathrm{grad}\, F_\alpha, \mathbf{v}_\alpha),$$

where θ is the temperature, F_α is the deformation gradient of the constituent α relative to a reference configuration, and

$$f = \{T_\alpha, \varepsilon, \mathbf{q}, \mathbf{m}_\alpha\}$$

are the constitutive quantities for the basic field variables $\{\rho_\alpha, \mathbf{v}_\alpha, \theta\}$ with governing equations consisting of (1) and (3). Constitutive theory of such a mixture has been considered by Bowen [1] in which consequences of the entropy principle have been obtained based on the entropy inequality of the form:

$$\frac{\partial \rho \eta}{\partial t} + \mathrm{div}(\rho \eta \mathbf{v} + \boldsymbol{\Phi}) - \rho s \geq 0, \tag{4}$$

where η is the entropy density of the mixture. The entropy flux and the entropy supply density are given by

$$\begin{aligned}
\boldsymbol{\Phi} &= \frac{1}{\theta}\left(\mathbf{q} - \sum_\alpha (\rho_\alpha(\psi_\alpha + \tfrac{1}{2}\mathbf{u}_\alpha^2)I - T_\alpha^T)\mathbf{u}_\alpha\right), \\
s &= \frac{1}{\theta}\left(r - \frac{1}{\rho}\sum_\alpha \rho_\alpha \mathbf{u}_\alpha \cdot \mathbf{b}_\alpha\right),
\end{aligned} \tag{5}$$

where ψ_α is the free energy density of the constituent α, and for the mixture we define

$$\psi = \frac{1}{\rho}\sum_\alpha \rho_\alpha \psi_\alpha,$$

and I is the identity tensor.

Remark

The specific form of the relations (5) can be derived provided that the Clausius-Duhem assumptions on the entropy flux and the entropy supply are valid for each constituent. However, only the entropy production of the mixture as a whole is postulated to be non-negative. It has been shown that such assumptions are appropriate (but not necessarily general enough) to account for the behavior of a mixture within the framework of continuum mechanics. On the other hand, the very expressions in (5) show that the Clausius-Duhem assumptions $\boldsymbol{\Phi} = \mathbf{q}/\theta$ and $s = r/\theta$ are not general enough to account for material bodies in general.

2.1 Summary of Results for Elastic Solid-Fluid Mixtures

For a mixture of a solid (with subindex s) and a fluid (with subindex f), the constitutive equations take the form:

$$f = \mathscr{F}(\theta, \operatorname{grad}\theta, \rho_f, F_s, \operatorname{grad}\rho_f, \operatorname{grad}F_s, V),$$

where $V = \mathbf{v}_f - \mathbf{v}_s$ is the relative velocity of the fluid through the solid.

We shall summarize the constitutive results from thermodynamic considerations obtained by Bowen ([1], see also [6] Appendix 5A) in the case of an elastic solid-fluid mixture:

$$\psi = \psi(\theta, \rho_f, F_s),$$

$$T_f = \rho_f \psi_f I - \frac{\partial \rho \psi}{\partial \rho_f} \rho_f I + \rho_f \frac{\partial \psi_f}{\partial V} \otimes V, \tag{6}$$

$$T_s = \rho_s \psi_s I + \frac{\partial \rho \psi}{\partial F_s} F_s^T + \rho_s \frac{\partial \psi_s}{\partial V} \otimes V.$$

Since there are only two constituents, we have $\mathbf{m}_f + \mathbf{m}_s = 0$ and we shall write $\mathbf{m} = \mathbf{m}_f = -\mathbf{m}_s$. Moreover, we have

$$\psi_f^0 = \widehat{\psi}_f(\theta, \rho_f, F_s),$$

$$\psi_s^0 = \widehat{\psi}_s(\theta, \rho_f, F_s), \tag{7}$$

$$\mathbf{m}^0 = \frac{\partial \rho_s \widehat{\psi}_s}{\partial \rho_f} \operatorname{grad}\rho_f - \frac{\partial \rho_f \widehat{\psi}_f}{\partial F_s}[\operatorname{grad}F_s],$$

where the superscript 0 indicates the value in equilibrium defined as processes for which $\operatorname{grad}\theta = 0$ and $V = 0$.

If we define the (equilibrium) chemical potential of the fluid and the (equilibrium) partial fluid pressure as

$$\mu_f = \frac{\partial \rho \psi}{\partial \rho_f}, \qquad p_f = \rho_f(\mu_f - \psi_f^0), \tag{8}$$

then we have

$$T_f^0 = -p_f I,$$

and the relation $(7)_3$ can be written as

$$\mathbf{m}^0 = \frac{p_f}{\rho_f} \operatorname{grad}\rho_f - \rho_f(\operatorname{grad}\widehat{\psi}_f)^0. \tag{9}$$

2.2 Jump Condition at Fluid-Permeable Surface

In [2] Liu considered an ideal fluid-permeable surface, i.e., a surface across which there is no jump of temperature and the solid constituent does not go through,

$$[\theta] = 0, \qquad \mathbf{v}_s = \mathbf{u}^*,$$

where [ ] denotes the jump across the surface and $\mathbf{u}^*$ is the surface velocity. It is proved that at such surfaces, the following jump condition holds,

$$\left[\!\!\left[\mu_f + \frac{1}{2}(\mathbf{v}_f - \mathbf{u}^*)^2 - (\mathbf{v}_f - \mathbf{u}^*) \cdot \frac{\partial \psi_f}{\partial \mathbf{v}_f} \right]\!\!\right] = 0. \tag{10}$$

In particular, if $\mathbf{v}_f = \mathbf{u}^*$, the the chemical potential of the fluid constituent is continuous over the surface.

3 Saturated Porous Media

The solid-fluid mixture considered in the previous section can be regarded as a model for saturated porous media provided that the concept of porosity, the volume fraction of the fluid constituent ϕ, is introduced:

$$\rho_f = \phi d_f, \qquad \rho_s = (1 - \phi)d_s, \tag{11}$$

where d_f and d_s are the true mass densities of the fluid and the solid constituents respectively.

3.1 Pore Fluid Pressure

We shall also regard the partial fluid pressure p_f in the mixture theory as the outcome of a "microscopic" pressure acting over the area fraction of surface actually occupied by the fluid in the pore. Thus we define the *pore fluid pressure P* as

$$P = \frac{p_f}{\phi}. \tag{12}$$

In this definition, we have tacitly assumed that the surface fraction is the same as the volume fraction ϕ defined in (11), which, of course, may not be true in general but is acceptable for most practical applications.

From (7), we have

$$\psi_f^0 = \widehat{\psi}_f(\theta, \phi d_f, F_s).$$

However, in most applications, it is reasonable to assume that in equilibrium the free energy of the fluid constituent is the same as the free energy of the pure fluid, i.e., we shall assume that

$$\psi_f^0 = \widehat{\psi}_f(\theta, d_f). \tag{13}$$

From (8), we have

$$\mu = \frac{P}{d_{\mathrm{f}}} + \psi_{\mathrm{f}}^0.$$

At an ideal fluid-permeable surface which allows the fluid to go through, since the true density does not change, $[d_{\mathrm{f}}] = 0$, it follows that

$$[\mu] = \left[\!\left[\frac{P}{d_{\mathrm{f}}}\right]\!\right] + [\widehat{\psi}_{\mathrm{f}}(\theta, d_{\mathrm{f}})] = \frac{1}{d_{\mathrm{f}}}[P].$$

Therefore the condition (10) implies that

$$[P] + d_{\mathrm{f}}\left[\!\left[\frac{1}{2}(\mathbf{v}_{\mathrm{f}} - \mathbf{u}^*)^2 - (\mathbf{v}_{\mathrm{f}} - \mathbf{u}^*)\cdot\frac{\partial\psi_{\mathrm{f}}}{\partial\mathbf{v}_{\mathrm{f}}}\right]\!\right] = 0.$$

In particular, if $\mathbf{v}_{\mathrm{f}} = \mathbf{u}^*$, then

$$[P] = 0. \tag{14}$$

This result agrees with our physical intuition that if the fluid does not flow from one side to the other side through the pores then the pressures in the pores on both sides must be equal. In [2], it has been shown by virtue of the jump condition (14) in equilibrium, that the fluid pressure measured from the manometer tube attached to the porous body is the pore fluid pressure.

3.2 Equations of Motion

The equations of motion for the fluid and the solid constituents in porous media can be written as

$$\phi d_{\mathrm{f}}\dot{\mathbf{v}}_{\mathrm{f}} = \operatorname{div} T_{\mathrm{f}} + \mathbf{m} + \phi d_{\mathrm{f}}\mathbf{g},$$
$$(1-\phi)d_{\mathrm{s}}\dot{\mathbf{v}}_{\mathrm{s}} = \operatorname{div} T_{\mathrm{s}} - \mathbf{m} + (1-\phi)d_{\mathrm{s}}\mathbf{g},$$

where $\dot{\mathbf{v}}_{\mathrm{f}}$ and $\dot{\mathbf{v}}_{\mathrm{s}}$ are the accelerations of the fluid and the solid respectively, and the external body force is the gravitational force $\mathbf{g}$.

Let us write the stresses in the following form,

$$T_{\mathrm{f}} = -\phi PI + \overline{T}_{\mathrm{f}},$$
$$T_{\mathrm{s}} = -(1-\phi)PI + \overline{T}_{\mathrm{s}}. \tag{15}$$

We call $\overline{T}_{\mathrm{f}}$ the *extra fluid stress* and $\overline{T}_{\mathrm{s}}$ the *effective solid stress*, since it reduces to the effective stress widely used in soil mechanics as we shall see later.

The equations of motion then become

$$\phi d_{\mathrm{f}}\dot{\mathbf{v}}_{\mathrm{f}} = -\phi\operatorname{grad}P - P\operatorname{grad}\phi + \operatorname{div}\overline{T}_{\mathrm{f}} + \mathbf{m} + \phi d_{\mathrm{f}}\mathbf{g},$$
$$(1-\phi)d_{\mathrm{s}}\dot{\mathbf{v}}_{\mathrm{s}} = -(1-\phi)\operatorname{grad}P + P\operatorname{grad}\phi + \operatorname{div}\overline{T}_{\mathrm{s}} - \mathbf{m} + (1-\phi)d_{\mathrm{s}}\mathbf{g}. \tag{16}$$

We obtain from (9), the interactive force $\mathbf{m}$ in equilibrium,

$$\mathbf{m}^0 = P\text{grad}\,\phi + \phi\left(\frac{P}{d_f} - d_f\frac{\partial\widehat{\psi_f}}{\partial d_f}\right)\text{grad}\,d_f, \tag{17}$$

by the use of (12) and (13). We shall call the non-equilibrium part of the interactive force

$$\mathbf{r} = \mathbf{m} - \mathbf{m}_0 \tag{18}$$

the *resistive force*, since it is the force against the flow of the fluid through the medium. We may also call $-\mathbf{r}$ the *drag force* acting upon the solid constituent.

3.3   Linear Theory and Darcy's Law

If we further assume that $|\text{grad}\,\rho_f|$, $|\text{grad}\,F_s|$, $|\text{grad}\,\theta|$, and $|V|$ are small quantities, then from (6), we have

$$\overline{T}_f = \rho_f\frac{\partial\psi_f}{\partial V}\otimes V \approx o(2)$$

is a second order quantity because the free energy of fluid constituent must be a scalar-valued isotropic function. Moreover, since the resistive force vanishes in equilibrium, i.e., $\text{grad}\,\theta = 0$ and $V = 0$, we can write

$$\mathbf{r} = -\phi\,RV - \phi\,G\text{grad}\,\theta + o(2), \tag{19}$$

where R and G are material parameters, which are tensor-valued functions of (θ,ϕ,d_f,F_s).

Therefore, by the use of (17) through (19), the equations of motion for the fluid constituent $(16)_1$ in the linear theory becomes

$$d_f\dot{\mathbf{v}}_f = -\text{grad}\,P - \mathbf{k} + d_f\mathbf{g}, \tag{20}$$

where

$$\mathbf{k} = RV + G\text{grad}\,\theta - \left(\frac{P}{d_f} - d_f\frac{\partial\widehat{\psi_f}}{\partial d_f}\right)\text{grad}\,d_f. \tag{21}$$

The equation of motion for the fluid constituent (20) is a generalization of *Darcy's law*. It is shown in [2], that for the case of classical Darcy's experiment (see [3]), the equation reduces to the original form of the Darcy's law, where the reciprocal of the resistivity constant R is call the *permeability tensor*.

Similarly, the equation $(16)_2$ becomes

$$(1-\phi)d_s\dot{\mathbf{v}}_s = -(1-\phi)\text{grad}\,P + \text{div}\,\overline{T}_s + \phi\mathbf{k} + (1-\phi)d_s\mathbf{g}, \tag{22}$$

which is the equation of motion of the solid constituent.

We can obtain an interesting equation for the solid constituent if we multiply the equation (20) by $-(1-\phi)$ and add it to the equation (22),

$$(1-\phi)(d_s\dot{\mathbf{v}}_s - d_f\dot{\mathbf{v}}_f) = \operatorname{div}\overline{T}_s + \mathbf{k} + (1-\phi)(d_s - d_f)\mathbf{g}. \tag{23}$$

In quasi-static case when the accelerations can be neglected, the two equations (20) and (23) for the fluid and the solid constituents can be written as

$$\begin{aligned}\operatorname{grad}P + \mathbf{k} &= d_f\mathbf{g},\\ -\operatorname{div}\overline{T}_s - \mathbf{k} &= (1-\phi)(d_s - d_f)\mathbf{g}.\end{aligned} \tag{24}$$

Note that the term $(1-\phi)(d_s - d_f)\mathbf{g}$, which represents the difference between the gravitational force of the solid and the fluid in the volume fraction of the solid, can be regarded as the *buoyancy force* on the solid constituent.

Furthermore, adding the two equations in (24) together, we obtain

$$-\operatorname{div}T = \rho\mathbf{g}. \tag{25}$$

Clearly this is the equilibrium equation for the mixture as a whole, where from (15)

$$T = T_f + T_s = -PI + \overline{T}_s + o(2), \qquad \rho = \phi\rho_f + (1-\phi)\rho_s, \tag{26}$$

are the total stress and the total mass density of the mixture respectively.

4 Incompressible Porous Media

We shall consider incompressible porous media, i.e., for which the true mass densities d_f and d_s are constants. For a linear theory, we have

$$\begin{aligned}T_f &= -\phi PI,\\ T_s &= -(1-\phi)PI + \overline{T}_s,\\ \mathbf{k} &= RV + G\operatorname{grad}\theta.\end{aligned} \tag{27}$$

where the pore pressure P is constitutively indeterminate and the effective solid stress is given by the constitutive relation, $\overline{T}_s = \overline{T}_s(\phi, F_s)$.

From (24), the equilibrium equations become

$$\begin{aligned}\operatorname{grad}P &= d_f\mathbf{g},\\ -\operatorname{div}\overline{T}_s &= (1-\phi)(d_s - d_f)\mathbf{g}.\end{aligned} \tag{28}$$

These two equations can be solved separately. Suppose that the x-coordinate is in the vertical downward direction, $\mathbf{g} = g\mathbf{e}_x$. Then the first equation can be integrated immediately to give

$$P = d_f gx + P_0, \qquad P_0 = P(0). \tag{29}$$

This result asserts that the equilibrium pore pressure is the hydrostatic pressure. It agrees with the observation in soil mechanics from experimental measurements that the manometer pressure in the soil is the pressure as if the medium were bulk fluid, unaffected by the presence of the solid constituent in the medium.

We remark that this result is sometimes overlooked in the mixture theory of porous media. It is mainly due to the fact that in the theory of *simple mixture*, which omits the second gradients of deformations as independent constitutive variables, the equilibrium interactive force $\mathbf{m}^0$ is identically zero by constitutive hypothesis. However from the relation (17), $\mathbf{m}_0$ is not a negligible quantity for a body with non-uniform porosity, $\mathbf{m}_0 = P\mathrm{grad}\,\phi$, and it is easy to see that in the absence of this term, the result (29) need not follow. This remark, we shall regard as a strong evidence that porous media must be treated as *non-simple mixtures* even for a linear equilibrium theory.

4.1 Effective Stress Principle

Now let us turn to the equation $(28)_2$ for the solid constituent, or equivalently the equilibrium equation for the whole mixture (25). After this equation is solved for the total stress T, the effective stress can be obtained from (26),

$$\overline{T}_s = T + PI,$$

where the pore pressure P is given by (29). This is the *effective stress principle* in soil mechanics first introduced by Terzaghi (see [4]).

4.2 An Equilibrium Solution

As an application of the equilibrium equation of the solid constituent to determine the porosity distribution in saturated soil, we shall consider a simple constitutive model that

$$\overline{T}_s = -\pi(\phi)I, \tag{30}$$

where π will be called the effective solid pressure. We have from $(28)_2$,

$$\mathrm{grad}\,\pi = (1-\phi)((d_s - d_f)\mathbf{g}.$$

First we note that if the effective pressure is constant, then either $\phi = 1$ or $d_s = d_f$. Both are uninteresting cases.

We shall further assume the constitutive relation for the effective pressure is given by a power law:

$$\pi(\phi) = b(1-\phi)^a, \tag{31}$$

where a and b are non-zero material constants. Then we have

$$-ab(1-\phi)^{a-2}\frac{d\phi}{dx} = (d_s - d_f)g.$$

For $a \neq 1$, integration of the equation leads to

$$(1-\phi)^{a-1} = \frac{a-1}{ab}(d_s - d_f)gx + (1-\phi_0)^{a-1}, \tag{32}$$

while for $a = 1$, it gives

$$(1 - \phi) = (1 - \phi_0) \exp\left(\frac{1}{b}(d_s - d_f)gx\right), \tag{33}$$

where $\phi_0 = \phi(0)$.

If experimental measurements of the porosity ϕ versus x in the soil are taken, then one can compare them with the solution (32) or (33) for the determination of the material constants a and b in the power law (31). It offers a check for the applicability of the present theory based on such a simple model.

References

1. Bowen, R.M.: The theory of Mixtures. In: Eringen, A.C. (ed.) Continuum Mechanics, vol. III. Academic Press, London (1976)
2. Liu, I.-S.: On chemical potential and incompressible porous media. Journal de Mécanique 19, 327–342 (1980)
3. Scheidegger, A.E.: Hydrodynamics in Porous Media. In: Flügge (ed.) Handbuch der Physik, vol. VIII/2. Springer, Heidelberg (1963)
4. Terzaghi, K.: Theoretical Soil Mechanics. Chapman & Hall, London (1951)
5. Truesdell, C.: Sulle basi della termomeccanica. Academia Nazionale dei Lincei, Rendiconti della Classe di Scienze Fisiche, Matematiche e Naturali 22(8), 33–38, 158–166 (1957)
6. Truesdell, C.: Rational Thermodynamics, 2nd edn. Springer, New York (1984)

An Extrapolation of Thermodynamics to Evolutionary Genetics

Ingo Müller

Dedicated to Krzyzstof Wilmanski on the occasion of his 70th birthday.

Abstract. Evolutionary genetics requires both: The randomness of mutations and the determination of selection. In that sense genetics is analogous to thermodynamics which requires the randomness of thermal motion and the purpose of energetic interaction. The analogy can be made explicit by defining an entropy of a population and a selective energy of interaction of the population with the environment, or a breeder.

1 Introduction

The extrapolation of thermodynamic ideas to other fields of knowledge – economy, sociology, genetics – is usually not appreciated; neither by thermodynamicists nor by the adherents of those fields *remote* from thermodynamics. Thermodynamicists deplore the simplicity of the mathematics that is involved while, on the other hand, the sociologists (say) complain about the use of mathematics – such as it is –, because according to them no mathematics is permissible that goes beyond the four basic elements of calculus, *e.g.* addition and multiplication.

And yet, in most fields of knowledge – upon reflection – we can detect a random influence on events that competes with the strife toward a definite and desirable goal, and prevents the full attainment of that goal. That competition between opposing tendencies was first identified in thermodynamics and it led to the doctrine of entropy and energy: Randomization under the influence of the thermal motion tends to increase the entropy, and energy tends to decrease, because it attracts the elements of a body toward the depth of definite potential wells. That theme dictates – again and again – physical properties and it has been identified for many circumstances in elasticity, chemistry, metallurgy, and meteorology.

Extrapolation of that competition into other fields requires some thought, and some insight. It is not a trivial exercise of *turning the crank*. In the present paper a

Ingo Müller
Technische Universität Berlin
e-mail: `ingo.mueller@alumni.tu-berlin.de`

case is made for the idea that evolutionary genetics may be understood and formulated in terms of entropy and energy.

Indeed, mutation and selection may act as opposing tendencies in the development of a population. Mutation is random so that – in the simplest case – it tends to be impartial to different genotypes in the population in the sense that they all have an equal chance of appearing. The intensity of mutation may be controlled – accelerated or slowed down – by radiation. Selection on the other hand is determined by the extant environment – possibly shaped by a breeder – and may prefer one phenotype in a population over others in the sense that the *fittest* has a more numerous progeny. In a manner of speaking mutation and selection *compete* and lead the population to a distribution of genotypes to which both tendencies contribute: Selection more when the intensity of mutation is weak, and mutation more, when its intensity is strong. Thus mutational *chance* may be balanced by selective *order*.

The situation is not unlike one frequently encountered in thermodynamics of a gas (say), where entropy and energy are often opposing each other in determining the shape of the gas. Entropy S measures the effect of the random thermal motion whose intensity is controlled by temperature T; therefore entropy is biggest for a homogeneous distribution of the atoms in the gas. The energy E on the other hand depends on the environmental situation, *e.g.* a gravitational field which pulls the atoms toward the state of minimum energy. A compromise of the two tendencies is found when the minimum of free energy $F = E - TS$ is reached. Thus the temperature determines to what extent energetic *order* dominates entropic *chance*. The theme was exploited and illustrated for many circumstances in [1].

The perceived analogy of thermodynamics and genetics was expounded in my paper [2] for simple cases of haploid and diploid populations. The present paper reviews the previous arguments – albeit only for haploids – and it emphasises the observation that evolutionary genetics can be interpreted in terms of the notions governing the thermodynamics of mixtures of reacting fluid constituents; that aspect was only touched on in [2] at the very end.

Mixtures of fluids may be seen as fluids with an internal structure, which is represented by the concentrations of the constituents. That interpretation is not a common one but it is all I have to offer for arguing that my contribution fits – loosely – into the scheme of the articles in this book.

2 Selective Free Energy of a Haploid Population

2.1 Model Population: Number of Realizations and Entropy

We consider a population of $2N$ haploid[1] cells. The cells are equal in all aspects except on one *locus* of one chromosome, which may carry alleles A or a. The A-cells and the a-cells are supposed to have the same molecular energy.

---

[1] Haploid cells are those with a single set of chromosomes. In [2] I have discussed haploid *and* diploid cells. The latter contain pairs of chromosomes.

When mutation occurs it is assumed to be of the type $A \leftrightarrow a$. Cells with alleles a may be selectively privileged by the environment. But, if there is no selective bias, a and A occur with equal probability, because they are energetically equal. The number of cells is characterized by N_A and N_a, respectively, – with $N_A + N_a = 2N$ – or by the relative frequencies

$$p = \frac{N_A}{2N} \qquad q = \frac{N_a}{2N}, \tag{1}$$

whichever is more convenient.

2.2   Entropy

By the rules of combinatorics the number R of realizations of a distribution $\{N_A, N_a\}$ is given by

$$\text{number of realizations } R = \frac{(2N)!}{N_A! N_a!}. \tag{2}$$

And by the rules of statistical thermodynamics the entropy of the population is defined by

$$\text{entropy } S = \ln R = \ln \frac{(2N)!}{N_A! N_a!}. \tag{3}$$

2.3   Mutation without Selection. Maximum Entropy

We assume that N_A and N_a are big enough that the Stirling formula can be applied and obtain from (3)

$$S = -N_A \ln N_A - N_a \ln N_a \quad \text{or, by } N_A + N_a = 2N \text{ and } (1)_2$$
$$S(q) = -2N [q \ln q + (1-q) \ln(1-q)]. \tag{4}$$

The statistical definition of entropy implies the expected growth-property of entropy by a simple argument. Basic is the reasonable assumption that all possible *realizations* of the distribution $\{N_A, N_a\}$ occur equally frequently in the course of the random mutational activity in the population. In all likelihood this means that a *distribution* with many realizations occurs more often than a distribution $\{N_A, N_a\}$ with few realizations. So, when we start out with a distribution $\{N_A, N_a\}$ with few realizations the mutation will *probably* lead to distributions with more realizations and eventually to equilibrium, the distribution with most realizations. By (2) through (4) this means that R and S will grow to a maximum. The speed of this growth is small in a physical sense, because time in genetics is measured in generations.

Thus the growth of entropy is purely probabilistic. It is entirely possible that the growth-process may be interrupted occasionally by a decrease of R and S and even in equilibrium there may be fluctuations away from the maxima of R and S.

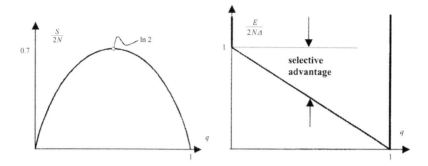

Fig. 1 a. Entropy as a function of q, b. Selective energy as a function of q.

$S(q)$ is plotted in Fig. 1a and inspection shows that equilibrium, – the maximum – occurs at $q = 1/2$ so that mutation without selection leads to an equi-distribution of A- and a-cells.

2.4 Selection without Mutation. Minimum of Selective Energy

It may happen that a-cells enjoy a selective bonus Δ. The bonus may consist of being able to find more adequate food and therefore have a more numerous progeny than the A-cells. In the end, – over many generations –, this may lead to dominance of the a-cells and the elimination of A-cells.

We consider the selective advantage to be given by the formula

$$E(q) = N_A\Delta = (2N - N_a)\Delta = 2N(1-q)\Delta. \tag{5}$$

It may be represented by a "potential well" of the type shown in Fig. 1b. The deepest point, at $q = 1$, corresponds to the selectively most preferred state and the highest point, at $q = 0$, occurs for the least preferred state. In analogy to physics we call E the selective energy of the population; its shortfall from $E(0) = 2N\Delta$ represents the selective advantage of the population, $cf.$ Fig. 2b.

> Thus we imagine that the environment, or the breeder, offers a *potential well* for the population. In the biological literature, *e.g.* see [3], the selectively preferred states are usually visualized as *peaks*. This is entirely equivalent, but we choose wells – instead of peaks – in order to emphasize the analogy with physics.

2.5 Mutation and Selection Together

When mutation and selection occur at the same time we do not expect the equi-distribution with maximum entropy to occur as the final state. Nor do we expect the selective energy to reach its minimum for $N_a = 2N$. The population will have to

Fig. 2 Selective free energies (bold curves). Their minima determine the values of q in constrained equilibrium for three given mutational intensities T.

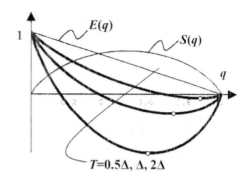

$T=0.5\Delta, \Delta, 2\Delta$

strike a compromise between the growth of entropy to a maximum and the decrease of selective energy to a minimum.

Suppose we start at some value of q close to one. Such a distribution has a small number of realizations and the stochastic tendency for the number of realizations to grow under the mutational activity will lead the population to smaller values of q. On the other hand, according to (5) and Fig. 1b, the selective energy is constrained to grow with decreasing q, so that the tendency of that energy to approach a minimum hampers the growth of entropy. And while the slope of $S(q)$ increases from $-\infty$ to 0 as q becomes smaller, the growth tendency of entropy becomes weaker. The slope of $E(q)$ on the other hand remains constant as q decreases.[2] It follows that eventually a given mutational intensity is no longer capable of pulling $E(q)$ upwards. If that happens, $S(q)$ – or the number R of realizations – has reached the maximum that is compatible with the constraint $E(q)$. This is the state of *constrained equilibrium*.

In analysis the most elegant manner to find a maximum under a constraint employs a Lagrange multiplier, here β. We have to determine

$$\text{the maximum of the expression } S_E(q) \equiv S(q) - \beta E(q). \tag{6}$$

Thus the number of realizations in constrained equilibrium equals

$$R_E(q) = \exp\left(\beta S_E(q)\right). \tag{7}$$

The Lagrange multiplier β is a measure for the mutational activity. Because, indeed, if β is small, the selective energy has little effect on $S_E(q)$ and the maximum of (6) occurs, because $S(q)$ reaches its unconstrained maximum, the equi-distribution for $N_a = N$, as if there were no selection. This means that a small β amounts to a strong mutational activity which overpowers selection. If on the other hand, β is large, the selective energy $E(q)$ dominates in $S_E(q)$ according to (6) and $S_E(q)$ becomes maximal because $E(q)$ is close to its minimum as if there were no mutation. Thus a large β implies a small mutational activity.

---

[2] Non-linear functions $E(q)$ with a non-constant slope will be considered later. Their effect may be drastic.

2.6   Mutational Intensity and Selective "Free Energy"

The situation just described is typical for many circumstances in statistical thermodynamics, although the jargon is different. In statistical thermodynamics we introduce a temperature

$$T = \frac{1}{\beta} \tag{8}$$

and – instead of maximizing $S_E = S - \beta E$ – we look for

$$\text{the minimum of the } \textit{free energy } F = E - TS. \tag{9}$$

We extrapolate the notions of temperature and free energy to the present case of
genetics of a population. The temperature thus corresponds to the *mutational intensity*. If T is small, E becomes a minimum and selection dominates with a target
state at $N_a \lesssim 2N$, and if T is large, S becomes a maximum and mutation dominates as N_a tends toward $N_a \gtrsim 2N$. For intermediate values of T we may most
easily determine the constrained equilibrium as given by *the* value of N_a that corresponds to the abscissa q_{min} of the minimum of the graph for the *selective free energy*
$F(q) = E(q) - TS(q)$ as shown in Fig. 2.

The requirements (6) and (9) are equivalent. The preference for (9) in thermodynamics is entirely traditional; we follow it here, because we wish to emphasize the
analogy of the reasoning in thermodynamics and genetics.

For the present case with a linear function $E(q)$, *cf.* (5), q_{min} may be calculated
analytically but for more complex selective energies this is impossible and we may
have to find the site of the minimum of the selective free energy graphically.

2.7   Other Forms of the Selective Energy

The "potential well" of the type shown in Fig. 1b is based on the assumption that
the privileged a-cell – privileged by selection – enjoys a bonus Δ, and that the bonus
for the population enjoys a selective advantage which is proportional to the number
of a-cells. This is a plausible ansatz and it may often be true.

However, more complex alternatives may be discussed. Thus both the a-cells
and the A-cells may be privileged, *if they are rare*. Thus, according to Haldane [4]
parasites will evolve so as to attack most effectively the *common* cells of the host
population; therefore the *rare* host cells will enjoy a selective advantage. Similarly
hosts will acquire immunity to the *common* type of parasites, so that the *rare* types of
parasites will be at an advantage. In such cases the potential well may be represented
graphically by a *double well*, *cf.* Fig. 3.

The function $E(q)$ may then be written in the form

$$E(q) = 2N\Delta q(1-q). \tag{10}$$

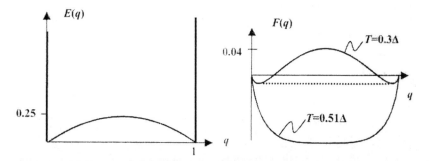

Fig. 3 a. Double well potential for q-dependent bonus, b. Free energies for $T = 0.3\Delta$ and $T = 0.51\Delta$.

Obviously in this case the *bonus* for the formation of a newly formed a-cell is given by $\Delta(1 - 2q)$; it depends on the relative frequency q. Actually it develops into a malus for $q < 1/2$. In that range the bonus is for A-cells.

The selective free energy appropriate for (10) reads

$$\frac{F(q)}{2N\Delta} = q(1-q) + \frac{T}{\Delta}\left[q\ln q + (1-q)\ln(1-q)\right]. \tag{11}$$

A simple calculation shows that for $T < 0.5\Delta$ this function has *two* minima, *cf.* Fig. 3b which shows the graphs of $F(q)$ for $T = 0.3\Delta$ and $T = 0.51\Delta$.

The interpretation is obvious: For the same (small) T the population may be rich in a-cells or rich in A-cells. Or more interestingly: Part of the population may be a-rich and the rest may be rich in A. Indeed, the selective free energy may be minimal on the convexification of the graph $F(q)$ – the horizontal dashed *common tangent* of the two convex parts of $F(q)$ – by splitting into parts, called *phases* in analogy to physics. The fractions of the cells in the two phases are x and $(1 - x)$ so that we have

$$q = (1-x)q^{L}\big|_{min} + xq^{R}\big|_{min}, \tag{12}$$

where $q^{L,R}\big|_{min}$ are the abscissae of the left and right minimum respectively. Thus a population may be meta-stable in a large range of q by consisting of two phases with different values of q in different proportions.

For $T = 0.5\Delta$ the two lateral minima coalesce and for $T > 0.5\Delta$ there is only a single minimum at $q_{min} = 1/2$. Inspection of Fig. 3b shows that for $T \gtrsim 0.5\Delta$ the minimum is flat and broad. The significance of the flatness of the selective free energy is discussed in [2]. It makes for large fluctuations.

An analogous situation occurs in a physical system which may decompose into two *phases*, a vapour and a liquid (say). Those phases disappear at a higher temperature which in thermodynamics is known as the *critical temperature*.

3 Analogy to Thermodynamics of Binary Chemically Reacting Mixtures

3.1 Chemical Potentials and Law of Mass Action

From the considerations of Section 2 above we conclude that the selective free energy of the population may be written in the form

$$F = E(N_A, N_a, \Delta) + T(N_A \ln N_A + N_a \ln N_a). \tag{13}$$

Formally this is identical to the free energy function of a binary mixture with constituents A and a and the selective energy $E(N_A, N_a, \Delta)$. We do not specify the form of $E(N_A, N_a, \Delta)$ at this stage, because it may take different forms as explained in Section 2, cf. (5), (10). In both cases it depends on a single variable, – apart from N_A, N_a –, the selective bonus Δ. Obviously the selective free energy F depends on T, Δ, N_A, N_a. Its total differential reads

$$dF = \underbrace{\left(N_A \ln \frac{N_A}{2N} + N_a \ln \frac{N_a}{2N} \right)}_{-S} dT + \frac{\partial E}{\partial \Delta} d\Delta$$
$$+ \underbrace{\left(\frac{\partial E}{\partial N_A} + T \ln \frac{N_A}{2N} \right)}_{\mu_A} dN_A + \underbrace{\left(\frac{\partial E}{\partial N_a} + T \ln \frac{N_a}{2N} \right)}_{\mu_a} dN_a. \tag{14}$$

The braces identify S as before – cf. (4) – and the chemical potentials μ_A and μ_a, used in the theory of mixtures to represent the effect of the changes of particle numbers of the constituents on the free energy. That notion is carried over now to the field of genetics.

3.2 "Chemical Equilibrium"

If the constraint $N_A + N_a = N$ is taken into account, (14) reads

$$dF = \underbrace{\left(N_A \ln \frac{N_A}{2N} + N_a \ln \frac{N_a}{2N} \right)}_{-S} dT + \frac{\partial E}{\partial \Delta} d\Delta + \underbrace{\left(\frac{\partial E}{\partial N_a} - \frac{\partial E}{\partial N_A} + T \ln \frac{N_a}{N_A} \right)}_{\mu_a - \mu_A} dN_a. \tag{15}$$

The constrained equilibrium discussed in Section 2 corresponds to the chemical equilibrium in the thermodynamic mixture theory. It is characterized by $\frac{\partial F}{\partial N_a} = 0$ – the law of mass action – and in the present case of the simple stoichiometric formula $A \leftrightarrow a$ that law, by (15), obviously requires the equality of the chemical potentials

$$\mu_A = \mu_a. \tag{16}$$

Fig. 4 Determination of the
constrained equilibria for
the selective energy (10).
The graphs represent the
two contributions to $\mu_A - \mu_a$
in (21) for two values of
T/Δ.

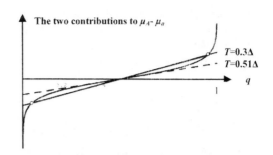

3.3 The Selective Energy Function (5)

Insertion of (5) into the definitions of μ_A and μ_a, the "chemical potentials" provides

$$\mu_A - \mu_a = -\Delta + \ln\frac{N_a}{N_A}. \tag{17}$$

They are equal for the constrained – or chemical – equilibrium and we obtain

$$N_A^{ce} = 2N\frac{1}{1+\exp\left(\frac{\Delta}{T}\right)} \quad \text{and} \quad N_a^{ce} = 2N\frac{\exp\left(\frac{\Delta}{T}\right)}{1+\exp\left(\frac{\Delta}{T}\right)}. \tag{18}$$

These equations represent the analytical forms of the numbers N_A and N_a appropriate to the minima of the selective free energy shown in Fig. 2.

3.4 The Selective Energy Function (10)

The previously considered case is the only one in which the equilibrium values of N_A and N_a can be expressed analytically. Thus for the selective energy (10), or

$$E = \frac{N_A N_a}{2N}\Delta, \tag{19}$$

the "chemical potentials" read

$$\mu_A = \frac{N_a}{2N}\Delta + T\ln\frac{N_A}{2N} \quad \text{and} \quad \mu_a = \frac{N_A}{2N}\Delta + T\ln\frac{N_a}{2N}. \tag{20}$$

and the equilibrium condition [16] requires

$$\mu_a - \mu_A = \frac{N_A - N_a}{2N}\Delta + T\ln\frac{N_a}{N_A} = 0 \tag{21}$$

which is impossible to solve for N_a (say) analytically. However, numerical or graphical methods are available. In the present case we refer to Fig. 4 which shows the functions $\ln\frac{q}{1-q}$ and $\frac{\Delta}{T}(1-2q)$, the latter for different values of the parameter Δ/T.

The abscissae of the lateral points of intersection – when they exist – are the equilibrium values q_{min}. They correspond, of course, to the sites of the minima in Fig. 3b. The central intersection corresponds to the maximum of the selective free energy and is unimportant for the discussion of stable equilibria.

References

1. Müller, I., Weiss, W.: Entropy and Energy – A Universal Competition. Springer, Heidelberg (2005)
2. Müller, I.: Thermodynamics and Evolutionary Genetics. Cont. Mech. & Thermodyn. (in press)
3. Dawkins, R.: Climbing mount improbable. Penguin Books (1997)
4. Haldane, J.B.S.: Disease and evolution. La Ricerca Scientifica, Suppl. A (1949)

Some Recent Results on Multi-temperature Mixture of Fluids

Tommaso Ruggeri

Abstract. We present a survey on some recent results concerning the different models of a mixture of compressible fluids. In particular we discuss the most realistic case of a mixture when each constituent has its own temperature (MT) and we first compare the solutions of this model with the one whit a unique common temperature (ST). In the case of Eulerian fluids it will be shown that the corresponding (ST) differential system is a *principal subsystem* of the (MT) one. Global behavior of smooth solutions for large time for both systems will also be discussed through the application of the Shizuta-Kawashima condition. Than we introduce the concept of the average temperature of mixture based upon the consideration that the internal energy of the mixture is the same as in the case of a single-temperature mixture. As a consequence, it is shown that the entropy of the mixture reaches a local maximum in equilibrium. Through the procedure of Maxwellian iteration a new constitutive equation for non-equilibrium temperatures of constituents is obtained in a classical limit, together with the Fick's law for the diffusion flux. To justify the Maxwellian iteration, we present for dissipative fluids a possible approach of a classical theory of mixture with multi-temperature and we prove that the differences of temperatures between the constituents imply the existence of a new *dynamical pressure* even if the fluids have a zero bulk viscosity. in the case of the one-dimensional steady heat conduction between two walls, we have verified that the main effect of multi-temperature is that the average temperature is not a linear function of the distance as in the case of the ST theory.

1 Mixtures in Rational Thermodynamics

The theory of homogeneous mixtures was developed, within the framework of rational thermodynamics by Truesdell [1], under the assumption that each constituent

Tommaso Ruggeri
Department of Mathematics and Research Center of Applied Mathematics (C.I.R.A.M.)
University of Bologna, Via Saragozza 8, 40123 Bologna, Italy
e-mail: ruggeri@ciram.unibo.it

obeys the same balance laws as a single fluid. A huge amount of literature appeared after that in the context of continuum approach, see e.g. [2, 3, 4, 5, 6, 7, 8]. The system of governing equations read:

$$
\begin{cases}
\dfrac{\partial \rho_\alpha}{\partial t} + \mathrm{div}\,(\rho_\alpha \mathbf{v}_\alpha) = \tau_\alpha, \\[2mm]
\dfrac{\partial (\rho_\alpha \mathbf{v}_\alpha)}{\partial t} + \mathrm{div}\,(\rho_\alpha \mathbf{v}_\alpha \otimes \mathbf{v}_\alpha - \mathbf{t}_\alpha) = \mathbf{m}_\alpha, \qquad (\alpha = 1,2,\ldots,n) \\[2mm]
\dfrac{\partial \left(\frac{1}{2}\rho_\alpha v_\alpha^2 + \rho_\alpha \varepsilon_\alpha \right)}{\partial t} \\[2mm]
\qquad + \mathrm{div}\, \left\{ \left(\tfrac{1}{2}\rho_\alpha v_\alpha^2 + \rho_\alpha \varepsilon_\alpha \right) \mathbf{v}_\alpha - \mathbf{t}_\alpha \mathbf{v}_\alpha + \mathbf{q}_\alpha \right\} = e_\alpha.
\end{cases}
\tag{1}
$$

On the left hand side, ρ_α is the density, $\mathbf{v}_\alpha$ is the velocity, ε_α is the internal energy, $\mathbf{q}_\alpha$ is the heat flux and $\mathbf{t}_\alpha$ is the stress tensor of the constituent α. The stress tensor $\mathbf{t}_\alpha$ can be decomposed into a pressure part $-p_\alpha \mathbf{I}$ and a viscous part $\boldsymbol{\sigma}_\alpha$ as

$$
\mathbf{t}_\alpha = -p_\alpha \mathbf{I} + \boldsymbol{\sigma}_\alpha.
$$

On the right hand sides τ_α, $\mathbf{m}_\alpha$ and e_α represent the production terms related to the interactions between constituents. Due to the total conservation of mass, momentum and energy of the mixture, the sum of production terms over all constituents must vanish

$$
\sum_{\alpha=1}^{n} \tau_\alpha = 0, \quad \sum_{\alpha=1}^{n} \mathbf{m}_\alpha = \mathbf{0}, \quad \sum_{\alpha=1}^{n} e_\alpha = 0.
$$

Mixture quantities $\rho, \mathbf{v}, \varepsilon, \mathbf{t}$ and $\mathbf{q}$ are defined as

$$
\rho = \sum_{\alpha=1}^{n} \rho_\alpha \qquad\qquad \text{total mass density,}
$$

$$
\mathbf{v} = \frac{1}{\rho} \sum_{\alpha=1}^{n} \rho_\alpha \mathbf{v}_\alpha \qquad\qquad \text{mixture velocity,}
$$

$$
\varepsilon = \varepsilon_I + \frac{1}{2\rho} \sum_{\alpha=1}^{n} \rho_\alpha u_\alpha^2 \qquad\qquad \text{internal energy,}
$$

$$
\left(\varepsilon_I = \tfrac{1}{\rho} \sum_{\alpha=1}^{n} \rho_\alpha \varepsilon_\alpha \right)
\tag{2}
$$

$$
\mathbf{t} = -p\mathbf{I} + \boldsymbol{\sigma}_I - \sum_{\alpha=1}^{n} (\rho_\alpha \mathbf{u}_\alpha \otimes \mathbf{u}_\alpha) \qquad \text{stress tensor,}
$$

$$
\mathbf{q} = \mathbf{q}_I + \sum_{\alpha=1}^{n} \rho_\alpha \left(\varepsilon_\alpha + \frac{p_\alpha}{\rho_\alpha} + \frac{1}{2} u_\alpha^2 \right) \mathbf{u}_\alpha \quad \text{flux of internal energy,}
$$

$$
(2)_{\text{cont.}}
$$

where $\mathbf{u}_\alpha = \mathbf{v}_\alpha - \mathbf{v}$ is the diffusion velocity of the component α, $p = \sum_{\alpha=1}^{n} p_\alpha$ is the total pressure, ε_I is the total intrinsic internal energy, $\mathbf{q}_I = \sum_{\alpha=1}^{n} \mathbf{q}_\alpha$ is the total intrinsic heat flux and $\boldsymbol{\sigma}_I = \sum_{\alpha=1}^{n} \boldsymbol{\sigma}_\alpha$ is the total intrinsic shear stress. By summation of equations (1), the conservation laws of mass, momentum and energy for the whole mixture appear in the same form as a single fluid. In such a way the balance laws for one constituent, say n, may be replaced by the conservation laws, leading to the following system of governing equations:

$$
\begin{cases}
\dfrac{\partial \rho}{\partial t} + \mathrm{div}\,(\rho \mathbf{v}) = 0, \\[2mm]
\dfrac{\partial (\rho \mathbf{v})}{\partial t} + \mathrm{div}\,(\rho \mathbf{v} \otimes \mathbf{v} - \mathbf{t}) = \mathbf{0}, \\[2mm]
\dfrac{\partial \left(\frac{1}{2}\rho v^2 + \rho \varepsilon\right)}{\partial t} \\[2mm]
\qquad + \mathrm{div}\,\left\{ \left(\tfrac{1}{2}\rho v^2 + \rho \varepsilon\right)\mathbf{v} - \mathbf{t}\mathbf{v} + \mathbf{q} \right\} = 0, \\[2mm]
\dfrac{\partial \rho_b}{\partial t} + \mathrm{div}\,(\rho_b \mathbf{v}_b) = \tau_b, \qquad (b = 1,\dots,n-1) \\[2mm]
\dfrac{\partial (\rho_b \mathbf{v}_b)}{\partial t} + \mathrm{div}\,(\rho_b \mathbf{v}_b \otimes \mathbf{v}_b - \mathbf{t}_b) = \mathbf{m}_b, \\[2mm]
\dfrac{\partial \left(\frac{1}{2}\rho_b v_b^2 + \rho_b \varepsilon_b\right)}{\partial t} \\[2mm]
\qquad + \mathrm{div}\,\left\{ \left(\tfrac{1}{2}\rho_b v_b^2 + \rho_b \varepsilon_b\right)\mathbf{v}_b - \mathbf{t}_b \mathbf{v}_b + \mathbf{q}_b \right\} = e_b,
\end{cases}
\tag{3}
$$

where the index b runs from 1 to $n-1$.

In this multi-temperature model (MT), used in particular in plasma physics [9], we have $5n$ independent field variables ρ_α, $\mathbf{v}_\alpha$ and T_α ($\alpha = 1,2,\dots,n$), where T_α is the temperature of constituent α. To close the system (3) of the field equations of the mixture process, we must write the constitutive equations for the quantities $p_\alpha, \varepsilon_\alpha, \mathbf{q}_\alpha, \boldsymbol{\sigma}_\alpha$ ($\alpha = 1,2,\dots,n$) and τ_b, $\mathbf{m}_b$, e_b ($b = 1,\dots,n-1$) in terms of the field variables $\rho_\alpha, \mathbf{v}_\alpha$ and T_α ($\alpha = 1,2,\dots,n$).

The system (3) is a particular case of the balance law one:

$$
\partial_t \mathbf{F}^0 + \partial_i \mathbf{F}^i = \mathbf{F},
\tag{4}
$$

where $\mathbf{F}^0$ (densities), $\mathbf{F}^i$ (fluxes) and $\mathbf{F}$ (productions) are $N-$vectors functions of the field variables and $\partial_t = \partial/\partial t$, $\partial_i = \partial/\partial x_i$. Principle of relativity requires that field equations are invariant with respect to Galilean transformations. Using a result of Ruggeri [10], it is possible to prove in the case of a mixture, the Galilean invariance dictates the velocity dependence in the source terms [11]:

$$\tau_b = \hat{\tau}_b;$$

$$\mathbf{m}_b = \hat{\tau}_b \mathbf{v} + \hat{\mathbf{m}}_b; \quad (b = 1,\ldots,n-1) \tag{5}$$

$$e_b = \hat{\tau}_b \frac{v^2}{2} + \hat{\mathbf{m}}_b \cdot \mathbf{v} + \hat{e}_b,$$

where $\hat{\tau}_b$, $\hat{\mathbf{m}}_b$ and $\hat{e}_b$ are independent of $\mathbf{v}$.

Another important restriction comes from the entropy inequality, i.e. there exists a supplementary balance law of entropy with an entropy production non negative::

$$\partial_t \rho S + \partial_i (\rho S v^i + \varphi^i) = \Sigma \geq 0,$$

where $\rho S = \sum_{\alpha=1}^{n} \rho_\alpha S_\alpha$, φ^i and Σ are the entropy density, the non-convective entropy flux and the entropy production, respectively. For example in the case of a mixture of Eulerian fluids, the entropy production becomes [11]:

$$\Sigma = \sum_{b=1}^{n-1} \left(\frac{\mu_n - \frac{1}{2}u_n^2}{T_n} - \frac{\mu_b - \frac{1}{2}u_b^2}{T_b} \right) \hat{\tau}_b + \left(\frac{\mathbf{u}_n}{T_n} - \frac{\mathbf{u}_b}{T_b} \right) \cdot \hat{\mathbf{m}}_b + \left(\frac{1}{T_b} - \frac{1}{T_n} \right) \hat{e}_b \geq 0. \tag{6}$$

This inequality allows to obtain the following structure of production terms. The internal parts of production terms (5) are chosen in such a way that the residual inequality (6) is actually a quadratic form. In particular in agreement with kinetic theory, we obtain:

$$\hat{\tau}_b = \sum_{c=1}^{n-1} \varphi_{bc} \left(\frac{\mu_n - \frac{1}{2}u_n^2}{T_n} - \frac{\mu_c - \frac{1}{2}u_c^2}{T_c} \right) + \sum_{c=1}^{n-1} g_{bc} \left(\frac{1}{T_c} - \frac{1}{T_n} \right);$$

$$\hat{\mathbf{m}}_b = \sum_{c=1}^{n-1} \psi_{bc} \left(\frac{\mathbf{u}_n}{T_n} - \frac{\mathbf{u}_c}{T_c} \right); \tag{7}$$

$$\hat{e}_b = \sum_{c=1}^{n-1} \theta_{bc} \left(\frac{1}{T_c} - \frac{1}{T_n} \right) + \sum_{c=1}^{n-1} g_{bc} \left(\frac{\mu_n - \frac{1}{2}u_n^2}{T_n} - \frac{\mu_c - \frac{1}{2}u_c^2}{T_c} \right),$$

where:

$$\mu_\alpha = \varepsilon_\alpha - T_\alpha S_\alpha + \frac{p_\alpha}{\rho_\alpha}$$

$(\alpha = 1,\ldots,n)$, are the chemical potentials of the constituents and

$$\begin{bmatrix} \varphi_{bc} & g_{bc} \\ g_{bc} & \theta_{bc} \end{bmatrix}, \quad \psi_{bc}$$

are phenomenological symmetric positive definite matrices $(b,c = 1,\ldots,n-1)$. In the sequel, our analysis will be restricted to a model of non-reacting mixtures, for which $\tau_b = 0$.

Due to the difficulties to measure the temperature of each component, a common practice among engineers and physicists is to consider only one temperature for the mixture. When we use a single temperature (*ST*), Eq. (3)$_6$ disappears and we get a unique global conservation of the total energy in the form (3)$_3$ (see for example [4]).

A further step of coarsening theory is the classical approach of mixtures (*CT*), in which the independent field variables are the density, the mixture velocity, the individual temperature of the mixture and the concentrations of constituents. In that case system reduces to the equations (3)$_1$-(3)$_4$.

2 Euler Fluids and Comparison between *MT* and *ST* Models

First we consider the case in which all the constituents of the mixture are Eulerian fluids, i.e. neither viscous nor heat-conducting, i.e.:

$$\mathbf{t}_\alpha = -p_\alpha \mathbf{I}, \quad \mathbf{q}_\alpha = 0 \; ; \quad (\alpha = 1,\ldots,n).$$

2.1 Symmetric Hyperbolic System and Principal Subsystems

For a generic hyperbolic system (4), when the entropy density $h^0 = -\rho S$ is a convex function of $\mathbf{u} \equiv \mathbf{F}^0$, it is possible to prove that there exists a privileged set of field variables, the *main field*

$$\mathbf{u}' = \frac{\partial h^0}{\partial \mathbf{u}}$$

such that original balance laws could be transformed in a symmetric form. In fact introducing the four potentials [12], [13]

$$h'^0 = \mathbf{u}' \cdot \mathbf{F}^0 - h^0; \qquad h'^i = \mathbf{u}' \cdot \mathbf{F}^i - h^i \quad (i = 1,2,3)$$

the original system can be put in the special symmetric form:

$$\partial_t \left(\frac{\partial h'^0}{\partial \mathbf{u}'} \right) + \partial_i \left(\frac{\partial h'^i}{\partial \mathbf{u}'} \right) = \mathbf{F}$$

$$\Longleftrightarrow \tag{8}$$

$$\frac{\partial^2 h'^0}{\partial \mathbf{u}' \partial \mathbf{u}'} \partial_t \mathbf{u}' + \frac{\partial^2 h'^i}{\partial \mathbf{u}' \partial \mathbf{u}'} \partial_i \mathbf{u}' = \mathbf{F} \; ; \quad (i = 1,2,3).$$

The Boillat proof [12, 14], has the advantage with respect to the symmetrization of Friedrichs and Lax [15] that the symmetric system is the original one. Moreover this includes, as a particular case, the example discovered first by Godunov for the fluid dynamics case and the Euler-Lagrange systems [16]. Symmetric structure of the system of balance laws is highly desirable due to hyperbolicity and local well-posedness of initial-value problems (see e.g. [17]). The main field components for the mixture of Euler fluids described by the system (3) have the form [11]:

$$\Lambda^\rho = \frac{1}{T_n} \left(\mu_n - \frac{1}{2} (\mathbf{u}_n + \mathbf{v})^2 \right)$$

$$\Lambda^{\mathbf{v}} = \frac{1}{T_n} (\mathbf{u}_n + \mathbf{v})$$

$$\Lambda^\varepsilon = -\frac{1}{T_n} \qquad\qquad (9)$$

$$\Lambda^{\rho_b} = \frac{1}{T_b} \left(\mu_b - \frac{1}{2} (\mathbf{u}_b + \mathbf{v})^2 \right) - \frac{1}{T_n} \left(\mu_n - \frac{1}{2} (\mathbf{u}_n + \mathbf{v})^2 \right)$$

$$\Lambda^{\mathbf{v}_b} = \frac{\mathbf{u}_b}{T_b} - \frac{\mathbf{u}_n}{T_n} - \left(\frac{1}{T_b} - \frac{1}{T_n} \right) \mathbf{v}$$

$$\Lambda^{\varepsilon_b} = \frac{1}{T_n} - \frac{1}{T_b}.$$

The use of the main field has still another advantage: the possibility of recognition of principal and equilibrium subsystems. Let us give a brief review of the results which can be found in [18]. Let us split the main field $\mathbf{u}' \in R^N$ into two parts $\mathbf{u}' \equiv (\mathbf{v}', \mathbf{w}')$, $\mathbf{v}' \in R^M$, $\mathbf{w}' \in R^{N-M}$, $(0 < M < N)$ and the system (8) with $\mathbf{F} \equiv (\mathbf{f}, \mathbf{g})$, reads:

$$\partial_t \left(\frac{\partial h'^0(\mathbf{v}', \mathbf{w}')}{\partial \mathbf{v}'} \right) + \partial_i \left(\frac{\partial h'^i(\mathbf{v}', \mathbf{w}')}{\partial \mathbf{v}'} \right) = \mathbf{f}(\mathbf{v}', \mathbf{w}'), \qquad (10)$$

$$\partial_t \left(\frac{\partial h'^0(\mathbf{v}', \mathbf{w}')}{\partial \mathbf{w}'} \right) + \partial_i \left(\frac{\partial h'^i(\mathbf{v}', \mathbf{w}')}{\partial \mathbf{w}'} \right) = \mathbf{g}(\mathbf{v}', \mathbf{w}'). \qquad (11)$$

Given an assigned constant value $\mathbf{w}'_*$ to $\mathbf{w}'$, we call principal subsystem of (8) the system:

$$\partial_t \left(\frac{\partial h'^0(\mathbf{v}', \mathbf{w}'_*)}{\partial \mathbf{v}'} \right) + \partial_i \left(\frac{\partial h'^i(\mathbf{v}', \mathbf{w}'_*)}{\partial \mathbf{v}'} \right) = \mathbf{f}(\mathbf{v}', \mathbf{w}'_*) \qquad (12)$$

In other words, a principal subsystem coincide with the first block of the system (10) putting $\mathbf{w}' = \mathbf{w}'_*$. In this case we have

Sub-Entropy Law: *The solutions of a principal subsystem* (12) *satisfy a supplementary subentropy law: The subentropy $\overline{h}^0$ is convex and therefore every principal subsystem is symmetric hyperbolic* [18].

Let $\lambda^{(k)}(\mathbf{v}', \mathbf{w}', \mathbf{n})$ and $\overline{\lambda}^{(\overline{k})}(\mathbf{v}', \mathbf{w}'_*, \mathbf{n})$, be the characteristic velocities of the total system (10), (11) and of the subsystem (12), respectively, $\mathbf{n} \in R^3$ being a unit vector. In general the solutions of the subsystem are not particular solutions of the system (for $\mathbf{w}' = \mathbf{w}'_*$) and the spectrum of $\overline{\lambda}$'s is not a part of the spectrum of λ's. However, if

$$\lambda_{\max} = \max_{k=1,2,\dots,N} \lambda^{(k)}, \quad \overline{\lambda}_{\max} = \max_{\overline{k}=1,2,\dots,M} \overline{\lambda}^{(\overline{k})},$$

and similarly for the minima, one obtains the following result

Sub-characteristic conditions: *Under the assumption that h^0 is a convex function, the following sub-characteristic conditions hold for every principal subsystem* [18]:

$$\lambda_{\max}(\mathbf{v}',\mathbf{w}'_*,\mathbf{n}) \geq \overline{\lambda}_{\max}(\mathbf{v}',\mathbf{w}'_*,\mathbf{n}),$$
$$\lambda_{\min}(\mathbf{v}',\mathbf{w}'_*,\mathbf{n}) \leq \overline{\lambda}_{\min}(\mathbf{v}',\mathbf{w}'_*,\mathbf{n}),$$

$\forall \mathbf{v}' \in R^M$ *and* $\forall \mathbf{n} \in R^3$, $\| \mathbf{n} \| = 1$.

Taking into account (9) and (3), we can recognize the following interesting principal subsystems:

Case 1 - *The single-temperature model is a principal subsystem of the multi-temperature.* Let us suppose that $\Lambda^{\varepsilon_b} = 0$ for $b = 1,\ldots,n-1$, then

$$T_1 = \ldots = T_n = T.$$

This principal subsystem contains only the energy conservation equation for the mixture, while energy balance equations for the constituents are dropped. Thus, one may conclude that single-temperature model naturally appears as a principal subsystem of the multi-temperature system.

Case 2 - *The equilibrium subsystem.* If we set

$$\Lambda^{\varepsilon_b} = \Lambda^{\mathbf{v}_b} = \Lambda^{\rho_b} = 0 \ \forall b = 1,...,n-1$$

i.e.:

$$T_b = T \ ; \ \mathbf{u}_b = 0 \ ; \ \mu_b = \mu \ \forall b = 1,...,n-1,$$

we have the equilibrium Euler subsystem (a single fluid system) with concentrations c_b being solutions of $\mu_1 = \mu_2 = \ldots = \mu_n$.

2.2 Qualitative Analysis

The system (3), is a particular case of a system of balance laws (4) and it is dissipative due to the presence of the productions that satisfy the entropy principle. Moreover we have verified that h^0 is a convex function of the densities $\mathbf{u} \equiv \mathbf{F}^0$. On the other hand the system (3) is of mixed type, some equations are conservation laws and the other ones are real balance laws, i.e., we are in the case in which

$$\mathbf{F}(\mathbf{u}) \equiv \begin{pmatrix} 0 \\ \mathbf{g}(\mathbf{u}) \end{pmatrix}; \qquad \mathbf{g} \in \mathbf{R}^{N-M}.$$

In this case the coupling condition discovered for the first time by Shizuta-Kawashima (K-condition) [19] plays a very important role in the analysis of global existence of smooth solutions. If it is satisfied the dissipation present in the second block of equations (balance laws) have effect also on the first block of equations (conservation laws). Hanouzet and Natalini [20] in one-space dimension and Wen-An Yong [21] in the multidimensional case, have proved the following theorem:

Assume that the system (4) is strictly dissipative and the K-condition is satisfied. Then there exists $\delta > 0$, such that, if $\|\mathbf{u}(x,0)\|_2 \leq \delta$, there is a unique global smooth solution, which verifies

$$\mathbf{u} \in \mathscr{C}^0\left([0,\infty); H^2\right)(\mathbf{R}) \cap \mathscr{C}^1\left([0,\infty); H^1(\mathbf{R}).\right)$$

Recently, Ruggeri and Serre [22] have proved in the one-dimensional case the stability of constant states:

Under natural hypotheses of strongly convex entropy, strict dissipativeness, genuine coupling and "zero mass" initial for the perturbation of the equilibrium variables, the constant solution stabilizes

$$\|\mathbf{u}(t)\|_2 = O\left(t^{-1/2}\right).$$

Lou and Ruggeri [23] have observed that exists a weaker form of K-condition, valid only for the genuine non-linear eigenvalues, that is a necessary (but, in general, not sufficient) condition for the global existence of smooth solutions.

2.3   The K-Condition in the Mixture Theories

It is important to observe that in general, under the same initial data, the solutions of the subsystem are not particular solutions of the full system! Therefore also if the *ST* appears a particular physical case of the *MT*, the solutions starting from the same initial data (with $T_\alpha(\mathbf{x},0) = T(\mathbf{x},0) = T_0(\mathbf{x})$) are different and with different regularity. In fact, for *ST* theory without chemical reactions it was proven [24], [25] that the K-condition is violated also for some genuinely non linear eigenvalues. Therefore, taking into account the results [23], in general, global smooth solutions do not exist, even if the initial data are small enough. Instead, for *MT* system it is possible to verify that the K-condition is satisfied for all eigenvalues. This means, roughly speaking, that the dissipation in *ST* is too weak with respect to the hyperbolicity and we do not have global smooth solutions for all the time; instead if we add the multi-temperature effect together with mechanical diffusion, the dissipation becomes enough to win the effect of hyperbolicity. Therefore we can conclude [11]:

If the initial data of the MT model are perturbations of equilibrium state, smooth solutions exist for all time and tends to the equilibrium constant state.

Also from this point of view the *MT* model provides a description more realistic than the *ST* model.

3   Average Temperature

The (*MT*) theory is of course the most realistic one and also in agreement with the kinetic theory [26] and it is a necessary theory in several physical situations, in particular in plasma physics [9]. Nevertheless, from the theoretical point of view, the main problem remains how it is possible to measure the temperatures of each constituent. Therefore, a question of definition of a macroscopic average temperature has to be posed. In this paper we reconsider the definition of average temperature recently proposed by Ruggeri and co-workers [27, 28, 29, 30, 31]. The main idea is to exploit the definition of internal energy to introduce (average) temperature T as a state variable for the mixture, and to do it in such a way that the intrinsic internal energy ε_I (see (2)$_3$) of the multi-temperature mixture resembles the structure of intrinsic internal energy of a single-temperature mixture. Therefore, *the following implicit definition of an average temperature is adopted*:

$$\rho \varepsilon_I(\rho_\beta, T) = \sum_{\alpha=1}^{n} \rho_\alpha \varepsilon_\alpha(\rho_\alpha, T) = \sum_{\alpha=1}^{n} \rho_\alpha \varepsilon_\alpha(\rho_\alpha, T_\alpha), \tag{13}$$

By expanding this relation in the neighborhood of the average temperature we have:

$$T = \frac{\sum_{\alpha=1}^{n} \rho_\alpha c_V^{(\alpha)} T_\alpha}{\sum_{\alpha=1}^{n} \rho_\alpha c_V^{(\alpha)}}, \tag{14}$$

where

$$c_V^{(\alpha)} = \left. \frac{\partial \varepsilon_\alpha(\rho_\alpha, T_\alpha)}{\partial T_\alpha} \right|_{T_\alpha = T}$$

is the specific heat at constant volume of constituent α. We observe that Eq. (14) gives the exact value of the average temperature in the case of the mixture of ideal gases for which the $c_V^{(\alpha)}$ are constant.

This definition of average temperature has several advantages with respect to usual ones used in the literature, as we can see in the following. First, as a consequence of the definition, the conservation law for the energy of mixture (3)$_3$ becomes an evolution equation for the average temperature T as in the case of (ST) and (CT). So, in the case of spatial homogeneous solution of the differential system (3) (whose solutions only depend on time), T is constant and all the non-equilibrium temperatures of each constituent T_α converge to T for large time, as we can see in the sequel.

The second advantage is related to the entropy of the whole mixture that, thanks to the introduction of this average temperature, reaches its maximum value when $T_\alpha = T$. In fact Ruggeri and Simić have proved in [28], using the Gibbs equations for each constituent, that the entropy density near the equilibrium becomes a negative definite quadratic form of the non-equilibrium variables $\Theta_\alpha = T_\alpha - T$ (*diffusion temperature flux*):

$$\rho S = \sum_{\alpha=1}^{n} \rho_\alpha S_\alpha(\rho_\alpha, T_\alpha) =$$
$$\sum_{\alpha=1}^{n} \rho_\alpha S_\alpha(\rho_\alpha, T) - \frac{1}{2T^2} \sum_{\alpha=1}^{n} \rho_\alpha c_V^{(\alpha)} \Theta_\alpha^2 + O(\Theta_\alpha^3).$$

4 Examples of Spatially Homogenous Mixture and Static Heat Conduction

In this section two simple examples will be provided in order to support previous theoretical considerations, and to stress the main features of multi temperature approach and the role of the average temperature.

4.1 Solution of a Spatially Homogenous Mixture

First we consider a non-reacting mixture of gases in the special case of spatially homogeneous fields, i.e. the case in which field variables depend solely on time [29], [28]. The governing equations (3) can be written in the following form:

$$\frac{d\rho}{dt} = 0; \quad \frac{d\mathbf{v}}{dt} = \mathbf{0}; \quad \frac{dT}{dt} = 0; \tag{15}$$

$$\frac{d\rho_b}{dt} = 0; \quad \rho_b \frac{d\mathbf{v}_b}{dt} = \hat{\mathbf{m}}_b; \quad \rho_b \frac{d\varepsilon_b}{dt} = \hat{e}_b. \tag{16}$$

where in the present case the material derivative $d/dt = \partial/\partial t$. From (15), (16), it is easy to conclude:

$$\rho = \text{const.}; \quad \mathbf{v} = \text{const.}; \quad T = \text{const.};$$
$$\rho_b = \text{const.}; \quad b = 1, \dots, n,$$

and due to Galilean invariance we may choose $\mathbf{v} = \mathbf{v}_0 = \mathbf{0}$ without loss of generality. It is also remarkable that the average temperature of the mixture remains constant during the process: $T(t) = T_0$.

In the sequel we shall regard only small perturbations of equilibrium state, $\mathbf{v}_\alpha = \mathbf{v}_0 = \mathbf{0}$, $T_\alpha = T_0$, $\alpha = 1, \dots, n$, and analyze their behavior. Therefore, the r.h.s. of $(16)_{2,3}$ could be linearized in the neighborhood of equilibrium and taking into account (7), we obtain:

$$\rho_b \frac{d\mathbf{v}_b}{dt} = -\sum_{c=1}^{n-1} \frac{\psi_{bc}^0}{T_0} (\mathbf{v}_c - \mathbf{v}_n); \tag{17}$$

$$\rho_b c_V^{(b)} \frac{dT_b}{dt} = -\sum_{c=1}^{n-1} \frac{\theta_{bc}^0}{T_0^2} (T_c - T_n), \tag{18}$$

where ψ_{bc}^0 and θ_{bc}^0 are entries of positive definite matrices evaluated in equilibrium. Note that $\mathbf{v}_b - \mathbf{v}_n = \mathbf{u}_b - \mathbf{u}_n$ and $T_b - T_n = \Theta_b - \Theta_n$. For what concerns the n-species v_n and T_n are obtained by the algebraic equations $(2)_2$ and (14).

In the particular case of a binary mixture the explicit solution of equations (17), (18), can be obtained and it reads:

$$\mathbf{v}_1(t) = \mathbf{v}_1(0) e^{-\frac{t}{\tau_V}}; \quad T_1(t) = T_0 + (T_1(0) - T_0) e^{-\frac{t}{\tau_T}},$$

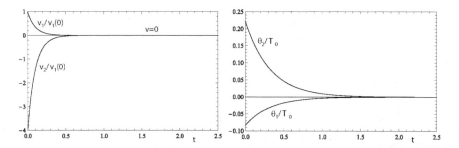

Fig. 1 Dimensionless velocities and diffusion temperature fluxes of the constituents versus time.

where τ_v and τ_T represent relaxation times that for ideal gas assume the expression

$$\tau_v = \frac{\rho_1 \rho_2 T_0}{\psi_{11}^0 \rho};$$

$$\tau_T = \frac{k \rho_1 \rho_2 T_0^2}{\theta_{11}^0 (\rho_1 m_2 (\gamma_2 - 1) + \rho_2 m_1 (\gamma_1 - 1))}.$$

Starting from these solutions, other field variables can be obtained by means of defining equations:

$$\rho_1 \mathbf{v}_1 + \rho_2 \mathbf{v}_2 = \rho \mathbf{v} = \mathbf{0}.$$
$$\rho_1 c_V^{(1)} T_1 + \rho_2 c_V^{(2)} T_2 = (\rho_1 c_V^{(1)} + \rho_2 c_V^{(2)}) T$$
$$= (\rho_1 c_V^{(1)} + \rho_2 c_V^{(2)}) T_0.$$

It is obvious that, due to dissipative character of the system, *all the non-equilibrium variables exponentially decay and converge to their equilibrium values*. In order to compare the values of τ_v and τ_T for ideal gases, and also to compute the actual values of variables in numerical example, the relations from kinetic theory has to be recalled [9]:

$$\theta_{11}^0 = \frac{3 m_1 m_2}{(m_1 + m_2)^2} k T_0^2 \Gamma_{12}'; \qquad \psi_{11}^0 = \frac{2 m_1 m_2}{m_1 + m_2} T_0 \Gamma_{12}',$$

where Γ_{12}' represents volumetric collision frequency, and the following estimate can be obtained:

$$\frac{\tau_T}{\tau_v} = \frac{2}{3} \frac{\rho (m_1 + m_2)}{\rho_1 m_2 (\gamma_2 - 1) + \rho_2 m_1 (\gamma_1 - 1)} > \frac{2}{3(\gamma_{max} - 1)} \geq 1, \qquad (19)$$

($\gamma_{max} = \max\{\gamma_1, \gamma_2\} \leq 5/3$).

In Figures 1, we present the graphs of normalized velocities, diffusion temperature fluxes [28]. It can be observed that, due to inequality (19), *the mechanical diffusion vanishes more rapidly than the thermal one.* This is in sharp contrast with widely adopted approach which ignores the influence of multiple temperature of each constituent of the mixture.

4.2 Static Heat Conduction Solution

Another simple example is the one-dimensional mixture of gas at rest ($\mathbf{v}_\alpha = 0$), without chemical reactions ($\tau_\alpha = 0$) between two walls $0 \le x \le L$, maintained at two different temperatures $T(0) = T_0, T(L) = T_L$ [30].

In both (CT) and (ST), the static field equation reduces to the global energy equation (3)$_3$ that reads div $\mathbf{q} = 0$. In the one-dimensional case, this equation combined with the Fourier law with constant heat conductivity, yields the classical result of a linear behavior temperature profile as for a single fluid:

$$T'' = 0 \iff T = (T_L - T_0)\xi + T_0$$

with $\xi = x/L$ and $'$ denotes $d/d\xi$. For what concerns the densities they are obtained by the conditions that the pressure of each constituent must be constant due to the momentum equations.

In the (MT) model, the situation is quite different. In fact if we consider the simple case of a binary mixture ($n = 2$), by taking into account of Eqs. (3), and (7), in the linear case, is reduced to

$$\begin{cases} \frac{dp_1}{dx} = 0, \ \frac{dp_2}{dx} = 0, \\ \frac{dq_1}{dx} = \beta \left(T_2 - T_1 \right), \\ \frac{dq_2}{dx} = \beta \left(T_1 - T_2 \right), \end{cases} \tag{20}$$

where $\beta = \theta_{11}/T_0^2$. By using the Fourier law, Eqs. (20)$_{2,3}$ can be rewritten as

$$\begin{cases} T_1'' = v_1(T_1 - T_2), \\ T_2'' = v_2(T_2 - T_1), \end{cases} \tag{21}$$

where we assume that the dimensionless quantities

$$v_1 = \frac{\beta L^2}{\chi_1}, \quad v_2 = \frac{\beta L^2}{\chi_2}, \tag{22}$$

are constant. The system (21) is equivalent to

$$\widehat{T}'' = 0, \qquad \Theta'' - \omega^2 \, \Theta = 0$$

with $\widehat{T} = vT_1 + (1 - v)T_2, \ \Theta = T_2 - T_1$ and

$$v = \frac{v_2}{v_1 + v_2} = \frac{\chi_1}{\chi_1 + \chi_2}, \quad \omega = \sqrt{v_1 + v_2}.$$

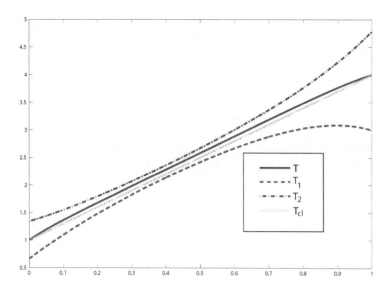

Fig. 2 Graphs of the average temperature T and constituent temperatures T_1, T_2 in terms of the dimensionless distance x/L. T_{cl} represents the classical straight line solution. T_0 is the temperature unit.

Consequently, we get the solution in the form

$$T_1 = \hat{T} - (1-v)\Theta, \quad T_2 = \hat{T} + v\Theta \tag{23}$$

with

$$\hat{T} = A\,\xi + B, \quad \Theta = \frac{1}{\sinh(\omega)}\{\Theta_L \sinh(\omega\,\xi) + \Theta_0\,\sinh(\omega(1-\xi))\}, \tag{24}$$

and A, B, Θ_0, Θ_L are constants of integration. In the case of ideal gas, equations $(20)_1$ and $(3)_2$ yield the constance of the internal energy densities of each constituent:

$$\rho_\alpha T_\alpha c_V^{(\alpha)} = P_\alpha = Const., \quad (\alpha = 1, 2). \tag{25}$$

and Eq. (14) yield the average temperature

$$\frac{1}{T} = \frac{\pi}{T_1} + \frac{1-\pi}{T_2}, \quad \text{with } \pi = \frac{P_1}{P_1 + P_2}. \tag{26}$$

The constant π belongs to $[0,1]$. It is interesting to observe that the *coldness* $1/T$ (inverse of the average temperature) belongs to the convex envelope of the component coldness $1/T_1$ and $1/T_2$. Equations (23), (24) and (26) give the explicit solution of T_1, T_2 and T as function of ξ and five constants of integration: $(A, B, \Theta_0, \Theta_L, \pi)$.

We observe the behavior of T is not a straight line as in the classic case of (CT) or (ST) theories; the multi-temperature effect is that the temperature is not a linear function of x (see Figure 2). Due to Eqs. (20)-(22), when $\varepsilon = 1/\beta$ tends towards zero, the solution of Eqs. (23), (24) and (26) converges towards the classical solution $T_1 = T_2 = T = \hat{T}$ for any $\xi \in]0,1[$. This result is true also at the boundary when Θ_0 and Θ_1 are of same order as ε.

Let $c = \rho_1/(\rho_1 + \rho_2)$ the concentration and $c(0) = c_0$. Equations (25), (23), (24), imply

$$c = \frac{c_0}{c_0 + \Omega(1 - c_0)}, \quad \text{with } \Omega = \frac{T_1}{T_2}\frac{T_{20}}{T_{10}}$$

and

$$T_{10} = B - (1 - v)\Theta_0, \quad T_{20} = B + v\Theta_0.$$

The concentration is function of the position x whereas in classical case $\Omega = 1$ and $c = c_0$. Ruggeri and Lou [30] have studied how is possible to determine in a unique way the constant of integration and they proved that for a mixture of n constituents the measure of the average temperature at $2(n-1)$ points allows to know the temperature behavior of each constituent in all points.

5 Maxwellian Iteration

To find a connection between the extended models of MT or ST with the classical theory CT it is necessary to use the procedure that is know as *Maxwellian iteration* (see [4]). In the present case the Maxwellian iterative procedure require to put in the left side of the system $(3)_{5,6}$ the zero$-th$ iterate, i.e the equilibrium state and in the right side the first iterate. Taking into account that in zeroth iteration $v_\alpha^{(0)} = v$ and consequently $d_b^{(0)}/dt = d/dt$, $J_b^{(0)} = u_b^{(0)} = 0$ and moreover $T_\alpha^{(0)} = T$, $q^{(0)} = q_b^{(0)} = 0$, $t^{(0)} = -p^{(0)}I = -p_0 I$, $t_b^{(0)} = -p_b^{(0)}I$, we obtain after some straightforward calculations, the Fick law for what concerns the momentum equations $(3)_5$, while for what concern the energy equations we obtain a new constitutive equations:

$$\Theta_a^{(1)} = -k_a \operatorname{div} v. \tag{27}$$

Equation (27), obtained by means of Maxwellian iteration gives the temperatures of each species as a constitutive equation in the same manner as the Fick law gives the velocities of each species.

For a mixture of ideal gases k_a is a linear combination of

$$\Omega_b = \rho_b T c_V^{(b)} \frac{\sum_{\alpha=1}^n \rho_\alpha c_V^{(\alpha)}(\gamma_b - \gamma_\alpha)}{\sum_{\alpha=1}^n \rho_\alpha c_V^{(\alpha)}}.$$

When all the constituents have the same ratio of specific heats we have $\Omega_b = 0$, and consequently $\Theta_b^{(1)} = 0$, $a = 1,\ldots,n-1$. In this case the diffusion temperature flux cannot be observed in the first approximation.

6 A Classical Approach of Multi-temperature Mixtures

Therefore the *ST* model using the Maxwellian iteration converge to the classic theory but when we start from the *MT* case we obtain a new constitutive equation that is not present in the classical approach of mixture theory.

To justify the results of the Maxwellian iteration, Gouin and Ruggeri [31] have constructed a classical theory of mixture with multi-temperature. The idea is to use the usual equation of the classical approach $(3)_{1-4}$, but now we suppose that each constituent has its an own temperature. In this approach the role of the average temperature previously defined by (13) is fundamental. In fact the multi-temperature effect appears through the pressure, that now near equilibrium has the form:

$$p = p_0 + \pi_\theta,$$

where

$$p_0 = \sum_{\alpha=1}^{n} p_\alpha(\rho_\alpha, T), \quad \pi_\theta = \sum_{b=1}^{n-1} r_b \Theta_b$$

and

$$r_b = \frac{1}{\rho_n c_V^{(n)}} \left\{ \rho_n c_V^{(n)} \frac{\partial p_b}{\partial T_b}(\rho_b, T) - \rho_b c_V^{(b)} \frac{\partial p_n}{\partial T_n}(\rho_n, T) \right\}. \tag{28}$$

Therefore, the total pressure p of the mixture is the sum of the equilibrium part p_0 depending on T and a new dynamical pressure part (as a non-equilibrium term) π_θ due to the difference of temperatures between the constituents.

Assuming that the internal energy $\varepsilon(\rho, T, c_b)$ and the equilibrium pressure $p_0(\rho, T, c_b)$ satisfy the Gibbs equation

$$T dS = d\varepsilon - \frac{p_0}{\rho^2} d\rho - \sum_{b=1}^{n-1} (\mu_b - \mu_n) \, dc_b,$$

in [31] was proved that hold the following equation:

$$\rho \frac{dS}{dt} + \text{div} \left\{ \frac{1}{T} \left(\mathbf{q} - \sum_{b=1}^{n-1} (\mu_b - \mu_n) \mathbf{J}_b \right) \right\} =$$

$$\mathbf{q} \cdot \text{grad} \left(\frac{1}{T} \right) - \sum_{b=1}^{n-1} \mathbf{J}_b \cdot \text{grad} \left(\frac{\mu_b - \mu_n}{T} \right) + \frac{1}{T} \text{tr} \left(\mathbf{J}_{mech} \mathbf{D} \right), \tag{29}$$

where the mechanical flux is

$$\mathbf{J}_{mech} = \boldsymbol{\sigma} - \pi_\theta \mathbf{I}.$$

Eq. (29) can be interpreted as a balance of entropy, if we accept

$$\Phi = \frac{1}{T} \left(\mathbf{q} - \sum_{b=1}^{n-1} (\mu_b - \mu_n) \mathbf{J}_b \right)$$

and

$$\Sigma = \mathbf{q} \cdot \text{grad} \left(\frac{1}{T} \right) - \sum_{b=1}^{n-1} \mathbf{J}_b \cdot \text{grad} \left(\frac{\mu_b - \mu_n}{T} \right) + \text{tr} \left(\mathbf{J}_{mech} \frac{\mathbf{D}}{T} \right) \tag{30}$$

as the entropy flux and the entropy production respectively.

We observe that the entropy production is the sum of products of the following quantities:

thermodynamic fluxes   **thermodynamic forces**
heat flux $\mathbf{q}$ temperature gradient grad $\left(\frac{1}{T} \right)$,
diffusion fluxes $\mathbf{J}_b$ chemical potential gradients $- \text{grad} \left(\frac{\mu_b - \mu_n}{T} \right)$,
mechanical flux $\mathbf{J}_{mech}$ velocity gradient $\frac{\mathbf{D}}{T}$.

In accordance with the case of a single temperature model [3] and [4] chapter 5, in *TIP* near equilibrium, the fluxes depend linearly on the associated forces (see also for the general methodology of *TIP* [32, 33, 34, 35]):

- For the heat flux and the diffusion fluxes, we obtain the constitutive equations in the form of Fourier and Fick,
- For Stokesian fluids, the last term of Eq. (30) corresponding to the mechanical production of entropy can be written in a separated form

$$\frac{1}{T} \text{tr} \left(\mathbf{J}_{mech} \mathbf{D} \right) = \frac{1}{T} \text{tr} \left(\sigma \mathbf{D}^D \right) - \frac{1}{T} \pi_\theta \, \text{div} \, \mathbf{v}.$$

We obtain the constitutive equation of the viscous stress tensor in the form of Navier Stokes, and the dynamical pressure part due to the difference of temperatures yields

$$\pi_\theta = \sum_{b=1}^{n-1} r_b \Theta_b = -L_\pi \, \text{div} \, \mathbf{v}, \tag{31}$$

where L_π is a scalar coefficient of proportionality.

The production of entropy must be non-negative (30) and therefore the *phenomenological coefficients* must satisfy the usual inequalities and moreover

$$L_\pi \geq 0.$$

Taking into account that terms r_b given by Eq. (28) depend on (ρ_b, T), from Eq. (31) we deduce that constitutive quantities Θ_a (depending *a priori* on $\nabla \mathbf{v}$) must be proportional to div $\mathbf{v}$:

$$\Theta_a = -k_a \, \text{div} \, \mathbf{v} \quad (a = 1, \cdots, n - 1).$$

This in perfect agreement with the Maxwellian iteration procedure presented in the previous section.

Let $\|M_{ab}\|$ be the matrix such that $k_a = \sum_{b=1}^{n-1} M_{ab} \, r_b$, we have

$$\Theta_a = -\sum_{b=1}^{n-1} M_{ab} \, r_b \operatorname{div} \mathbf{v} \quad (a = 1, \cdots, n-1). \tag{32}$$

Introducing expression (32) into Eq. (31), we obtain

$$L_\pi = \sum_{a,b=1}^{n-1} M_{ab} \, r_a r_b \geq 0,$$

and assuming the Onsager symmetry property, $M_{ab} = M_{ba}$ $(a, b = 1, \cdots, n-1)$, we deduce that coefficients M_{ab} are *associated with a positive definite quadratic form*.

Finally, the results are the same as in the classical theory, but moreover we get new constitutive equations (32) for the difference of temperatures.

We have considered the simple case of Stokes fluids. If the fluid is non Stokesian, the Navier-Stokes stress tensor of viscosity contains also the bulk viscosity λ and the stress tensor $\mathbf{t}$ becomes

$$\mathbf{t} = -(p_0 + \pi_\theta)\mathbf{I} + \sigma = -p\,\mathbf{I} + 2\nu\,\mathbf{D}^D,$$

with

$$p = p_0 + \pi_\theta + \pi_\sigma.$$

The non-equilibrium pressure $p - p_0$ is separated in two different parts. The first one $\pi_\sigma = -\lambda \operatorname{div} \mathbf{v}$ is related to the bulk viscosity and the second one $\pi_\theta = -L_\pi \operatorname{div} \mathbf{v}$ is related to the multi-temperature effects between components.

Due to a non-zero dynamical pressure even for Stokes fluids, we conclude that multi-temperature mixtures of fluids have a great importance. Perhaps such a model may be used to analyze the evolution of the early universe in which the dynamical pressure seems essential [36, 37].

7 Conclusions

In this survey we have reconsidered the definition of an average temperature in the context of multi-temperature approach to the theory of mixtures of fluids. It was based upon the assumption that internal energy of the mixture should retain the same form as in a single-temperature approach. The supremacy of this definition is supported by a simple derivation of entropy maximization result in equilibrium as its consequence and the result that the average temperature remains constant for spatially homogenous mixture. Furthermore, by means of Maxwellian first iterative procedure we have derived constitutive equations for non-equilibrium variables, mechanical diffusion flux $\mathbf{J}_a$ and diffusion temperature flux Θ_a, in the neighborhood of equilibrium. It was shown that the first iterate of diffusion flux $\mathbf{J}_a^{(1)}$ coincides with the classical generalized Fick's laws which can be obtained in TIP. However, diffusion temperature flux $\Theta_a^{(1)}$ is found to be proportional to $\operatorname{div} \mathbf{v}$ – a new result which is in accordance with recent observations within classical TIP framework of multi-temperature mixture [31].

Moreover in the case of the one-dimensional steady heat conduction between two walls, we have verified that the main effect of multi-temperature is that the average temperature is not a linear function of the distance as in the case of the *ST* theory. These last result could be used during some experiments to show the order of magnitude of the difference between *ST* and *MT*.

Maybe some ideas of the present work can be used in the theory of porus media or granular materials, fields in which Krzysztof Wilmanski gave fundamental contributions (see [8] and references therein).

Acknowledgment. This paper is dedicated to Krzysztof Wilmanski one of my dearest friends on the occasion of his 70th birthday.

References

1. Truesdell, C.: Rational Thermodynamics. McGraw-Hill, New York (1969)
2. Müller, I.: Arch. Ration. Mech. Anal. 28, 1 (1968)
3. Müller, I.: Thermodynamics. Pitman, London (1985)
4. Müller, I., Ruggeri, T.: Rational Extended Thermodynamics. Springer, New York (1998)
5. Hütter, K.: Continuum Methods of Physical Modeling. Springer, New York (2004)
6. Rajagopal, K.R., Tao, L.: Mechanics of Mixtures. World Scientific, Singapore (1995)
7. Atkin, R.J., Craine, R.E.: Quart. J. Mech. Appl. Math. 29, 209 (1976)
8. Wilmanski, K.: Continuum Thermodynamics - Part I: Foundations. World Scientific, Singapore (2008)
9. Bose, T.K.: High Temperature Gas Dynamics. Springer, Berlin (2003)
10. Ruggeri, T.: Galilean Invariance and Entropy Principle for Systems of Balance Laws. The Structure of the Extended Thermodynamics. Contin. Mech. Thermodyn. 1, 3 (1989)
11. Ruggeri, T., Simić, S.: On the Hyperbolic System of a Mixture of Eulerian Fluids: A Comparison Between Single- and Multi-Temperature Models. Math. Meth. Appl. Sci. 30, 827 (2007)
12. Boillat, G.: Sur l'existence et la recherche d'équations de conservation supplémentaires pour les Systémes Hyperboliques. C.R. Acad. Sc. Paris 278A, 909 (1974)
13. Ruggeri, T., Strumia, A.: Main field and convex covariant density for quasi-linear hyperbolic systems. Relativistic fluid dynamics. Ann. Inst. H. Poincaré 34A, 65–84 (1981)
14. Boillat, G.: In CIME Course. In: Ruggeri, T. (ed.) Recent Mathematical Methods in Nonlinear Wave Propagation. Lecture Notes in Mathematics, vol. 1640, pp. 103–152. Springer, Heidelberg (1995)
15. Friedrichs, K.O., Lax, P.D.: Systems of conservation laws with a convex extension. Proc. Nat. Acad. Sci. USA 68, 1686–1688 (1971)
16. Godunov, S.K.: An interesting class of quasilinear systems. Sov. Math. 2, 947–948 (1961)
17. Dafermos, C.: Hyperbolic Conservation Laws in Continuum Physics. Springer, Berlin (2001)
18. Boillat, G., Ruggeri, T.: Hyperbolic Principal Subsystems: Entropy Convexity and Sub characteristic Conditions. Arch. Rat. Mech. Anal. 137, 305–320 (1997)
19. Shizuta, Y., Kawashima, S.: Systems of equations of hyperbolic-parabolic type with applications to the discrete Boltzmann equation. Hokkaido Math. J. 14, 249–275 (1985)

20. Hanouzet, B., Natalini, R.: Global existence of smooth solutions for partially dissipative hyperbolic systems with a convex entropy. Arch. Rat. Mech. Anal. 169, 89 (2003)
21. Yong, W.A.: Entropy and global existence for hyperbolic balance laws. Arch. Rat. Mech. Anal. 172(2), 247 (2004)
22. Ruggeri, T., Serre, D.: Stability of constant equilibrium state for dissipative balance laws system with a convex entropy. Quart. Appl. Math. 62(1), 163–179 (2004)
23. Lou, J., Ruggeri, T.: Acceleration waves and weaker Kawashima Shizuta condition. Rendiconti del Circolo Matematico di Palermo, Serie II 78, 187–200 (2006)
24. Ruggeri, T.: Global Existence, Stability and Non Linear Wave Propagation in Binary Mixtures. In: Fergola, P., Capone, F., Gentile, M., Guerriero, G. (eds.) Proceedings of the International Meeting in honour of the Salvatore Rionero 70th Birthday, Napoli 2003, pp. 205–214. World Scientific, Singapore (2004)
25. Ruggeri, T.: Some Recent Mathematical Results in Mixtures Theory of Euler Fluids. In: Monaco, R., Pennisi, S., Rionero, S., Ruggeri, T. (eds.) Proceedings WASCOM 2003, pp. 441–454. World Scientific, Singapore (2004)
26. Ikenberry, E., Truesdell, C.: On the pressures and the flux of energy in a gas according to Maxwell's kinetic theory. J. Rational Mech. Anal. 5, 1 (1956)
27. Ruggeri, T., Simić, S.: Mixture of Gases with Multi-temperature: Identification of a macroscopic average temperature. In: Proceedings Mathematical Physics Models and Engineering Sciences, Liguori Editore, Napoli, p. 455 (2008)
28. Ruggeri, T., Simić, S.: Average temperature and Maxwellian iteration in multitemperature mixtures of fluids. Phys. Rev. E 80, 026317 (2009)
29. Ruggeri, T., Simić, S.: Asymptotic Methods in Non Linear Wave Phenomena. In: Ruggeri, T., Sammartino, M. (eds.), p. 186. World Scientific, Singapore (2007)
30. Ruggeri, T., Lou, J.: Heat Conduction in multi-temperature mixtures of fluids: the role of the average temperature. Physics Letters A 373, 3052 (2009)
31. Gouin, H., Ruggeri, T.: Identification of an average temperature and a dynamical pressure in a multi-temperature mixture of fluids. Phys. Rev. E 78, 01630 (2008)
32. Eckart, C.: The Thermodynamics of Irreversible Processes. II. Fluid Mixtures. Phys. Rev. 58, 269 (1940)
33. Onsager, L.: Reciprocal Relations in Irreversible Processes. I. Phys. Rev. 37, 405 (1931); Reciprocal Relations in Irreversible Processes. II. Phys. Rev. 38, 2265 (1931)
34. de Groot, S.R., Mazur, P.: Non-Equilibrium Thermodynamics. North-Holland, Amsterdam (1962)
35. Gyarmati, I.: Non-Equilibrium Thermodynamics. Field Theory and Variational Principles. Springer, Berlin (1970)
36. de Groot, S.R., van Leeuwen, W.A., van Weert, C.G.: Relativistic Kinetic Theory. North-Holland, Amsterdam (1980)
37. Weinberg, S.: Entropy generation and the survival of protogalaxies in an expanding universe. Astrophys. J. 168, 175 (1971)

Part III
EXTENSIONS OF CONSTITUTIVE LAWS

Hypocontinua

Gianfranco Capriz

Abstract. The unusual title is justified, mainly, by two remarks: (i) the background is provided by (and the balance laws derive from within) an attempt to model, inside the scheme of the continuum, the behaviour of a drove of profuse minute granules and thus to overcome, to an extent, the dilemma discrete/continuous [1]; (ii) the terms hypoelasticity and hypoplasticity are already in common use for theories that are close to one proposed here [2, 3].

1 Preamble

Here bodies are imagined to span in three dimensions over fit regions $\mathscr{B}$ of ordinary Euclidean space (e.g., to fill the interior of an appropriate container or the whole exterior of a vessel, as usually obtains for gases). Consequently the aim is a doctrine based on some concepts and methods of the theory of continua; however, contrary to the attendant instinctive perception, each body is figured as an assembly of relatively very small flowing granules. Thus a conflict ensues as the invoked theory relies strictly on a primary, though by and large surreptitiously condoned, axiom of perfect identifiability of material elements and on consequent corollaries built on bijections from a preferred, possibly only ideal, reference stance. The axiom is obviously fallacious within the imagined context. Think of the extreme case of gas dynamics as derived from the kinetic theory, though, luckily in that instance, a related default is met, actually, only when persevering to deeper estimates. A multifield theory must be proposed for such ephemeral bodies where bare knowledge of instantaneous and local circumstances and paces is presumed. The ensuing balance laws mimic those appropriate to micromorphic bodies, though 'coshaping' time-derivatives must be engaged and mass loss and gain by suffusion of granules must

Gianfranco Capriz
Dipartimento di Matematica, Università di Pisa, via F. Buonarroti 2, I-56127 Pisa, Italy
and
Accademia dei Lincei, via della Lungara 10, I-00165 Roma, Italy
e-mail: gianfrancocapriz@nettare.net

be accounted for. Beside the requests for mass balance and for balance of tensorial inertia, there is an equation analogous to Cauchy's, one for tensorial moment of momentum and one for Reynolds' kinetic tensor (which may be variously intended to measure balanced cross-over of granules or the intensity of their collisions). The last balance is the decisive ingredient in branding hypocontinua.

The next arduous task pertains the choice of constitutive laws; arduous owing to the spread of events surveyed and the score of items to be chosen (three 'stress' tensors and two 'self agents'). At least some general norms need be invoked even to identify the guise mentioned above. To assist in speculations a very special case is explored here, when a bold constraint is enforced, coercing congruence between macro and micro stretching. Surprisingly, under some side conditions, seemingly mild, even appropriate within the context, one is led to balance equations which echo those originally proposed by C. Truesdell for hypoelasticity and, more recently, studied by many for hypoplasticity; coincidence may even be forced, but by a quaint revision of the role and meaning of some fields, a revision causing consequent radical contrast with proposed accessory thermodynamic traits.

2 Essential Ingredients

In its most complete format [1], the scheme sees the body (of mass density ρ) subject to macrodeformations as in the standard narrations (x, place; $v(x, \tau)$, velocity at x and at time τ; $L = \text{grad } v$), but, besides, affected, at the level of each loculus $\mathbf{e}(x)$ (i.e., of a representative volume element centered at x), by an affine mesodistortion, with rate B, with causes changes in a local mesoinertia tensor Y and in a tensor moment of mesomomentum K (both per unit mass), while safeguarding the standard affine link between the two: $K = YB^T$. Furthermore, suffusion of granules from/to loculi might be present at a rate σ, due to discrepancy between the two disfigurements

$$\sigma = \text{tr}\,(L - B),\tag{1}$$

beyond an otherwise equilibrated cross-transit of granules at a rate measured by the correlation tensor of peculiar velocities: the Reynolds' tensor H (notice that this name is used, rather than 'Maxwell pressure tensor', to promote its kinetic rather than its dynamic role here). Actually, in view of the fact that granules are supposed to be undistinguishable (all assumed here to have the same mass μ), the tensor H could provide, rather, an appraisal of the frequency and direction of collisions among granules.

Sadly, in [1], the balance equations were printed as copied from earlier papers (where they were derived on the basis of slightly different and not fully consistent premises), rather than checked for specific consistency. Properly the equations involve the 'coshaping' time-derivatives of Y, K and H; thus they should read as follows:

$$\frac{\partial \rho}{\partial \tau} + \text{div}\,(\rho v) = \sigma\rho \quad [\text{or } \dot{\rho} + \rho\,\text{tr}\,B = 0],\tag{2}$$

$$\frac{\partial Y}{\partial \tau} + (\operatorname{grad} Y)\,v + \sigma Y - YB^T - BY = 0 \quad \left[\text{or } \dot{Y} + \sigma Y = K + K^T\right], \tag{3}$$

$$\rho\left[\frac{\partial v}{\partial \tau} + (\operatorname{grad} v)\,v + \sigma v\right] = \rho b + \operatorname{div} T, \tag{4}$$

$$\rho\left[\frac{\partial K}{\partial \tau} + (\operatorname{grad} K)\,v + \sigma K - KB^T - BK - H\right] = \rho M - A + \operatorname{div} \mathbf{m}$$

$$\left[\text{or } \rho\left(Y\dot{B}^T - H\right) = \rho M - A + \operatorname{div} \mathbf{m}\right], \tag{5}$$

$$\rho\left[\frac{\partial H}{\partial \tau} + (\operatorname{grad} H)\,v + \sigma H - HB^T - BH\right] = \rho J - Z + \operatorname{div} \mathbf{j}; \tag{6}$$

$$\operatorname{skw} T = \operatorname{skw} A \quad \left[\text{sometimes, more deeply, } T = -A^T\right]. \tag{7}$$

The discrepancies (an addendum BYB^T in the right-hand side of the fourth equation (5) and $2(BH + HB^T)$ in the fifth (6)) could, formally, have been thought as buried within the, contextually largely unspecified, tensors A and Z. However, such lame justification would have left to A and Z the task of completing the delivery of the apparent actions consequent on the effects of entrainment. As written above, the mentioned 'coshaping' time-derivatives

$$\dot{K} - KB^T - BK = G\left(G^{-1}KG^{-T}\right)^{\cdot} G^T$$

and

$$\dot{H} - HB^T - BH = G\left(G^{-1}HG^{-T}\right)^{\cdot} G^T$$

of K and H put directly in due evidence the apparent inertia, largely controlled by the tensor G generated as recalled below.

Surely, in (4), (7), T is the usual Cauchy stress, whereas $\mathbf{m}$ and $\mathbf{j}$ are third-order hyperstresses (the first as is current in theories of micromorphic continua); $-A$ and $-Z$ count the effect of internal actions and b, M, J count that of external agencies.

The last equation substitutes, within the context, the classical requirement of symmetry for T.

In the process that led to the proposal of (2)-(6), the appropriate form of the left-hand sides was sought with careful concern for the context, calling also attention on some disregarded minor side effects. The formulation of the right-hand sides, instead, was based on no more that a tautology (as Truesdell and Toupin declare it to be in Sect. 157 of [4]). A deeper scrutiny should still be pursued and might lead to amendments, chiefly because it might intimate a better model of physical occurrences consequent on an 'Euler cut', which, as a rule, is imagined as acting only at the macro level but is, in fact, operating also through each loculus.

To make the set (2)-(7) into a forecasting tool, constitutive laws must be devised, appropriate for a any special material in hand, for T, $\mathbf{m}$, $\mathbf{j}$ and for Z and A (the last one obeying rule (7)), so that they all become pertinent functions of shape and kinetic circumstances, possibly of L, B, H and of their gradients.

When, exceptionally, a preferred shape (or even placement) exists or can be conveniently imagined without injury, the placement gradient F becomes available and, consequently, Green's deformation tensor C and the corresponding rotation tensor R:

$$F = RC^{\frac{1}{2}}, \quad C = F^T F, \quad R = F(F^T F)^{-\frac{1}{2}}.$$

The tensor L, already mentioned, measures the time-rate of F and its symmetric part D that of C:

$$L = \dot{F} F^{-1}, \quad D = \mathrm{sym}\, L = \frac{1}{2} F^{-T} \dot{C} F^{-1}.$$

As a match, one is led to introduce, perhaps artificially, similar tensors for the affine mesodistortion:

$$G = R' N^{\frac{1}{2}}, \quad N = G^T G, \quad R' = GN^{-\frac{1}{2}}.$$

Actually G is generated through Y, declaring the latter to coincide with GG^T a constant factor δ^2 apart (δ^3 being the volume of $\mathbf{e}$), assuming besides the expected relation of $\dot{G}$ with B, corrected, however, for the effects of suffusion:

$$B = \dot{G} G^{-1} + \frac{1}{2} \sigma I \quad (I, \text{ the identity tensor});$$

explicitly:

$$B = \dot{G} G^{-1} + \frac{1}{2} I \,\mathrm{tr}\left(L - \dot{G} G^{-1}\right).$$

Consequently one finds the condition which identifies R', initial value apart:

$$\dot{R}' R'^T = \mathrm{skw}\left[\left(Y^{\frac{1}{2}}\right)^{\cdot} Y^{-\frac{1}{2}}\right] - Y^{-\frac{1}{2}} (\mathrm{skw}\, K) Y^{-\frac{1}{2}}. \tag{8}$$

Thus, consistency with the second balance equation (3) and with the standard affine expression of K in terms of Y and B ($K = YB^T$) is assured. At the same time it asseverated that all relevant fields are derived from present local patterns and their paces of change even when they mimic fields elsewhere shown as typically material.

3 Reminder of Definitions. Comments

The fields ρ, v, Y, K, H arise, actually, from the, possibly disorderly, setting and stirring of granules within each loculus $\mathbf{e}$ via averages with a mesodensity $\mu\tilde{\theta}$, if $\tilde{\theta}$ is number density of granules at y (a subplace within $\mathbf{e}$) and μ is the mass of one granule (presuming that all granules have the same mass, as already pledged).

In fact, $\tilde{\theta}$ itself is a derived quantity, the fundamental entity being the distribution θ which, at time τ within $\mathbf{e}(x)$, expresses the numerosity of the crowd of granules transiting through the subplace y with velocity w. Verbatim:

$$\tilde{\theta}(\tau, x; y) = \int_{\mathcal{V}} \theta(\tau, x; y; w)\, dw,$$

where $\mathcal{V}$ is the full vector space of all velocities.

Notice that the distribution $\hat{\theta}(\tau,x;w)$, which enters the standard account of phenomena by the kinetic theory, is the total distribution over $\mathbf{e}(x)$

$$\hat{\theta}(\tau,x;w) = \int_{\mathbf{e}} \theta(\tau,x;y;w)\ dy.$$

Thus, whereas θ declares a property at the microlevel involving only granules crossing instantaneously the same subplace y, $\hat{\theta}$ lifts that property at the mesolevel merging the whole lot of granules within $\mathbf{e}(x)$ into the spot x, though still keeping them arranged into batches in accord with their velocity; hence it preserves the statistical character of θ, contrary to what occurs of $\tilde{\theta}$ which takes rather, as we said, the attributes of a mesodensity, so that the value of the field ρ at x is given by

$$\mu\omega\delta^{-3}, \quad \omega = \int_{\mathbf{e}} \tilde{\theta}(\tau,x;y)\ dy.$$

The average velocity at y, $\tilde{w}$, is given obviously by

$$\tilde{w} = \tilde{\theta}^{-1} \int_{\mathscr{V}} \theta w\ dw$$

and the value of the field v at x by

$$v(\tau,x) = \omega^{-1} \int_{\mathbf{e}} \tilde{\theta}\tilde{w}\ dy.$$

Similarly

$$Y = \omega^{-1} \int_{\mathbf{e}} \tilde{\theta}y \otimes y\ dy, \quad K = \omega^{-1} \int_{\mathbf{e}} \tilde{\theta}y \otimes \tilde{w}\ dy$$

and

$$H = \omega^{-1} \int_{\mathbf{e}} \tilde{\theta}\tilde{c} \otimes \tilde{c}\ dy, \quad \text{with } \tilde{c} = \tilde{w} - v - By.$$

There is a cascade of peculiar velocities, including $\tilde{c}$. The deepest one is the peculiar random velocity c

$$c = w - \tilde{w} \quad \text{or} \quad c = \bar{c} - \tilde{c}$$

with $\bar{c} = w - v - By$.

Kinetic energy tensors (per unit mass) ensue: an exhaustive one, W, involving all details of motion

$$W = \frac{1}{2}\omega^{-1} \int_{\mathbf{e}} \int_{\mathscr{V}} \theta w \otimes w\ dydw = \frac{1}{2}\omega^{-1} \int_{\mathscr{V}} \hat{\theta}w \otimes w\ dw$$

and a mesoscopic 'reduced' one $\tilde{W}$:

$$\tilde{W} = \frac{1}{2}\omega^{-1} \int_{\mathbf{e}} \tilde{\theta}\tilde{w} \otimes \tilde{w}\ dy,$$

for which an explicit expression is easily derived

$$\tilde{W} = \frac{1}{2}\left(v \otimes v + BYB^T + H\right).$$ (9)

The difference

$$\hat{W} = W - \tilde{W} = \frac{1}{2}\omega^{-1}\int_{\mathscr{V}}\hat{\theta}c \otimes c\,dc$$ (10)

is not accessible via the density $\tilde{\theta}$ and the ensuing mesoscopic multifield theory alone; thermodynamic postulations must be brought to bear.

To put the matter into appropriate evidence, we must premise a theorem of kinetic energy associated with the balance equations (2)-(6). For any fit region $\mathfrak{b}$ belonging to $\mathscr{B}$ and under appropriate smoothness conditions, it reads, in tensorial form (i.e., as a virial theorem), as follows

$$\int_{\mathfrak{b}}\rho\left\{\dot{\tilde{W}} - \text{sym}\left[B\left(BYB^T\right) + H\right] + \sigma\tilde{W}\right\}dx =$$

$$= \int_{\mathfrak{b}}\rho\left[\text{sym}\left(v \otimes b + BM\right) + \frac{1}{2}J\right]dx +$$

$$+ \int_{\partial\mathfrak{b}}\left\{\text{sym}\left[v \otimes Tn + B\left(\mathbf{m}n\right)\right] + \frac{1}{2}\mathbf{j}n\right\}d\mathscr{H}^2 -$$

$$- \int_{\mathfrak{b}}\left[\frac{1}{2}Z + \text{sym}\left(LT^T + BA + \mathbf{bm}^t\right)\right]dx.$$ (11)

The sum of the first two addenda under the integral sign within the left-hand term express the coshaping time derivative of $\tilde{W}$

$$\overset{\triangle}{\tilde{W}} = \frac{1}{2}\left(v \otimes v\right)^{\cdot} + G\left[G^{-1}\left(\tilde{W} - v \otimes v\right)G^{-T}\right]^{\cdot}G^T;$$

besides: n is the unit vector normal to $\partial\mathfrak{b}$, the exponent T means transposition, the exponent t means minor right transposition when applied to a third order tensor, namely $(\mathbf{bm}^t)_{ij} = \mathbf{b}_{ihk}\mathbf{m}_{khj}$.

The tensor power of internal actions has density

$$\text{sym}\left(\frac{1}{2}Z + LT^T + BA + \mathbf{bm}^t\right);$$ (12)

it is invariant under observer motion if and only if

$$T = -A^T,$$ (13)

under the restriction for Z to depend on kinetic quantities only if they are intrinsic (such as σ, for instance).

The scalar power, i.e. the trace of (12) is similarly invariant under the already mentioned looser condition (7), again on the proviso that trace of Z be independently invariant. It is invariably agreed that either (7) or (13) (depending on the depth of the choice) should be satisfied identically by the constitutive laws for T and A, rather than only verified along actual flows, as an interpretation of either as balance laws would directly suggest.

Peculiarly, the tensor $\mathbf{j}$ does not affect the power (12); consequently we will assume later that it does not react to the enforcement of any internal constraint. Similarly, the absence of the deviator of Z persuades us that $\mathrm{dev}\,Z$ cannot contain, under internally constrained circumstances a reactive additive component. Of course, the matter may be contentious and, again, some later developments, based on the mentioned convictions, may possibly need amendment.

Formally one can write a further equation for the sole portion

$$W^* = \tilde{W} - \frac{1}{2}v \otimes v = \frac{1}{2}\left(BYB^T + H\right)$$

of the kinetic energy W, portion due only to the deeper motions accounted for merely by the multifield theory. Under conditions of sufficient regularity and in view of the arbitrariness in the choice of $\mathfrak{b}$, that equation can be written in differential form

$$\rho\left\{\dot{W}^* - 2\,\mathrm{sym}\left[\left(B - \frac{1}{2}\sigma I\right)W^*\right]\right\} = \tag{14}$$

$$= \rho\left(\mathrm{sym}\,BM + \frac{1}{2}J\right) + \mathrm{sym}\,\mathrm{div}\,(B\mathbf{m}) + \frac{1}{2}\,\mathrm{div}\,\mathbf{j} - \left[\frac{1}{2}Z + \mathrm{sym}\left(BA + \mathbf{bm}^t\right)\right].$$

Taking the trace of both sides and using the notation

$$\kappa^* = \mathrm{tr}\,W^*, \quad \lambda = \rho\,\mathrm{tr}\left[B\left(2W^* + M\right)\right] + \frac{1}{2}\,\mathrm{tr}\left(\rho J - Z\right), \quad h_i = -B_{jk}\mathbf{m}_{kji} - \frac{1}{2}\mathbf{j}_{jji} \tag{15}$$

we obtain a scalar equation in a format convenient for thermodynamic speculations

$$\rho\dot{\kappa}^* + \rho\kappa^*\sigma + B\cdot A^T + \mathbf{b}^t\cdot\mathbf{m} = \rho\lambda - \mathrm{div}\,h. \tag{16}$$

4 Balance Laws for Hypocontinua

The generality (but, also, the consequent, perhaps disturbing, residual vagueness) of the proposed theory for ephemeral continua must be counterbalanced by descending to some specific issues or by offering a link with at least one known set of concepts and results. The choice below exhibits a route towards (and an alternative interpretation of) rules of hypoelastic and hypoplastic behaviour.

For our purpose the most obvious artifice is to choose a significant, possibly simple, constraint and to examine its consequences. Take the strong restriction of exact local coincidence of macro and meso placement gradients: $F = G$ or, rather, of the corresponding rates: $L = B$; a coincidence meant to be assured by a 'perfect' internal constraint. The accepted tenet (appropriately extended here) is that, then, the stresses T and A, hyperstress and power $\operatorname{tr} Z$ split into the sum of active and reactive components: $T = \overset{a}{T} + \overset{r}{T}$, etc., of which only the first ones can be specified constitutively and the second must be eliminated to obtain 'pure' equations of balance.

The process may be implemented readily when the 'perfection' of the constraint is intended to imply that the reactive agents (performing, as already figured, only through T, A and $\operatorname{tr} Z$) operate altogether powerlessly during any allowed process (i.e., here, a process within which

$$B = L, \quad \mathbf{b} = \operatorname{grad} L = \operatorname{grad} \operatorname{grad} v$$

and, of course, also $\sigma = 0$):

$$\frac{1}{2} \operatorname{tr} \overset{r}{Z} + L \cdot \left(\overset{r}{A} + \overset{r}{T}{}^T \right) + {}^t\overset{r}{\mathbf{m}} \cdot \operatorname{grad} \operatorname{grad} v = 0$$

for any choice of the fields v and $L = \operatorname{grad} v$. If $\operatorname{tr} Z$ is not affected by the choice of L, as already presumed, and accounting for the absence of reactive components for $\operatorname{dev} Z$, again as already evidenced, we conclude that

$$\overset{r}{Z} = 0, \quad \overset{r}{A} = -\overset{r}{T}{}^T, \quad \overset{r}{\mathbf{m}}_{ijk} + \overset{r}{\mathbf{m}}_{kji} = 0.$$

In a large class of bodies (see [1], Sect. 8) $\mathbf{m}$ is symmetric in the first two indices and we accept that the property hold also here and transit to $\overset{r}{\mathbf{m}}$; we conclude that

$$\overset{r}{\mathbf{m}}_{jik} + \overset{r}{\mathbf{m}}_{jki} = 0,$$

therefore $\overset{r}{\mathbf{m}}$ turns out to be skew in the two last indices and, finally, its double divergence vanishes

$$\overset{r}{\mathbf{m}}_{ijk,jk} = 0.$$

Now, by taking the transpose of the divergence of each member of the fourth equation (5) and then subtracting it, thus modified and hence devoid of $\overset{r}{\mathbf{m}}$, member by member from (4), we can eliminate also the reactive components of T and A.

Thus we come to the set of balance laws we were seeking for a hypocontinuum:

$$\frac{\partial \rho}{\partial \tau} + \operatorname{div} (\rho v) = 0, \tag{17}$$

$$\frac{\partial Y}{\partial \tau} + (\operatorname{grad} Y) v - Y L^T - LY = 0, \tag{18}$$

$$\rho \left(\frac{\partial v}{\partial \tau} + Lv \right) - \operatorname{div} \left\{ \rho \left[\frac{\partial L}{\partial \tau} + (\operatorname{grad} L) v \right] Y \right\} =$$

$$= \rho b - \operatorname{div} \left(\rho H^T \right) + \operatorname{div} \left[\operatorname{sym} \left(\overset{a}{T} + \overset{a}{A} \right) - \rho H - \operatorname{div} \overset{a}{\mathbf{m}} \right], \tag{19}$$

$$\rho \left[\frac{\partial H}{\partial \tau} + (\operatorname{grad} H) v - HL^T - LH \right] = \rho J - Z + \operatorname{div} \mathbf{j}. \qquad (20)$$

The complexity even of this 'reduced' system is obvious at first glance: blatantly it involves third-order partial derivatives of v. Thus, in general, a deep analytical study is required, even beyond physical interpretations of events depicted. For instance, boundary conditions to be associated with the system cannot be expected to mimic the standard model strictly; even a trivial recourse to the divergence theorem shows that inertial effects might be relevant at the frontier. Granularity and permeability of restraining walls may play an important role; their influence on the inner flows must be identified and portrayed mathematically and that portrayal demands information on the 'substructure' of the boundary. In any case the variety of continua for which the balance laws may be presumed to apply makes general statements unfeasible: constitutive decisions must be taken also in regards of boundary (and initial?) conditions. When extremely loose matter (such as a gas) is studied, free boundaries have no meaning; rather, containers with rigid (possibly moving) walls are usually involved or exterior domains beyond rigid vessels. For loose (though not so loose) matter, free boundaries might imply that ρ tend to zero there; at least ρH tend to zero or, minimally, $\rho H n = 0$, n, unit normal vector.

For the purposes of this section, it seems advisable to restrict further attention to still more particular situations; the proposals advanced and successfully pursued in hypoelasticity and hypoplasticity suggest consideration of the case when Y is relatively very small; in other words, when stirring of the grains (broadly measured by H) has heavier effects than whirling (broadly measured by moments of momentum). If terms involving Y, M and $\mathbf{m}$ can be disregarded, the equation (18) drops out of the system and (19) reads

$$\rho \left(\frac{\partial v}{\partial \tau} + Lv \right) = \rho b + \operatorname{div} \left[\operatorname{sym} \left(\overset{a}{T} + \overset{a}{A} \right) - \rho H \right] \qquad (21)$$

whereas the first and the fourth remain nominally the same. The formal similarity with well-known equations appears evident (compare, for instance, with (5), (6), (7) of [2]). The only glaring conflict is in the option selected for the unknown function in the last equation: convincingly it is here the kinetic quantity H, as is v and was originally K in the other balance laws, rather than a jarring dynamic quantity as the stress. However, multiplied by ρ, H achieves the physical dimensions of a stress; as such it appears under the divergence operator in the right-hand side of (21) as a legitimate representative of the pressure generated by collisions. There is no term corresponding to our active components of T and A in (7) nor hyperstress in (5) of [2], but the absences are constitutive choices, admissible within our context.

Another instance where there are significant, though not so radical, simplifications occurs when the constraint calls on B to differ from L by a skew addendum (i.e., a mesorotation is compounded with the macrostretching) and Y is supposed to be proportional to the identity by a factor α. Then σ vanishes again, the equation (3) provides simply a differential equation for α

$$\frac{\partial \alpha}{\partial \tau} + (\operatorname{grad} \alpha) \cdot v = \frac{2}{3}\alpha \operatorname{tr} L.$$

The reactive component of A must be symmetric and must coincide with the similar component of T, so that they can be both be eliminated by combining appropriately the equation (5) with (4). Finally the equation obtained from (5) by taking the skew components of both sides becomes also 'pure' and conditions the choice of the skew addendum in B.

Many questions, of course, remain still open even leaving aside demands regarding thermodynamic phenomena associated with the mechanical decisions suggested above.

References

1. Capriz, G.: On ephemeral continua. Physical Mesomechanics 11, 285–298 (2008)
2. Truesdell, C.A.: Hypoelasticity. J. Rational Mech. Anal. 4, 83–133 (1955)
3. Kolymbas, D.: An outline of hypoplasticity. Arch. Appl. Mech. 61, 143–154 (1991)
4. Truesdell, C.A., Toupin, R.A.: The classical field theories. In: Handbuch der Physik, III/1. Springer, Berlin (1960)

On Constitutive Choices for Smectic Elastomers

Pasquale Giovine

Abstract. A thermo–mechanical continuum model for smectic-C elastomers is developed within the setting of multifield theories describing material substructures. Smectic elastomers are layered materials exhibiting a solid-like elastic response along the layer normal and a rubbery one in the plane, hence possess microstructure both of the material and local type (the nematic microstructure and the lamellae, respectively). The balance equations are derived from the general theory of continua with constrained microstructure [2] and, after, the appropriate constitutive relations are proposed along with the thermodynamic restrictions and the invariance principles. At the end we compare our theory with two previous proposals which are recovered to be particular cases of this one.

1 Introduction

Liquid crystal elastomers [16] have attracted much attention in recent years because they uniquely combine the rubber elasticity of polymer networks with the anisotropic properties of liquid crystals [10, 15] and, therefore, provide exciting challenges for fundamental research (experimental and theoretical) and open new possibilities for novel device applications, for instance in sensors and actuators, and for new materials utilized, particularly, at microscale; we refer, for example, to shape memory alloys, shear-thickening fluids, metallic foams and rubber elastomers.

Essentially, any phase known from conventional liquid crystals can be made in elastomeric form, such as, *e.g.*, nematic, smectic-A and smectic-C. Among the various phases, nematic elastomers have been studied most extensively to date, so we focus our attention by particularly restricting the interest on a specific sub-set of the elastomers, namely the smectic-C elastomers, which display a number of interesting

Pasquale Giovine
Dipartimento di Meccanica e Materiali, Facoltà di Ingegneria, Università Mediterranea,
Via Graziella n.1, Località Feo di Vito, I-89122, Reggio Calabria, Italy
e-mail: giovine@unirc.it

mechanical and optical properties, with potential applications ranging from nonlinear optics to artificial muscles, due to the coupling between liquid crystal ordering transition and rubber elasticity [16]. Experimental interest in smectic-C elastomers is also strongly motivated by the existence of a chiral ferroelectric phase.

The present paper is organized as follows: the geometric model and the kinematic descriptors are discussed in §2; balance equations for classical and microstructural interactions are described in §3; §4 deals with constraints and §5 with constitutive prescriptions; concluding remarks and comparisons are presented in §6.

2   The Smectic Elastomers

Smectic-C liquid crystals elastomers are rubbery materials that have the macroscopic symmetry properties of smectics liquid crystals and are characterized by rod-like molecules (mesogen) assembled in a layered structure, and tilted at a fixed angle with respect to the layer normal [13]. The mesogens are attached to polymer chains, which are cross-linked to obtain a rubber-like solid; the coupling to liquid crystal ordering leads to rubbery response in the tangential directions, and solid-like along the layer normal.

In the smectic-A phase, the director $\mathbf{n}$ describing the average orientation of constituent mesogens is parallel to the normal $\mathbf{m}$ of the smectic layers, whereas in the smectic-C phase, it has a component in the plane of the layers.

The new techniques has produced elastic networks that stabilize the liquid crystalline order so that mono-domain or single crystal samples result and they leave at same time sufficient mobility for the mesogenic component to reorient when mechanical or electrical fields are applied.

As representing in Fig. 1, to describe smectic ordering we employ the unit layer normal $\mathbf{m}$ and the unit mesogen director $\mathbf{n}$ which describe the local orientation of the mesogens. Clearly $\mathbf{n}$ can be decomposed into its components parallel and orthogonal to $\mathbf{m}$, defining, in this manner, the vector $\mathbf{t}$, usually called as $\mathbf{t}$-director.

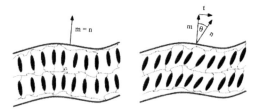

Fig. 1 The cross-section, before (left) and after (right), of a deformed smectic-C elastomers.

Molecules of smectics are organized in layers, the lamellae, delimited by surfaces Σ_ξ, ξ an integer, defined by

$$\omega(\mathbf{x}, \tau, \xi \lambda_*) = 0 , \tag{1}$$

where λ_* is an appropriate lenght-scale and τ the time. We do not presume here (as it was done in [4, 1]) that surfaces are necessarily equidistant: the layers are allowed to compress or expand paying in energy, even if we exclude any permeation of molecules from one layer to another in our developments. Thus a larger class of motions is possible.

To fix ω uniquely, one may exploit a property which stems from the quasi-solid behaviour of the smectics, that is the existence of a reference placement, at an instant τ_*, where the surfaces Σ_ξ^* are equidistant planes for which

$$\omega(\mathbf{x}_*, \tau_*, \xi\lambda_*) = \mathbf{m}_* \cdot \mathbf{x}_* - \xi\lambda_* \,, \tag{2}$$

where $\mathbf{m}_*$ is a constant unit vector normal to the planes and λ_* the distance between two adjacent planes. But we also suppose that the surfaces are entrained by the gross motion, say $\mathbf{x} = \tilde{\mathbf{x}}(\mathbf{x}_*, \tau)$, thus, by using the inverse image $\mathbf{x}_* = \bar{\mathbf{x}}_*(\mathbf{x}, \tau)$ of the motion, we can write Eq. (1) as

$$\left[\omega(\mathbf{x}, \tau, \xi\lambda_*) = \right] \mathbf{m}_* \cdot \bar{\mathbf{x}}_*(\mathbf{x}, \tau) - \xi\lambda_* = 0 \,. \tag{3}$$

We observe that ω should be defined over a discrete set of values $\{\xi\lambda_*\}$, but the cardinality of $\{\xi\lambda_*\}$ is so high, while the thickness of layers so minute, that we can take $\omega(\mathbf{x}, \tau, \cdot)$ as defined over a whole interval of real numbers, within the continuum scheme, so that, at each instant τ, the collection of all surfaces Σ_ξ completely fills the smectic cell under consideration, whereas $\Sigma_\xi \cap \Sigma_{\xi'} = \emptyset$, if $\xi \neq \xi'$.

Now let λ be the distance between two adjacent smectic surfaces, hence, if $\mathbf{x}$ is a point of the ξth surface, for Eq. (1) it satisfies the relation $\omega(\mathbf{x}, \tau, \xi\lambda_*) = 0$, then $(\mathbf{x} + \lambda\mathbf{m})$ is a point of the $(\xi + 1)^{th}$ surface and therefore satisfies the relation $\omega(\mathbf{x} + \lambda\mathbf{m}, \tau, (\xi + 1)\lambda_*) = 0$. Moreover, for Eq. (3), ω depends linearly on ξ and we obtain:

$$\omega(\mathbf{x} + \lambda\mathbf{m}, \tau, (\xi + 1)\lambda_*) \approx \omega(\mathbf{x}, \tau, \xi\lambda_*) + (\text{grad}\,\omega) \cdot \lambda\mathbf{m} - \lambda_* = 0. \tag{4}$$

Hence, always for Eq. (3),

$$|\text{grad}\,\omega|\lambda = \lambda_* \tag{5}$$

and $|\text{grad}\,\omega|^{-1}$ represents the thickness of the lamellae measured in multiples of λ_*, for which there is a dilatation of the layer if $|\text{grad}\,\omega| < 1$ and a compression if $|\text{grad}\,\omega| > 1$.

Besides, as the lamellae follow the macromotion, if we introduce the gross deformation gradient $\mathbf{F} := \frac{\partial \mathbf{x}}{\partial \mathbf{x}_*}$, a second order tensor with positive determinant, the unit layer normal $\mathbf{m}$ is given by

$$\mathbf{m} = \frac{\mathbf{F}^{-T}\mathbf{m}_*}{|\mathbf{F}^{-T}\mathbf{m}_*|} \,, \tag{6}$$

where the exponent $-T$ indicates the transpose of the inverse (see, *e.g.*, Eq. (1.24) of [12]).

3   Balance Equations for Continua with Vectorial Microstructure

We obtain the thermo-dynamics equations for smectic elastomers within the setting of the general multifield theory of continua with microstructure depicted in [2], by using the developments pursued in [3, 7] and adapted to our broader context.

Were the microstructural variable a privileged material direction characterized by a vector $\mathbf{n}$, hence a rotation $\mathbf{Q} = e^{-\mathscr{E}\mathbf{s}}$ of the observer of associated axial vector $\mathbf{s}$, where $\mathscr{E}$ is Ricci's permutation tensor and e the basis of natural logarithms, induces the vector $\mathbf{n}$ to change into $\mathbf{n}_s = \mathbf{Q}\mathbf{n}$; moreover, the infinitesimal generator $\mathscr{A}$ of the group of rotations on the microstructure in the translation space $\mathscr{V}$, i.e., the operator describing the effect of a rotation of the observer on the value $\mathbf{n}_s$ of the microstructure to the first order in $\mathbf{s}$, is given by

$$\mathscr{A}(\mathbf{n}) := \left.\frac{d\mathbf{n}_s}{d\mathbf{s}}\right|_{\mathbf{s}=\mathbf{0}} = \mathscr{E}\mathbf{n}. \tag{7}$$

The total kinetic energy density per unit mass is here the sum of the classical translational term $\frac{1}{2}\dot{\mathbf{x}}^2$ and of the microstructural one $\kappa(\mathbf{n},\dot{\mathbf{n}})$, which expresses the inertia related to the admissible mesogenic micromotions (the superposed dot denotes the material time derivative); this extra term is a non-negative scalar function, homogeneous in $\mathbf{n}$, such that $\kappa(\mathbf{n},\mathbf{0}) = 0$ and $\frac{\partial^2 \kappa}{\partial \dot{\mathbf{n}}^2} \neq \mathbf{0}$, and it is related to the kinetic co-energy $\chi(\mathbf{n},\dot{\mathbf{n}})$ by the Legendre transform (see, also, [5]):

$$\frac{\partial \chi}{\partial \dot{\mathbf{n}}} \cdot \dot{\mathbf{n}} - \chi = \kappa. \tag{8}$$

The equations of balance governing an admissible thermomechanical process for a continuum with vectorial microstructure in absence of constraints are then

$$\dot{\rho} + \rho \operatorname{div}\dot{\mathbf{x}} = 0, \tag{9}$$

$$\rho\ddot{\mathbf{x}} = \rho\mathbf{f} + \operatorname{div}\mathbf{T}, \tag{10}$$

$$\rho\left[\frac{d}{d\tau}\left(\frac{\partial \chi}{\partial \dot{\mathbf{n}}}\right) - \frac{\partial \chi}{\partial \mathbf{n}}\right] = \rho\mathbf{b} - \mathbf{z} + \operatorname{div}\mathbf{S}, \tag{11}$$

$$\operatorname{skw}\mathbf{T} = \frac{1}{2}\mathscr{E}\left[\mathscr{A}^T\mathbf{z} + \left(\operatorname{grad}\mathscr{A}^T\right)\mathbf{S}\right] = \mathbf{0}, \tag{12}$$

$$\rho\dot{\varepsilon} = \mathbf{T}\cdot\mathbf{L} + \mathbf{z}\cdot\dot{\mathbf{n}} + \mathbf{S}\cdot\operatorname{grad}\dot{\mathbf{n}} + \rho\lambda - \operatorname{div}\mathbf{q}, \tag{13}$$

$$\rho\dot{\eta} \geq \rho\lambda\theta^{-1} - \operatorname{div}\left(\theta^{-1}\mathbf{q}\right), \tag{14}$$

where $\operatorname{div}(\cdot)$ denotes the divergence field and $\operatorname{skw}(\cdot)$ the skew part of a 2nd–order tensor.

Eq. (9) is the conservation law of mass, with ρ the mass density and $\mathbf{L}$ the usual velocity gradient so defined: $\mathbf{L} := \operatorname{grad}\dot{\mathbf{x}}\,(=\dot{\mathbf{F}}\mathbf{F}^{-1})$; Eq. (10) is the standard law of balance of Cauchy, where $\rho\mathbf{f}$ is the vector density per unit volume of external bulk forces and $\mathbf{T}$ the Cauchy's stress tensor; Eq. (11) is the balance of microstructural interactions, in which $\rho\mathbf{b}$ and $-\mathbf{z}$ are the resultant vector densities per unit volume of external bulk micro-interactions and of internal self-force, respectively, while $\mathbf{S}$ is the 2nd–order microstress tensor that, in general, is not necessarily completely related to a sort of boundary microtractions, unless it is possible to define a physically significant connection on the manifold of values of the microstructure by which the gradient on it may be evaluated in covariant manner (see [6]); Eq. (12) is the balance law of moment of momentum in which we can observe that the Cauchy's tensor $\mathbf{T}$ is a not necessarily symmetric–valued tensor field, unless microstructural contributes are null; Eq. (13) is the balance of internal energy of specific density per unit mass ε, where λ is the scalar rate of heat generation per unit mass due to irradiation and $\mathbf{q}$ the heat flux vector; the principle of entropy (14), whose specific density per unit mass is η, is here accepted as it applies in its classical purely thermal form, with θ the (positive) absolute temperature.

The choice for an explicit expression of the kinetic co-energy χ is further improved by the fact to must have, as κ, the same value for all observers at rest, that is, it must be invariant under the Galilean group and hence satisfy the condition

$$\dot{\mathscr{A}}^T\frac{\partial\chi}{\partial\dot{\mathbf{n}}}+\mathscr{A}^T\frac{\partial\chi}{\partial\mathbf{n}}=\mathbf{0};\tag{15}$$

by inserting the expression (7) of $\mathscr{A}$ into Eq. (15), one obtains the relation

$$\dot{\mathbf{n}}\times\frac{\partial\chi}{\partial\dot{\mathbf{n}}}+\mathbf{n}\times\frac{\partial\chi}{\partial\mathbf{n}}=\mathbf{0}:\tag{16}$$

this condition requires χ to depend on $\dot{\mathbf{n}}$ and $\mathbf{n}$ only through the scalar products $|\dot{\mathbf{n}}|^2$, $\dot{\mathbf{n}}\cdot\mathbf{n}$ and $|\mathbf{n}|^2$. Now we observe that $\mathbf{n}$ is a unit vector, thus $\dot{\mathbf{n}}\cdot\mathbf{n}$ is null and $|\mathbf{n}|^2=1$ and hence χ depends only on $|\dot{\mathbf{n}}|^2$; usually one thinks of the kinetic co-energy as quadratic in $\dot{\mathbf{n}}$, that is $\chi=\frac{1}{2}\beta|\dot{\mathbf{n}}|^2$, with β a scalar coefficient of mesogenic inertia.

By inserting the kinetic co-energy into Eq. (11), by using again the expression (7) of $\mathscr{A}$ now into Eq. (12) and by introducing the Helmholtz free energy per unit mass $\psi(:=\varepsilon-\theta\eta)$ into the inequality of Clausius–Duhem (13), we obtain the following equivalent form of previous relations:

$$\rho\beta\ddot{\mathbf{n}}=\rho\mathbf{b}-\mathbf{z}+\mathrm{div}\,\mathbf{S},\tag{17}$$

$$\mathrm{skw}\left[\mathbf{T}^T+\mathbf{n}\otimes\mathbf{z}+(\mathrm{grad}\,\mathbf{n})\mathbf{S}^T\right]=\mathbf{0},\tag{18}$$

$$\rho\left(\dot{\psi}+\dot{\theta}\eta\right)\leq\mathbf{T}\cdot\mathbf{L}+\mathbf{z}\cdot\dot{\mathbf{n}}+\mathbf{S}\cdot\mathrm{grad}\,\dot{\mathbf{n}}-\theta^{-1}\mathbf{q}\cdot\mathrm{grad}\,\theta,\tag{19}$$

where the symbol $\otimes$ denotes the usual diadic tensor product between vectors.

4   Constraints for Smectic Elastomers

Here $\mathbf{n}$ represents the unit mesogen director of the optical axis in the current placement and, hence, all time-rates of change of $\mathbf{n}$ must satisfy the property

$$\dot{\mathbf{n}} = \mathbf{P}\mathbf{u}, \tag{20}$$

where $\mathbf{P}$ is the projector in the plane orthogonal to $\mathbf{n}$, that is $\mathbf{P} := \mathbf{I} - \mathbf{n} \otimes \mathbf{n}$, and $\mathbf{u}$ an arbitrary vector in $\mathscr{V}$.

Therefore we are in presence of the *internal mechanical constraints* (6) and (20), in addition to the fact that the continuum is incompressible and so the symmetric part of $\mathbf{L}$, $\mathbf{D}$ ($:= \mathrm{sym}\,\mathbf{L}$), must be taken as traceless: thus one of its eigenvalues vanishes, while the other two are opposite.

By following classical theories (see [11] and [8]), we suppose that each quantity which, in absence of the constraints, is ruled by a constitutive prescription (that is $\mathbf{T}, \mathbf{z}, \mathbf{S}, \mathbf{q}, \varepsilon, \eta, \psi$) is now the sum of one *active* and one *reactive* component

$$\mathbf{T} = \mathbf{T}_a + \mathbf{T}_r, \quad \mathbf{z} = \mathbf{z}_a + \mathbf{z}_r, \quad \text{etc.} \tag{21}$$

and only the active component is given in term of the independent thermokinetic variables through suitable constitutive relations.

In the case of *internally frictionless* constraints, the reactive part is specified, in this wider thermomechanic rather than purely mechanical context, by the property that the entropy production due to the reaction is null, that is the contribution of the reactions to the inequality (19) are identically zero for every process allowed by the constraint (see, also, §27 of [2]):

$$\rho \left(\dot{\psi}_r + \dot{\theta}\eta_r \right) + \theta^{-1}\mathbf{q}_r \cdot \mathrm{grad}\,\theta = \mathbf{T}_r \cdot \mathbf{L} + \mathbf{z}_r \cdot \dot{\mathbf{n}} + \mathbf{S}_r \cdot \mathrm{grad}\,\dot{\mathbf{n}}. \tag{22}$$

By placing the constraint (20) into Eq. (22), we obtain the relation

$$\rho \left(\dot{\psi}_r + \dot{\theta}\eta_r \right) + \theta^{-1}\mathbf{q}_r \cdot \mathrm{grad}\,\theta = \tag{23}$$
$$= \mathbf{T}_r \cdot \mathbf{L} + \left\{ \mathbf{P}^T \mathbf{z}_r + \left[\mathrm{grad}\left(\mathbf{P}^T \right) \right] \mathbf{S}_r \right\} \cdot \mathbf{u} + \left(\mathbf{S}_r^T \mathbf{P} \right) \cdot \mathrm{grad}\,\mathbf{u},$$

for every totally free choice of an arbitrary scalar $\dot{\theta}$, two arbitrary vectors $\mathbf{u}$ and $(\mathrm{grad}\,\theta)$, an arbitrary tensor $(\mathrm{grad}\,\mathbf{u})$ and an arbitrary traceless tensor $\mathbf{L}$, due to incompressibility constraint of the smectic elastomers.

Hence we are led to the following conditions

$$\psi_r = \mathrm{const.}, \quad \eta_r = 0, \quad \mathbf{q}_r = \mathbf{0}, \quad \mathbf{T}_r = -\pi \mathbf{I},$$
$$\mathbf{P}^T \mathbf{z}_r + \left[\mathrm{grad}\left(\mathbf{P}^T \right) \right] \mathbf{S}_r = \mathbf{0}, \quad \mathbf{S}_r^T \mathbf{P} = \mathbf{0} \tag{24}$$

($\mathbf{I}$ being the identity tensor). The transpose of last condition assures us that

$$\left[\mathrm{grad}\left(\mathbf{P}^T \right) \right] \mathbf{S}_r = \mathrm{grad}\left(\mathbf{P}^T \mathbf{S}_r \right) - \mathbf{P}^T \,\mathrm{div}\,\mathbf{S}_r = -\mathbf{P}^T \,\mathrm{div}\,\mathbf{S}_r \tag{25}$$

and therefore, from Eq. (24)$_5$ and the symmetry of P, we have that

$$\mathbf{P} \left(\mathbf{z}_r - \mathrm{div}\,\mathbf{S}_r \right) = \mathbf{0}, \tag{26}$$

that is, the reactive part of the total distress of smectic-C elastomers has null projection in the plane orthogonal to the mesogen director $\mathbf{n}$; hence we are able to project the micromomentum balance (17) on that plane in order to obtain a first pure field equation governing all thermo–mechanical processes of smectic elasomers:

$$\mathbf{P}\left(\rho\beta\ddot{\mathbf{n}} - \rho\mathbf{b} + \mathbf{z}_a - \operatorname{div}\mathbf{S}_a\right) = \mathbf{0}. \tag{27}$$

We obtain a second pure field equation from the balance of angular momentum (18), if we use Eq. (17) and observe that $\mathbf{T}_r$ is symmetric and relation (24)$_6$ can be written in this alternative way:

$$\mathbf{S}_r = \mathbf{n} \otimes \left(\mathbf{S}_r^T \mathbf{n}\right); \tag{28}$$

therefore we have:

$$
\begin{aligned}
\operatorname{skw}\mathbf{T}_a &= \operatorname{skw}\left[\mathbf{n}\otimes\mathbf{z} + (\operatorname{grad}\mathbf{n})\mathbf{S}^T\right] = \\
&= \operatorname{skw}\left[\mathbf{n}\otimes(\mathbf{z} - \operatorname{div}\mathbf{S}) + \operatorname{div}(\mathbf{n}\otimes\mathbf{S})\right] = \\
&= \operatorname{skw}\left[\mathbf{n}\otimes\rho\,(\mathbf{b} - \beta\ddot{\mathbf{n}}) + \operatorname{div}(\mathbf{n}\otimes\mathbf{S}_a) + \operatorname{div}(\mathbf{n}\otimes\mathbf{S}_r)\right] = \\
&= \operatorname{skw}\left\{\mathbf{n}\otimes\rho\,(\mathbf{b} - \beta\ddot{\mathbf{n}}) + \operatorname{div}(\mathbf{n}\otimes\mathbf{S}_a) + \operatorname{div}\left[\mathbf{n}\otimes\mathbf{n}\otimes\left(\mathbf{S}_r^T\mathbf{n}\right)\right]\right\} = \\
&= \operatorname{skw}\left[\mathbf{n}\otimes\rho\,(\mathbf{b} - \beta\ddot{\mathbf{n}}) + \operatorname{div}(\mathbf{n}\otimes\mathbf{S}_a)\right],
\end{aligned} \tag{29}
$$

since the tensor $\left\{\operatorname{div}\left[\mathbf{n}\otimes\mathbf{n}\otimes\left(\mathbf{S}_r^T\mathbf{n}\right)\right]\right\}$ is symmetric.

Moreover, once a micromotion of mesogen directors is ensued from Eq. (27), for Eq. (26) we are able to express the corresponding microstructural reactions to constraints by projecting Eq. (17) along $\mathbf{n}$, within the intrinsic indeterminacy derived from a partial identity of effects of the fields $\mathbf{z}_r$ and $\mathbf{S}_r$, as pointed out in §§205 and 227 of [14] or in Remark 1, §3 of [9]: it is

$$\mathbf{z}_r - \operatorname{div}\mathbf{S}_r = \left\{\left[\rho\,(\mathbf{b} - \beta\ddot{\mathbf{n}}) - \mathbf{z}_a + \operatorname{div}\mathbf{S}_a\right]\cdot\mathbf{n}\right\}\mathbf{n}. \tag{30}$$

By inserting the expressions of reactive part of Cauchy stress (24)$_4$ and of skew part of active one (29) in the Cauchy's Eq. (10), we obtain:

$$\rho\ddot{\mathbf{x}} = \rho\mathbf{f} - \operatorname{grad}\pi + \operatorname{div}\left\{\operatorname{sym}\mathbf{T}_a + \operatorname{skw}\left[\mathbf{n}\otimes\rho\,(\mathbf{b} - \beta\ddot{\mathbf{n}}) + \operatorname{div}(\mathbf{n}\otimes\mathbf{S}_a)\right]\right\}. \tag{31}$$

Instead the effects due to constraints disappear from the equation of evolution for the temperature of the smectic-C elastomer, in fact by using the definition of the Helmholtz free energy ψ and results (24) and (27), other than Eqs. (20) and (21), in the energy balance (13), we get the following equation at once

$$\rho\overline{(\dot{\psi}_a + \theta\eta_a)} = \mathbf{T}_a\cdot\mathbf{L} + \rho\left[\mathbf{P}(\mathbf{b} - \beta\ddot{\mathbf{n}})\right]\cdot\mathbf{u} + \operatorname{div}\left[\mathbf{S}_a^T(\mathbf{Pu}) - \mathbf{q}_a\right] + \rho\lambda \tag{32}$$

as the term $(-\pi\operatorname{tr}\mathbf{L})$ is null for the incompressibility of the continuum.

In Eq. (32) the term $\left[\mathbf{S}_a^T(\mathbf{Pu})\right]$ is the so-called interstitial work flux of mechanical nature in excess of the usual flux due to surface tractions and owing to interactions between lamellae, $[-\rho\beta(\mathbf{P\ddot{n}})\cdot\mathbf{u}]$ is the rate of working of the mesogen microinertia and $[\rho(\mathbf{Pb})\cdot\mathbf{u}]$ is an elastomer heat source due to external microactions.

5 Restrictions on the Constitutive Equations

Here we consider the simple thermoelastic case for which all the active components of dependent constitutive fields are given by smooth functions of the set of thermokinetic variables $\mathscr{S} \equiv \{\mathbf{F}, \mathbf{n}, \mathbf{N} := \mathrm{grad}\,\mathbf{n}, \theta, \mathbf{g} := \mathrm{grad}\,\theta\}$, i.e.,

$$\{\psi_a, \eta_a, \mathbf{T}_a, \mathbf{z}_a, \mathbf{S}_a, \mathbf{q}_a\} = \{\hat{\psi}, \hat{\eta}, \hat{\mathbf{T}}, \hat{\mathbf{z}}, \hat{\mathbf{S}}, \hat{\mathbf{q}}\}(\mathscr{S}); \tag{33}$$

hence we check the compatibility of these prescriptions with the Clausius–Duhem inequality, in its reduced version (19), by introducing the constraint (20), the condition (22) of perfect constraints, the functional dependence of the free energy ψ_a, by using the chain rule, and the identity: $\overline{\mathrm{grad}\,\mathbf{n}} = \mathrm{grad}\,\dot{\mathbf{n}} - (\mathrm{grad}\,\mathbf{n})\mathbf{L}$. It is

$$\left(\mathbf{T}_a - \rho\frac{\partial\psi}{\partial\mathbf{F}}\mathbf{F}^T + \rho\mathbf{N}^T\frac{\partial\psi}{\partial\mathbf{N}}\right)\cdot\mathbf{L} + \left[\mathbf{P}\left(\mathbf{S}_a - \rho\frac{\partial\psi}{\partial\mathbf{N}}\right)\right]\cdot\mathrm{grad}\,\mathbf{u} +$$
$$+ \left\{\mathrm{grad}\left[\mathbf{P}\left(\mathbf{S}_a - \rho\frac{\partial\psi}{\partial\mathbf{N}}\right)\right] + \mathbf{P}\left[\mathbf{z}_a - \rho\frac{\partial\psi}{\partial\mathbf{n}} - \mathrm{div}\left(\mathbf{S}_a - \rho\frac{\partial\psi}{\partial\mathbf{N}}\right)\right]\right\}\cdot\mathbf{u} -$$
$$- \rho\frac{\partial\psi}{\partial\mathbf{g}}\cdot\dot{\mathbf{g}} - \rho\dot{\theta}\left(\eta_a + \frac{\partial\psi}{\partial\theta}\right)\dot{\theta} - \theta^{-1}\mathbf{q}_a\cdot\mathbf{g} \geq 0, \tag{34}$$

for any traceless tensor $\mathbf{L}$, tensor $(\mathrm{grad}\,\mathbf{u})$, vectors $\mathbf{u}$ and $\dot{\mathbf{g}}$ and scalar $\dot{\theta}$, where the index a for the free energy ψ is implied.

Therefore the following constitutive equations hold for smectic elastomers:

$$\left(\mathbf{T}_a - \rho\frac{\partial\psi}{\partial\mathbf{F}}\mathbf{F}^T + \rho\mathbf{N}^T\frac{\partial\psi}{\partial\mathbf{N}}\right) \in \mathrm{Sph}, \quad \mathbf{P}\left(\mathbf{S}_a - \rho\frac{\partial\psi}{\partial\mathbf{N}}\right) = \mathbf{0}, \quad \frac{\partial\psi}{\partial\mathbf{g}} = 0$$
$$\mathbf{P}\left[\mathbf{z}_a - \rho\frac{\partial\psi}{\partial\mathbf{n}} - \mathrm{div}\left(\mathbf{S}_a - \rho\frac{\partial\psi}{\partial\mathbf{N}}\right)\right] = \mathbf{0}, \quad \eta_a = -\frac{\partial\psi}{\partial\theta}, \tag{35}$$
$$\mathbf{q}_a\cdot\mathbf{g} \leq 0,$$

where the usual Gibbs relation for the entropy density η_a and Fourier inequality for the heat flux $\mathbf{q}_a$ hold, while from Eqs. $(35)_{3,5}$ it follows that η_a and ψ_a are independent on $\mathbf{g}$; Sph is the vectorial space of spherical tensor.

Moreover, for relations $(35)_{1,2,4}$, they must exist two scalar quantities ϖ and α and a vector $\mathbf{a}$, all functions of the set $\mathscr{S}$, such that

$$\mathbf{T}_a = \rho\frac{\partial\psi}{\partial\mathbf{F}}\mathbf{F}^T - \rho\mathbf{N}^T\frac{\partial\psi}{\partial\mathbf{N}} - \varpi\mathbf{I}, \quad \mathbf{S}_a = \rho\frac{\partial\psi}{\partial\mathbf{N}} - \mathbf{n}\otimes\mathbf{a},$$
$$\mathbf{z}_a = \rho\frac{\partial\psi}{\partial\mathbf{n}} + \mathrm{div}\left(\mathbf{S}_a - \rho\frac{\partial\psi}{\partial\mathbf{N}}\right) + \alpha\mathbf{n}; \tag{36}$$

the term $\left(-\rho \mathbf{N}^T \frac{\partial \psi}{\partial \mathbf{N}}\right)$ in the relation $(36)_1$ plays the analogous rôle of Ericksen's stress in liquid crystals, while the pressure $(-\varpi \mathbf{I})$ is in addition to reactive one.

By using Eq. $(36)_2$ in $(36)_3$, we have:

$$\mathbf{z}_a = \rho \frac{\partial \psi}{\partial \mathbf{n}} - \mathbf{N}\mathbf{a} + (\alpha - \operatorname{div}\mathbf{a})\,\mathbf{n}. \tag{37}$$

Remark. The constitutive restrictions (36) permit us to clarify better the rôle represented by the unusual moment of momentum balance (29); in fact it reduces to a compatibility condition on the free energy ψ, which comes out with the use of Eq. (27) also:

$$\operatorname{skw}\left[\mathbf{F}\frac{\partial \psi}{\partial \mathbf{F}}^T + \mathbf{n} \otimes \frac{\partial \psi}{\partial \mathbf{n}} + \mathbf{N}^T\frac{\partial \psi}{\partial \mathbf{N}} + \mathbf{N}\left(\frac{\partial \psi}{\partial \mathbf{N}}\right)^T\right] = \mathbf{0} \tag{38}$$

and it simply expresses the condition of frame–indifference for $\hat{\psi}$, namely,

$$\hat{\psi}(\mathbf{QF}, \mathbf{Qn}, \mathbf{QNQ}^T, \theta) = \hat{\psi}(\mathbf{F}, \mathbf{n}, \mathbf{N}, \theta), \tag{39}$$

for each rotation $\mathbf{Q} \in SO(3)$.

The field equations which rules the thermo–mechanical processes in a thermo–elastic smectic elastomer are hence obtained by inserting the constitutive results (36) in Eqs. (27), (31) and (32):

$$\mathbf{P}\left[\beta \ddot{\mathbf{n}} - \mathbf{b} + \frac{\partial \psi}{\partial \mathbf{n}} - \operatorname{div}\left(\frac{\partial \psi}{\partial \mathbf{N}}\right)\right] = \mathbf{0}, \tag{40}$$

$$\rho \ddot{\mathbf{x}} = \rho \mathbf{f} - \operatorname{grad}\tilde{\pi} + \rho \operatorname{div}\left(\frac{\partial \psi}{\partial \mathbf{F}}\mathbf{F}^T - \mathbf{N}^T\frac{\partial \psi}{\partial \mathbf{N}}\right), \tag{41}$$

$$\rho \theta \left(\frac{\partial \psi}{\partial \theta}\right)^{\cdot} + \rho \lambda - \operatorname{div}\mathbf{q}_a = 0, \tag{42}$$

as ρ is constant, for the constraint of incompressibility $\det \mathbf{F} = 1$, and where $\tilde{\pi} := \pi + \varpi$. Last equation is the pure equation of evolution for the temperature.

6 Final Remarks and Conclusions

Some special cases are of prominent interest when the free energy ψ is additively decomposed in four basic terms, namely, a nematic potential, two elastic energies of compression of layers and of tilt of molecules with respect to the normal $\mathbf{m}$ to the layers and a thermal energy, that is,

$$\psi = \frac{1}{2}\mu|\mathbf{N}|^2 + \frac{1}{2}\gamma(|\operatorname{grad}\omega| - 1)^2 + \frac{1}{2}\nu|\mathbf{n} \times \mathbf{m}|^2 + \frac{1}{2}\sigma(\theta - \theta_*)^2, \tag{43}$$

respectively, where μ, γ, ν and σ are thermo–elastic constants, while θ_* is the referential value of the temperature.

The balance Eqs. (40)–(42) have now the following form:

$$\mathbf{P}\left(\beta\ddot{\mathbf{n}} - \mathbf{b} + \nu\,\mathbf{Qn} - \mu\,\Delta\mathbf{n}\right) = \mathbf{0},$$

$$\rho\ddot{\mathbf{x}} = \rho\mathbf{f} - \operatorname{grad}\tilde{\pi} +$$

$$+\rho\operatorname{div}\left\{\gamma[(\mathbf{m} - \operatorname{grad}\omega)\otimes\operatorname{grad}\omega] + \nu[(\mathbf{n}\cdot\mathbf{m})\mathbf{m}\otimes\mathbf{Qn}] - \mu(\operatorname{grad}\mathbf{n})^T\operatorname{grad}\mathbf{n}\right\},$$

$$\rho\sigma\theta\dot{\theta} + \rho\lambda + \zeta\Delta\theta = 0,$$

$$(44)$$

where we used the identity $\frac{\partial(\mathbf{F}^{-1})}{\partial\mathbf{F}} = \mathbf{F}^{-1}\otimes\mathbf{F}^{-1}$ and the Fourier's law of constant ζ for the heat flux vector $\mathbf{q}_a$, while $\mathbf{Q}\,(:=\mathbf{I} - \mathbf{m}\otimes\mathbf{m})$ is the projector on the plane tangent to the layers.

In statics, when the body forces can be neglected, the first two balance equations reduce to the form

$$\mathbf{P}\left(\nu\,\mathbf{Qn} - \mu\,\Delta\mathbf{n}\right) = \mathbf{0},$$

$$\rho\operatorname{div}\left\{\gamma[(\mathbf{m} - \operatorname{grad}\omega)\otimes\operatorname{grad}\omega] + \nu[(\mathbf{n}\cdot\mathbf{m})\mathbf{m}\otimes\mathbf{Qn}] - \mu(\operatorname{grad}\mathbf{n})^T\operatorname{grad}\mathbf{n}\right\} = \operatorname{grad}\tilde{\pi},$$

$$(45)$$

similar to those proposed in [7] for smectic liquid crystals and resolved in two cases in the linear approximation.

Another particular example of our theory was presented in [1] in the elastic case by considering contributions to the free energy due to nematic potential and tilting effects only and a minimum problem was resolved by minimizing over the tilt rotation to obtain the engineering stress as relationship to the shear strain.

In conclusion, here we presented a continuum model for smectic-C elastomers where a unit vector $\mathbf{n}$, the mesogen director, models additional microstructural degrees of freedom associated with polymer-chain shape. We obtained the balance equations within a general theory of contrained vectorial microstructure and then specialized them to the thermo-elastic case. We presented the pure field equations for the micro-motion and the temperature, and the Cauchy's equation for the macro-motion in which the Cauchy's stress is no more symmetric, while a spherical tensor of reactive pressure is also present. The model is consistent with previous theories.

Acknowledgements. This research was supported by the Department of Mechanics and Materials, Faculty of Engineering of the "Mediterranean" University of Reggio Calabria, Italy.

References

1. Buonsanti, M., Giovine, P.: On a Minimum Problem in Smectic Elastomers. In: Santini, A., Moraci, N. (eds.) 2008 Seismic Engng. Conf. commem. the 1908 Messina and Reggio Calabria Earthquake (MERCEA 2008). Amer. Inst. Physics Conference Proc., vol. 1020, pp. 1350–1357. AIP, New York (2008)
2. Capriz, G.: Continua with Microstructure. Springer Tracts in Natural Philosophy, vol. 35. Springer, New York (1989)

3. Capriz, G.: Smectic Liquid Crystals as Continua with Latent Microstructure. Meccanica 30, 621–627 (1995)
4. Capriz, G.: Smectic Elasticity. In: Batra, R.C., Beatty, M.F. (eds.) Contemporary Research in Mechanics and Mathematics of Materials, pp. 199–204. CIMNE, Barcelona (1996)
5. Capriz, G., Giovine, P.: On Microstructural Inertia. Math. Mod. Meth. Appl. Sc. 7, 211–216 (1997)
6. Capriz, G., Giovine, P.: Remedy to Omissions in a Tract on Continua with Microstructure. In: Atti XIII Congresso AIMETA 1997, Siena, Meccanica Generale, vol. I, pp. 1–6 (1997)
7. Capriz, G., Napoli, G.: Swelling and Tilting in Smectic Layers. Appl. Math. Letters 14, 673–678 (2001)
8. Capriz, G., Podio–Guidugli, P.: Internal Constraint. In: Truesdell, C. (ed.) Rational Thermodynamics, Appendix 3A, 2nd edn., pp. 159–170. Springer, New York (1984)
9. Capriz, G., Podio–Guidugli, P.: Formal Structure and Classification of Theories of Oriented Materials. Annali Mat. Pura Appl. (IV) CXV, 17–39 (1977)
10. de Gennes, P.G., Prost, J.: The Physics of Liquid Crystals, 3rd edn. Oxford Science Publ., Oxford (2003)
11. Gurtin, M.E., Podio–Guidugli, P.: The Thermodynamics of Constrained Materials. Arch. Rational Mech. Anal. 51, 192–208 (1973)
12. Manacorda, T.: Introduzione alla Termomeccanica dei Continui. Pitagora Editrice, Bologna (1979)
13. Stenull, O.: Smectic Elastomers Membranes. Phys. Rev. E 75, 051702 (2007)
14. Truesdell, C., Toupin, R.A.: The Classical Field Theories. In: Flügge, S. (ed.) Handbuch der Physik, III/1. Springer, Berlin (1960)
15. Virga, E.G.: Variational Theories for Liquid Crystals. Chapman & Hall, London (1994)
16. Warner, W., Terentjev, E.M.: Liquid Crystal Elastomers. Clarendon Press, Oxford (2003)

A Note on the Representation of Cosserat Rotation

J.D. Goddard

Abstract. This brief article provides an independent derivation of a formula given by Kafadar and Eringen (1971) connecting two distinct Cosserat spins. The first of these, the *logarithmic spin* represents the time rate of change of the vector defining finite Cosserat rotation, whereas the second, the *instantaneous spin*, gives the local angular velocity representing the infinitesimal generator of that rotation. While the formula of Kadafar and Eringen has since been identified by Iserles et al. (2000) as the differential of the Lie-group exponential, the present work provides an independent derivation based on quaternions. As such, it serves to bring together certain scattered results on quaternionic algebra, which is currently employed as a computational tool for representing rigid-body rotation in various branches of physics, structural and robotic dynamics, and computer graphics.

1 Background: Cosserat Rotations

From the conventional continuum-mechanical viewpoint, a Cosserat continuum[1] is defined via a differentiable map assigning spatial position $\mathbf{x}(\mathbf{x}^\circ, t)$ and microstructural rotation $\mathbf{P}(\mathbf{x}^\circ, t)$ to each material particle, with $\mathbf{x} = \mathbf{x}^\circ$, $\mathbf{P} = \mathbf{I}$ in a given reference configuration, where $\mathbf{P} \in \mathbf{SO}(3)$ denotes a real, proper orthogonal tensor.

We can express the kinematics concisely in terms of the map $\mathbb{R}^3 \to \mathbb{R}^3 \times \mathbf{SO}(3)$ given by

$$\mathbf{x}^\circ \to \{\mathbf{x}, \boldsymbol{\theta}\}, \quad \text{where } \boldsymbol{\theta} = -\frac{1}{2}\boldsymbol{\varepsilon} : \boldsymbol{\Theta}, \text{ and } \boldsymbol{\Theta} = -\boldsymbol{\varepsilon} \cdot \boldsymbol{\theta},$$

$$\text{i.e.} \quad \theta_i = -\frac{1}{2}\varepsilon_{ijk}\Theta^{jk}, \text{ and } \Theta_{ij} = -\varepsilon_{ijk}\theta^k, \tag{1}$$

J.D. Goddard

Department of Mechanical and Aerospace Engineering 0411, University of California, San Diego, 9500 Gilman Dr., La Jolla, CA, 92093-0411, USA

e-mail: jgoddard@ucsd.edu

---

[1] As noted by numerous authors, e.g. [1, 2, 3, 4], this is a special case of Eringen's *microstretch* continuum [5], which involves both rotation and dilatation.

and the *Cayley-Gibbs-Rodrigues* relation [2, 6, 7, 8, 9, 10]

$$\mathbf{P} = \exp\mathbf{\Theta} = \mathbf{I} + \left(\frac{\sin\vartheta}{\vartheta}\right)\mathbf{\Theta} + \left(\frac{1-\cos\vartheta}{\vartheta^2}\right)\mathbf{\Theta}^2, \ \vartheta = \{-\mathrm{tr}(\mathbf{\Theta}^2)/2\}^{1/2} \qquad (2)$$

Here $\hat{\boldsymbol{\theta}} = \boldsymbol{\theta}/|\boldsymbol{\theta}|$ represents the axis of rotation and $\vartheta = |\boldsymbol{\theta}| = (\theta^i\theta_i)^{1/2}$ the angle of rotation about the axis.

If $\mathbf{P}$ is taken as primary variable, the skew-symmetric tensor $\mathbf{\Theta} = \log\mathbf{P} \in \mathfrak{so}(3)$ (Lie algebra) represents an inverse of the map $\mathfrak{so}(3) \rightarrow \mathbf{SO}(3)$ (Lie group), and it can be defined uniquely and computed by various methods [11]. Alternatively, and more conveniently, we may regard $\boldsymbol{\theta}$ as the primary variable, with (1) defining the associated map or *Cosserat placement* $\mathbb{R}^3 \rightarrow \mathbb{R}^6$.

To connect the vector of the *logarithmic spin* $\boldsymbol{\Omega} = d\mathbf{\Theta}/dt$ to that of the *instantaneous spin* $\mathbf{N} = (d\mathbf{P}/dt)\mathbf{P}^\mathsf{T}$, where

$$\frac{d}{dt} := \left(\frac{\partial}{\partial t}\right)_{\mathbf{x}^\circ},$$

we recall the rather remarkable result of Kafadar and Eringen [5](Eqs.(2)-(9)), which can be expressed in the present notation as:

$$\boldsymbol{v} := -\frac{1}{2}\boldsymbol{\varepsilon} : \mathbf{N} = \boldsymbol{\Lambda}\boldsymbol{\omega}, \text{ where } \boldsymbol{\omega} = -\frac{1}{2}\boldsymbol{\varepsilon} : \boldsymbol{\Omega},$$

with

$$\boldsymbol{\Lambda} = \mathbf{I} + \left(\frac{1-\cos\vartheta}{\vartheta^2}\right)\mathbf{\Theta} + \left(\frac{\vartheta-\sin\vartheta}{\vartheta^3}\right)\mathbf{\Theta}^2, \qquad (3)$$

and

$$\boldsymbol{\Lambda}^{-1} = \mathbf{I} - \frac{1}{2}\mathbf{\Theta} + \frac{1}{\vartheta^2}\left(1 - \frac{\vartheta}{2}\cot\frac{\vartheta}{2}\right)\mathbf{\Theta}^2$$

Either definition of spin is acceptable, and this relation makes it easy to relate their conjugate stresses.

Kafadar and Eringen [5] derive a formulas equivalent to (3), and Iserle et al. [8](Eqs. B.10-B.11) later have given them as differentials of the Lie-group exponential, cf. [7](Eqs. 17-19). The purpose of the present article is to give an independent derivation by means of the quaternionic representation of the Lie-algebra/Lie-group connection $\mathbf{SO}(3) = \exp\{\mathfrak{so}(3)\}$. It is hoped that this derivation will clarify certain relations between quaternions, matrix algebra and Lie groups.

2 Quaternions as Tensors

With no claim to originality, the object here is to provide a concise summary of scattered results from numerous treatises on quaternions, many of which are presented under a different guise elsewhere, e.g. in the much more comprehensive journal article [12]. The knowledgeable reader can skip to the following section for the derivation of the main result.

Hamilton's [13] quaternions represent a special case of the non-commutative hypercomplex (Clifford) algebras, which are known to be isomorphic to matrix algebras [14][2]. With this in mind, it is convenient for the present purposes to adopt the tensorial representation of quaternions:

$$
\mathbf{Z} = z^i \mathbf{E}_i = z^0 \mathbf{E}_0 + z^1 \mathbf{E}_1 + z^2 \mathbf{E}_2 + z^3 \mathbf{E}_3 \doteq
\begin{pmatrix}
z^0 & -z^1 & -z^2 & -z^3 \\
z^1 & z^0 & z^3 & -z^2 \\
z^2 & -z^3 & z^0 & z^1 \\
z^3 & z^2 & -z^1 & z^0
\end{pmatrix}
$$

$$
= \begin{pmatrix}
z^0 \mathbb{I} + z^1 \mathbb{J} & -z^2 \mathbb{I} + z^3 \mathbb{J} \\
z^2 \mathbb{I} + z^3 \mathbb{J} & z^0 \mathbb{I} - z^1 \mathbb{J}
\end{pmatrix}, \quad (4)
$$

$$
\text{with} \quad \mathbb{I} := \begin{pmatrix} 1 & 0 \\ 0 & 1 \end{pmatrix}, \quad \mathbb{J} := \begin{pmatrix} 0 & -1 \\ 1 & 0 \end{pmatrix}, \quad \text{and} \quad \mathbb{J}^2 = -\mathbb{I},
$$

which represents a 4-dimensional subspace $\mathfrak{Q}$ of the 7-dimensional vector space consisting of the sum of skew-symmetric and isotropic 4-tensors. The z^i are real, and $\doteq$ indicates the matrix of tensor components relative to given basis $\mathbf{e}_i$, $i = 0 \dots, 3$. Superscripts on z^i allow for curvilinear tensors, and the Einstein summation convention is observed here and in the following[3]. The 2×2 *identity* matrix $\mathbb{I}$ and *symplectic* matrix $\mathbb{J}$ are to be interpreted as belonging to the appropriate blocks of the 4×4 matrix, and the replacement $\mathbb{I} \to 1$, $\mathbb{J} \to \iota$ establishes an isomorphism with the algebra of 2×2 complex matrices[4]. The basis elements $\mathbf{E}_i$, which are obtained from (4) by taking $z^j = \delta_i^j$, satisfy

$$
\mathbf{E}_i^2 = \begin{cases} \mathbf{E}_0, & i = 0, \\ -\mathbf{E}_0, & i \neq 0 \end{cases} \quad \text{and} \quad \mathbf{E}_i \mathbf{E}_j = \begin{cases} \varepsilon_{ij}{}^k \mathbf{E}_k, & \text{for } i, j, k \neq 0, \\ \mathbf{E}_i, & \text{for } j = 0, \\ \mathbf{E}_j, & \text{for } i = 0, \end{cases} \quad (5)
$$

which yields the well-known product rules for general quaternions. The *deviator* (or "vector part") $\mathbf{Z}'$, *conjugate* $\mathbf{Z}^*$, *modulus* $|\mathbf{Z}|$, and *inverse* of a general quaternion $\mathbf{Z}$ are defined respectively by

$$
\mathbf{Z}' := \text{Dev}(\mathbf{Z}) = \mathbf{Z}|_{z^0 = 0}, \quad \text{with} \quad \mathbf{Z} = z^0 \mathbf{E}_0 + \mathbf{Z}', \quad \mathbf{Z}^* = z^0 \mathbf{E}_0 - \mathbf{Z}',
$$

$$
|\mathbf{Z}| = \frac{1}{2} [\text{tr}(\mathbf{Z} \mathbf{Z}^*)]^{1/2}, \quad \text{and} \quad \mathbf{Z}^{-1} = \mathbf{Z}^* / |\mathbf{Z}|^2, \quad (6)
$$

As discussed below, *unitary* quaternions, defined by $\mathbf{Q}^{-1} = \mathbf{Q}^*$, represent spatial rotations.

---

[2] Given this fact, the pure mathematician might wish to be excused from further reading of the present work.

[3] However, the z^i in (4) do not obey the standard tensor-transformation rules unless interpreted in terms of 4-vectors.

[4] In that case, the $\mathbf{E}_i$, $i = 1, 2, 3$, are identical with the well-known *Pauli spin matrices* [12], up to permutation and multiplication by $\pm \iota$.

With matrix scalar product for matrices $\mathbf{A}, \mathbf{B}$ defined by $\mathbf{A} : \mathbf{B} = \mathrm{tr}(\mathbf{A}^*\mathbf{B})$, it is clear from the preceding relations that the $\mathbb{E}_i$ represent an orthogonal basis, whose reciprocal basis is $\mathbb{E}^i = \frac{1}{4}\mathbb{E}_i$. Therefore, the transformation of $\mathfrak{Q}$ into the space of 4-vectors with basis $\mathbf{e}_i$ is given by

$$\boldsymbol{\gamma} = \mathbf{e}_i \otimes \mathbb{E}^i \text{ with } \mathbf{z} = \boldsymbol{\gamma} : \mathbf{Z}, \text{ i.e. } \gamma^j_{\ kl} = (\mathbb{E}^j)_{kl} \text{ with } z^j = \gamma^{jkl}(\mathbf{Z})_{kl}, \tag{7}$$

and the inverse $\mathbf{Z} = \boldsymbol{\gamma}^{-1} \cdot \mathbf{z}$ is obviously given by $(\boldsymbol{\gamma}^{-1})_{klj} = (\mathbb{E}_j)_{kl}$.

The projection $\mathfrak{Q} \to \mathfrak{so}(3)$

$$\mathbf{S}' = \mathrm{dev}(\mathbf{S}), \text{ with } \mathbf{S} = \boldsymbol{\Pi}\mathbf{S}, \ \boldsymbol{\Pi} = \mathbb{E}_0 - \mathbf{e}_0 \otimes \mathbf{e}_0 \tag{8}$$

provides a connection between the Lie algebra $\mathfrak{so}(3)$ and the algebra of quaternions, as defined by the preceding matrix representations. Thus, given a 4-vector $\mathbf{x}$ with quaternion $\mathbf{X} = \boldsymbol{\gamma}^{-1} \cdot \mathbf{x}$, one obtains from it the 3-vector ("physical-space" vector) $\mathbf{x}'$:

$$\mathbf{x}' = \boldsymbol{\gamma} : \mathbf{X}', \text{with } \mathbf{x} = x^0 \mathbf{e}_0 + \mathbf{x}', \text{ and } \mathbf{x}^* = x^0 \mathbf{e}_0 - \mathbf{x}', \tag{9}$$

Then, the quaterionic product[5] is given in terms of 3-vector operations as:

$$\mathbf{xy} := \boldsymbol{\gamma} : (\mathbf{X}\mathbf{Y}) = x^0 \mathbf{y}' + y^0 \mathbf{x}' + (x^0 y^0 - \mathbf{x}' \cdot \mathbf{y}')\mathbf{e}_0 + \mathbf{x}' \times \mathbf{y}',$$

$$\text{with } \mathbf{x}' \times \mathbf{y}' = \frac{1}{2}(\mathbf{x}'\mathbf{y}' - \mathbf{y}'\mathbf{x}'), \tag{10}$$

which, with the proviso that the vector space be enlarged to 4-vectors, adds a new operation to the usual 3-vector operations.

Without loss of generality, we employ the usual quaterionic convention $\mathbf{e}_0 \equiv 1$, and we make a distinction between a quaternion and its 4-vector only when necessary to clarify tensor-transformation formulae. Hence, letting lower-case bold Greek refer either to 3-vectors or 3rd-rank tensors, we have the well-known polar or exponential representation (cf. e.g. [12])

$$\mathbf{z} = \rho e^{\boldsymbol{\phi}} = \rho(\cos\varphi + \hat{\boldsymbol{\phi}}\sin\varphi),$$

$$\text{where } \varphi = |\boldsymbol{\phi}|, \ \rho = |\mathbf{z}|, \ \boldsymbol{\phi} = \boldsymbol{\phi}', \ \hat{\boldsymbol{\phi}} = \boldsymbol{\phi}/\varphi = -\hat{\boldsymbol{\phi}}^*, \tag{11}$$

which remains valid when the pair $\mathbf{z}, \boldsymbol{\phi}$ is replaced by the corresponding quaternions $\mathbf{Z} = \boldsymbol{\gamma}^{-1} \cdot \mathbf{z}$ and $\mathbf{F} = \boldsymbol{\gamma}^{-1} \cdot \boldsymbol{\phi}$. The same formula serves to define the logarithm and its various branches.

3 Application to Cosserat Rotations

The important special case $\rho = 1$ of (11) gives a unitary quaternion $\mathbf{q} = \boldsymbol{\gamma} : \mathbf{Q}$, which represents an orthogonal transformation $\mathbf{P} \in \mathbf{SO}(3)$ of space-vectors, according to

$$\mathbf{y}' = \boldsymbol{\gamma} : (\mathbf{Q}\mathbf{X}'\mathbf{Q}^*) = \mathbf{q}\mathbf{x}'\mathbf{q}^* = \mathbf{P}\mathbf{x}' = \mathbf{P}^{1/2}(\mathbf{x}'\mathbf{P}^{-1/2}) \tag{12}$$

---

[5] Not to be confused with the Gibbs dyadic notation for the tensor product $\mathbf{x} \otimes \mathbf{y}$.

It is easy to show [9] by (10)-(12) that

$$\hat{\boldsymbol{\theta}} = \hat{\boldsymbol{\phi}}, \ |\boldsymbol{\theta}| = 2|\boldsymbol{\phi}| \ (\mathrm{mod}\, 2\pi), \ \text{hence } \mathbf{p} = \mathbf{q}^2 \ \text{ or } \ \mathbf{q} = \pm \mathbf{p}^{1/2},$$

$$\text{where } \mathbf{q} = e^{\boldsymbol{\phi}} \text{ and } \mathbf{p} = e^{\boldsymbol{\theta}},$$

(13)

which involves the map between skew-Hermitian quaternions $\mathbf{Z} = -\mathbf{Z}^*$ and rotations:

$$e^{\zeta} = \cos \zeta + \sin \zeta \, \hat{\boldsymbol{\zeta}} = \boldsymbol{\gamma} : e^{\mathbf{Z}},$$

$$\text{with } \ e^{\mathbf{Z}} = \mathbf{I} + \sin \zeta \, \hat{\mathbf{Z}} + (1 - \cos \zeta) \, \hat{\mathbf{Z}}^2, \ \mathbf{Z} = \boldsymbol{\Pi}\mathbf{Z},$$

(14)

where the unit skew-Hermitian vector $\hat{\boldsymbol{\zeta}} = -\hat{\boldsymbol{\zeta}}^*$ represents the axis of rotation, and ζ the angle of rotation about the axis.

The relations (13)-(14) allow for an easy verification of certain results of Kafadar and Eringen [5] involving the derivatives of the one-parameter Lie group represented by orthogonal transformations $\mathbf{P}(t) = \exp\boldsymbol{\Theta}(t)$. In particular, given the relations,

$$\boldsymbol{v} = -\frac{1}{2}\boldsymbol{\varepsilon} : \mathbf{N}, \ \mathbf{N} = \frac{d\mathbf{P}}{dt}\mathbf{P}^{\mathrm{T}}, \ \text{with } \boldsymbol{v} = \frac{d\mathbf{p}}{dt}\mathbf{p}^* = 2\frac{d\mathbf{q}}{dt}\mathbf{q}^*$$

(15)

we can now derive the desired relation between $\mathbf{N}$ or $\boldsymbol{v}$ and $\boldsymbol{\omega} = -\frac{1}{2}\boldsymbol{\varepsilon} : \boldsymbol{\Omega}$, where $\boldsymbol{\Omega} := d\boldsymbol{\Theta}/dt$.

Letting $(\dot{\ }) = d(\ )/dt$ and employing the representation $\mathbf{q} = \exp\boldsymbol{\phi} = \cos\varphi + \hat{\boldsymbol{\phi}}\sin\varphi$, one finds readily by (13) and the relation $\hat{\boldsymbol{\theta}}^2 = -1$ that

$$\boldsymbol{v} = 2\dot{\mathbf{q}}\mathbf{q}^* = \dot{\vartheta}\hat{\boldsymbol{\theta}} + \sin\vartheta\dot{\hat{\boldsymbol{\theta}}} + (1 - \cos\vartheta)\hat{\boldsymbol{\theta}}\dot{\hat{\boldsymbol{\theta}}},$$

(16)

which corresponds to Eq. (7) of [5]. However, in view of the relation $\boldsymbol{\theta}^2 = -\vartheta^2$, it follows that

$$\dot{\vartheta} = -\frac{1}{2}\left(\hat{\boldsymbol{\theta}}\dot{\boldsymbol{\theta}} + \dot{\boldsymbol{\theta}}\hat{\boldsymbol{\theta}}\right) \ \text{and} \ \hat{\boldsymbol{\theta}}\dot{\vartheta} = \frac{1}{2}\left(\dot{\boldsymbol{\theta}} - \hat{\boldsymbol{\theta}}\dot{\boldsymbol{\theta}}\hat{\boldsymbol{\theta}}\right)$$

(17)

Substitution of these expressions into (16) and application of (10) gives after some algebra

$$\boldsymbol{v} = \dot{\boldsymbol{\theta}} + \left(\frac{1 - \cos\vartheta}{\vartheta^2}\right)\boldsymbol{\theta}\times\dot{\boldsymbol{\theta}} + \left(\frac{\vartheta - \sin\vartheta}{\vartheta^3}\right)\boldsymbol{\theta}\times(\boldsymbol{\theta}\times\dot{\boldsymbol{\theta}})$$

$$\equiv \left\{\mathbf{I} + \left(\frac{1 - \cos\vartheta}{\vartheta^2}\right)\boldsymbol{\Theta} + \left(\frac{\vartheta - \sin\vartheta}{\vartheta^3}\right)\boldsymbol{\Theta}^2\right\}\boldsymbol{\omega}$$

(18)

representing Eq. (8) of [5], as cited above in (3).

4 Conclusions

The connection between Cosserat spins given by Kafadar and Eringen [5] is rather easily established by means of the quaternionic representation of rotations. While the present work is intended mainly to establish that fact, Section 2 provides tensor representations that may be useful in a broader range of applications. In particular, the formula (11) with $\rho \neq 1$ describes superposed Cosserat rotation and dilatation, suggesting a convenient of representation of Eringen's microstretch continuum. In particular, one sees that a complex quaternion of the form

$$\mathbf{Z}(\mathbf{x}^\circ) = \mathbf{x} + \imath \log \mathbf{z}$$

where $\mathbf{z}$ is defined by (11), defines a more general *microstretch placement*, a subject to be considered in a future work.

Acknowledgement

This is offered as a token of friendship to Professor Krzysztof Wilmanski, on the occasion of his seventieth birthday, and to acknowledge his early recognition of the relevance of structured-continuum models to the mechanics of heterogeneous materials [15]. I also thank Professor Reuven Segev for clarifying certain algebraic details in the derivation of Kafadar and Eringen [5].

References

1. Ericksen, J., Truesdell, C.: Exact theory of stress and strain in rods and shells. Arch. Rat. Mech. Anal. 1(1), 295–323 (1957)
2. Truesdell, C., Toupin, R.A.: Principles of classical mechanics and field theory. In: Flügge, S. (ed.) Encyclopedia of Physics (Handbuch der Physik), vol. 3/1. Springer, Berlin (1960)
3. Mindlin, R.D.: Micro-structure in linear elasticity. Arch. Ratl. Mech. Anal. 16(1), 51–78 (1964)
4. Cowin, S.C.: Stress functions for cosserat elasticity. Int. J. Solids Struct. 6(4), 389–398 (1970)
5. Kafadar, C.B., Eringen, A.C.: Micropolar media. i. the classical theory. Int. J. Eng. Sci. 9(3), 271–305 (1971)
6. Mehrabadi, M.M., Cowin, S.C., Jaric, J.: Six-dimensional orthogonal tensor representation of the rotation about an axis in three dimensions. Int. J. Solids Struct. 32(3-4), 439–449 (1995)
7. Park, J., Chung, W.K.: Geometric integration on euclidean group with application to articulated multibody systems. IEEE Trans. Robotics 21(5), 850–863 (2005)
8. Iserles, A., Munthe-Kaas, H.Z., Nørsett, S.P., Zanna, A.: Lie-group methods. Acta numerica 9(1), 215–365 (2000)
9. Spring, K.W.: Euler parameters and the use of quaternion algebra in the manipulation of finite rotations: A review. Mech. Mach. Theo. 21(5), 365–373 (1986)
10. Bauchau, O.A., Choi, J.Y.: The vector parameterization of motion. Nonlin. Dyn. 33(2), 165–188 (2003)
11. Gallier, J., Xu, D.: Computing exponentials of skew symmetric matrices and logarithms of orthogonal matrices. Int. J. Robot. Autom. 17, 1–11 (2002)

12. Lüdkovsky, S.V., van Oystaeyen, F.: Differentiable functions of quaternion variables. Bull. Sci. math. 127(9), 755–796 (2003)
13. Hamilton, W.R.S.: Elements of quaternions. Longmans, Green & Co., London (1866)
14. Morris, A.O.: On a generalized clifford algebra (ii). Q J. Math 19(1), 289–299 (1968)
15. Wilmanski, K.: Dynamics of bodies with microstructure. Arch. Mech. Stos. 20(6), 705–744 (1968)

Material Uniformity and the Concept of the Stress Space

Serge Preston and Marek Elżanowski

The authors would like to dedicate this paper to Krzysztof Wilmański on the occasion of his 70th birthday.

Abstract. The notion of the stress space, introduced by Schaefer [14], and further developed by Kröner [7] in the context of materials free of defects, is revisited. The comparison between the Geometric Theory of Material Inhomogeneities and the Stress Space approach is discussed. It is shown how to extend Kröner's approach to the case of the material body with inhomogeneities (defects).

1 Introduction

The work presented in this note is a continuation of the earlier work by Ciancio *et al* [3]. Its main objective is to investigate the relation between the Geometric Theory of Material Inhomogeneities (Epstein and Elżanowski [4], Wang and Truesdell [16]) and the description of the continuous distribution of defects based on the concepts of the intermediate configuration and the stress space (Bilby [2], Kröner [6], [7], Lee [9], Stojanovic [15]).

We are particularly interested in describing effectively the residual stresses associated with the presence of material inhomogeneities (defects). To this end, we employ the Bilby-Kröner-Lee multiplicative decomposition of the deformation gradient (and the concept of the intermediate configuration) as well as the notions of the stress and strain spaces of Schaefer [14] and Kröner [7]. Using the language of modern Differential Geometry we show that the multiplicative decomposition of the deformation gradient, exemplifying the elasto-plastic material behavior, leads to the introduction of the uniformity tensor which plays the role similar to that of the material isomorphism of the Geometric Theory of Material Inhomogeneities.

Serge Preston · Marek Elżanowski
Portland State University, Portland Oregon, USA
e-mail: {serge,elzanowskim}@pdx.edu

When discussing the role of the uniformity tensor, and its uniqueness, we show the importance of the concept of the intermediate configuration.

In employing the notion of the stress space we follow Kröner's idea (see Kröner [7]) of the non-holonomic transformation between the spaces of strain and stress. This allows us to introduce the residual stress metric, the Ricci tensor of which is interpreted as the residual stress tensor. The said non-holonomic transformation, known as the residual stress function, represents a constitutive law relating the residual stress to the material strain of the intermediate configuration of the inhomogeneous material body. We show how to reconcile the introduction of the residual stress function with the existence of the uniformity tensor.

The paper has the following layout. In the next section we introduce the basic notions of Continuum Mechanics and the Geometric Theory of Material Inhomogeneities. In Section 3 we discuss the Bilby-Kröner-Lee multiplicative decomposition of the deformation gradient introducing the concept of the uniformity tensor and the notion of the material strain. In Section 4 the construction of the stress space is presented. The paper is concluded by a couple of examples in Section 5.

2 Hyperelastic Unifomity

We start by reviewing some basic concepts of Continuum Mechanics and the Geometric Theory of Material Inhomogeneities restricting our presentation to hyperelastic materials only.

2.1 Configurations and the Cauchy Metric

In Continuum Mechanics the **material body** is usually represented by a connected 3-dimensional smooth oriented manifold M with a piece-wise smooth boundary ∂M. However, as the issues discussed in this paper are of the local nature only, it is sufficient to consider M as a connected, open domain in $\mathbb{R}^3$ with coordinates $\{X^I\}$, $I = 1, 2, 3$. We assume that the **physical space** our body is placed in is the 3-dimensional Euclidean vector space E^3 equipped with the (flat) Euclidean metric **h**. Given a global Cartesian coordinate system $\{x^i\}$, $i = 1, 2, 3$, in E^3, de facto allowing us to identify E^3 with $\mathbb{R}^3$, the metric **h** takes the form $h_{ij}dx^i dx^j$ where the standard summation convention is enforced. A **configuration** of the body M, often called its **placement**, is an (differentiable) embedding $\phi : M \to \mathbb{R}^3$. Its **deformation gradient** at a point $X \in M$ is a linear isomorphism from the tangent space $T_X M$ into the tangent space $T_{\phi(X)}\mathbb{R}^3$. Namely,

$$\mathbf{F}(X) \equiv \phi_*(X) : T_X M \to T_{\phi(X)}\mathbb{R}^3, \tag{1}$$

where ϕ_* denotes the tangent map of ϕ. The deformation gradient at a material point, say $Y \in M$, is represented (in the given coordinates systems on M and E^3) by the non-singular matrix of partial derivatives of ϕ, that is,

$$F_I^i(Y) = \frac{\partial \phi^i}{\partial X^I}(Y) \equiv \phi_{,I}^i(Y), \tag{2}$$

where $\phi^i(X^1, X^2, X^3) = x^i$, $i = 1, 2, 3$. The material equivalence of the special metric **h**, relative to the placement ϕ the body is at, is the right **Cauchy-Green deformation tensor C** obtained by the pull-back of the Euclidean metric **h** to the body manifold M. That is,

$$\mathbf{C} \equiv \phi^* \mathbf{h}, \tag{3}$$

where ϕ^* denotes the pull-back map. The matrix $C_{IJ} = h_{ij}\phi_{,I}^i \phi_{,J}^j$ evaluated at the point X is the coordinate representation of the tensor **C**.

2.2 Material Uniformity

Recall, that the material is called **hyperelastic** if its constitutive response is completely determined by a single scalar-valued function, say W, called the **elastic energy density** (per unit reference volume υ_0[1]). We assume that W is a function of a material point and the deformation gradient at this point, that is, $W = W(X, \mathbf{F}(X))$.

The material body M is considered **uniform** if it is made of the same material at all points. In mathematical terms, this means that for any pair of material points, say X and Y, there exists a linear isomorphism, referred to as a **material isomorphism**

$$K_X^Y : T_X M \rightarrow T_Y M, \tag{4}$$

between the corresponding tangent spaces such that

$$W(Y, \mathbf{F}(Y))K_X^{Y*} d\upsilon_0(Y) = W(X, \mathbf{F}(Y)K_X^Y)d\upsilon_0(X) \tag{5}$$

holds for all possible deformation gradients **F**, where K_X^{Y*} denotes the pullback of a 3-form by the mapping K_X^Y. Equivalently, the material body is considered uniform, if there exist material isomorphisms

$$P_X : T_{X_0} M \rightarrow T_X M, \tag{6}$$

called the **implants**, from a fixed point $X_0 \in M$ to every point $X \in M$ such that $K_X^Y = P_Y \circ P_X^{-1}$ and the equation (5) holds. The said fixed point X_0 can be arbitrarily chosen. In fact, it is often convenient to think about its tangent space $T_{X_0}M$ as a standing alone vector space V with the orthogonal frame $\{\mathbf{e}_i\}$, $i = 1, 2, 3$, and its own metric being the the Euclidean (flat) metric **h** at the origin.

Having the implants P_X available we can "pull-back" the Euclidean volume element of the **archetype** V to the material manifold M. Indeed, let

$$\upsilon_P(X) \equiv P_X^{-1}(\mathbf{e}_1 \wedge \mathbf{e}_2 \wedge \mathbf{e}_3) \tag{7}$$

---

[1] Although we do not explicitly utilize here the concept of the reference configuration, we assume that assigning coordinates to the body manifold $M \subset \mathbb{R}^3$ is equivalent to selecting its reference configuration; see Epstein and Elżanowski [4].

and let $J_P : M \to \mathbb{R}$ be a real-valued function such that

$$\upsilon_P(X) = J_P(X)\upsilon_0(X) \tag{8}$$

at every $X \in M$. Let $GL(V)$ be the group of all linear automorphism of the archetype V and define a real-valued function $\widehat{W} : M \times GL(V) \to \mathbb{R}$ by

$$\widehat{W}(X,\mathbf{A}) \equiv J_P^{-1}(X)W(X,\mathbf{A}P_X^{-1}), \tag{9}$$

for all $X \in M$ and any $\mathbf{A} \in GL(V)$. It should now be easy to see that the uniformity condition (5) is equivalent to

$$\widehat{W}(X,\mathbf{A}) = \widehat{W}(Y,\mathbf{A}) \tag{10}$$

for all $\mathbf{A} \in GL(V)$ and any pair of material points X and Y. In other words, the material body M is uniform if its **archetypical energy density** function $\widehat{W}$ is material point independent. Consequently, the strain energy density function of the uniform material body M is such that

$$W(X,\mathbf{F}(X)) = J_P(X)\widehat{W}(\mathbf{F}(X)P_X) \tag{11}$$

for any (non-singular) deformation gradient $\mathbf{F}$ and some archetypical energy function $\widehat{W}$ obeying the relation (10). For the clarity and the simplicity of our presentation we assume here that that archetypical energy $\widehat{W}$ has the trivial isotropy group[2].

2.3 Material Connections

It is normally assumed that the material isomorphisms, consequently the implants, are locally smoothly distributed on M. This implies that the materially uniform body M can be equipped with the (locally smooth) global material **uniform frame field**

$$\mathbf{p}_J(X) \equiv P_X(\mathbf{e}_J) = P_J^I \frac{\partial}{\partial X^I}, \quad I = 1,2,3. \tag{12}$$

The material isomorphisms K_X^Y, or equivalently the material implants P_X, establish a long distance parallelism on M as $K_X^Y(\mathbf{p}_J(X)) = \mathbf{p}_J(Y)$, $j = 1,2,3$. Such a parallelism defines a **material connection**, say ω, the curvature of which vanishes identically. Indeed, as evident from the definition of the global uniform frame field (12), the corresponding parallel transport is curve independent. The torsion of the connection ω provides the measure of the non-integrability of the material frame field $\mathbf{p}_J$, $J = 1,2,3$. This, in turn, is accepted as a "measure" of the local **non-homogeneity** of the given material body. More precisely, the hyperelastic body M, as defined by

---

[2] Note, that if the isotropy group of $\widehat{W}$ is nontrivial the material implants P_X are not necessarily uniquely defined. Indeed, suppose that the archetypical energy function $\widehat{W}$ has a continuous isotropy group, say $G \subset GL(V)$. Then, given an implant P_X, $P_X G$ is also an implant as long as $\mathbf{G} \in G$, Epstein and Elżanowski [4].

the strain energy density function W, is considered locally **homogeneous** provided there exists[3] a material connection ω such that its torsion vanishes identically.

As we have mentioned earlier, the implant maps induce the uniform material frame field (12). They also induce the corresponding **uniform material metric g** defined by the pull-back of the Euclidean metric of the archetype. That is,

$$\mathbf{g} \equiv P_X^{-1*} \mathbf{h} \tag{13}$$

or

$$g_{IJ} = (P_X^{-1})_I^K (P_X^{-1})_J^L h_{KL} \tag{14}$$

in the corresponding local coordinate systems. The availability of the metric $\mathbf{g}$ allows one to consider the corresponding Levi-Civita connection ω_g, that is the connection in which the material frame field $\mathbf{p}_I$, $I = 1, 2, 3$, is parallel and $\mathbf{g}$-orthonormal. It can be shown that the curvature of the connection ω_g is defined by the torsion of the material connection ω, Wang and Truesdell [16].

3 The Multiplicative Decomposition of the Deformation Gradient

In this section we will look at the uniformity and the homogeneity of a material body from a somewhat different perspective, that is, using the concept of an intermediate configuration. To this end, suppose that we are given the material body M in a configuration $\phi : M \rightarrow \mathbb{R}^3$. Its deformation gradient $\mathbf{F}$ can be viewed as a two-point tensor field on M, i.e., the tangent bundle mapping $\phi_* : TM \rightarrow T\mathbb{R}^3$ over the base mapping (configuration) $\phi : M \rightarrow \mathbb{R}^3$. Let

$$\mathbf{F} = \mathbf{F}^e \mathbf{F}^p \tag{15}$$

be the Bilby-Kröner-Lee multiplicative decomposition (**BKL-decomposition**, in short) of the deformation gradient where $\mathbf{F}^e$ is understood as the elastic part of the deformation gradient while $\mathbf{F}^p$ is its inelastic (plastic) component (see for example Bilby [2], Kröner [6] and Lee [9]). Assume, that every time the deformation gradient of a material configuration is available[4] we have means of identifying its BKL-decomposition. Interpreting the relation (15) as the composition of (tangent) bundle maps it is only natural (at least from the mathematical stand point) to consider the **intermediate configuration** $\widehat{\phi} : M \rightarrow \mathbb{R}^3$ as the base map for the bundle map $\mathbf{F}^p$. Indeed, if $\mathbf{F}^e$ and $\mathbf{F}^p$ are to represent bundle maps one is required to

---

[3] If the isotropy group G of the archetypical energy function $\widehat{W}$ is nontrivial and continuous, different material parallelisms, and different material connections are possible, all gauged by the isotropy group G. However, if the isotropy group G is discrete the corresponding material connection is unique, as implied by the local smoothness of the distribution of implants. In this instance and, in particular, when the group G is trivial, the torsion of ω may be considered the true measure of the local non-homogeneity, Epstein and Elżanowski [4].

[4] The strain tensor may be a better measure of the deformation, see Remark 1.

introduce a configuration $\widehat{\phi}$, which we assume to be a (differential) embedding, such that the composition

$$\widehat{\psi} \equiv \phi \circ \widehat{\phi}^{-1} \tag{16}$$

is well defined and the bundle maps $\mathbf{F}^e$ and $\mathbf{F}^p$ are based over $\widehat{\psi}$ and $\widehat{\phi}$, as illustrated by the following diagram

$$
\begin{array}{ccccc}
TM & \xrightarrow{\mathbf{F}^p} & T\widehat{\phi}(M) & \xrightarrow{\mathbf{F}^e} & T\phi(M) \\
\downarrow & & \downarrow & & \downarrow \\
M & \xrightarrow{\widehat{\phi}} & \widehat{\phi}(M) & \xrightarrow{\widehat{\psi}} & \phi(M)
\end{array}
$$

Thus, given the material point $X \in M$, $\mathbf{F}^p(X) \in T_{\widehat{\phi}(X)}\mathbb{R}^3$ while $\mathbf{F}^e(\widehat{\phi}(X)) \in T_{\phi(X)}\mathbb{R}^3$. Realize that despite the fact that the deformation gradient $\mathbf{F}$ is the tangent map of its base mapping (configuration) ϕ, the elements of the BKL-decomposition (15) are, in general, nonintegrable. That is, the inelastic part $\mathbf{F}^p$ is not necessarily the tangent map of the intermediate configuration[5] $\widehat{\phi}$ and the elastic part $\mathbf{F}^e$ is not the gradient of its base map $\widehat{\psi}$. Still, both elements of the BKL-decomposition, viewed as the bundle maps, are based over the corresponding base maps $\widehat{\phi}$ and $\widehat{\psi}$. All that implies that there exists a tangent bundle map $\mathbf{D} : TM \to TM$, over the identity map of M, such that

$$\mathbf{F}^p = \widehat{\phi}_* \circ \mathbf{D} \tag{17}$$

and

$$\mathbf{F}^e = \phi_* \circ (\widehat{\phi}_* \circ \mathbf{D})^{-1}. \tag{18}$$

We shall call the tensor $\mathbf{D}$ the **uniformity tensor**. Note that $\mathbf{D}$, when evaluated at a material point, say X, is effectively the implant map (6). Indeed, given the material point $X \in M$ the BKL-decomposition can be presented, using the language of the Geometric Theory of Material Inhomogeneities, as $\mathbf{F}^e = \mathbf{F}P_X$ and $\mathbf{F}^p = P_X^{-1}$, where P_X denotes the corresponding material isomorphism from the fixed material point X_0 (or the archetype V) to X; see for example Maugin and Epstein [12]. Invoking the Euclidean parallelism in $\mathbb{R}^3$, which allows us to view P_X as a map from T_XM to itself with $P_{X_0} = \mathbf{I}$, we can equate the uniformity tensor $\mathbf{D}(X)$ with P_X^{-1}.

It seems that the BKL-decomposition of the deformation gradient $\mathbf{F}$ and the intermediate configuration $\widehat{\phi}$ define completely[6] the uniformity structure of the material. Note however that given the deformation gradient $\mathbf{F}$ its BKL-decomposition is not necessarily uniquely defined. Indeed, replace the configuration $\widehat{\phi}$ with $\widetilde{\phi} = \beta \circ \widehat{\phi}$, where β is a diffeomorphism of the physical space E^3 (or simply a change of its

---

[5] It is often argued that the (inelastic) intermediate configuration $\widehat{\phi}$ is defined uniquely by the material and the history of the deformation leading to the current configuration ϕ. See Lee and Agah-Tehrani [10] where the relaxation (unloading) of the material to the intermediate configuration is discussed.

[6] See also Ciancio *et al.* [3].

coordinate system). Then, the plastic part of the deformation gradient corresponding to the new intermediate configuration $\widetilde{\phi}$ is given by

$$\mathbf{F}_{\beta}^{p} = \beta_{*} \circ \mathbf{F}^{p} \tag{19}$$

subsequently changing the form of the elastic part. The tensor $\mathbf{D}$ is not affected, however, by such a change of the intermediate configuration. On the other hand, if the intermediate configuration $\widehat{\phi}$ gets replaced by $\widehat{\phi} \circ \alpha$, where α can be interpreted, using the language of the Geometric Theory of Material Inhomogeneities, as the change of the archetype V^{7}, then all the elements of the BKL-decomposition do change. Indeed, the new uniformity tensor

$$\mathbf{D}_{\alpha} = \alpha_{*}^{-1} \circ \mathbf{D} \circ \alpha_{*} \tag{20}$$

and the new inelastic part of the deformation gradient is

$$\mathbf{F}_{\alpha}^{p} = \mathbf{F}^{p} \circ \alpha_{*}. \tag{21}$$

Once the uniformity tensor $\mathbf{D}$ is available the uniform material metric $\mathbf{g}$, (13), can be represented as

$$g_{KM} = D_{K}^{I} D_{M}^{J} h_{IJ}. \tag{22}$$

The metric $\mathbf{g}$ is, in general, not flat. Hence, the corresponding material Levi-Civita connection $\omega\mathbf{g}$ has non-vanishing curvature. On the other hand, the vanishing of the curvature tensor, say $\mathbf{Rg}$, of the connection $\omega\mathbf{g}$, or equivalently its Ricci tensor $\mathbf{Rcg}$, implies the flatness of the metric $\mathbf{g}$, Kobayashi and Nomizu [5]. The flatness of the material metric should be viewed as the indication of the local material homogeneity. In fact, when the metric $\mathbf{g}$ is flat, one is allowed to select the uniformity tensor $\mathbf{D}$ as the identity $\mathbf{I}$ rendering the choice of the intermediate configuration $\widehat{\phi}$ arbitrary and the BKL-decomposition integrable.

Remark 1. The commonly used measure of the deformation of a material body, particularly well suited for the theory of the small deformations, is the **strain tensor**

$$\mathbf{E} = \frac{1}{2} \ln(\mathbf{h}^{-1}\mathbf{C}) \cong \frac{1}{2}\mathbf{h}^{-1}(\mathbf{C} - \mathbf{h}) \tag{23}$$

comparing the metric of the deformed state ϕ and the metric of the undeformed (reference) state, Marsden and Hughes [11]. Utilizing this measure of the deformation the **inelastic strain** of the BKL-decomposition takes the form

$$\mathbf{E}^{p} = \frac{1}{2} \ln(\mathbf{h}^{-1}\mathbf{D}\widehat{\mathbf{C}}\mathbf{D}^{T}) \cong \frac{1}{2}\mathbf{h}^{-1}(\mathbf{D}\widehat{\mathbf{C}}\mathbf{D}^{T} - \mathbf{h}) \tag{24}$$

where

$$\mathbf{E}_{in}^{p} = \frac{1}{2} \ln(\mathbf{h}^{-1}\widehat{\mathbf{C}}) \cong \frac{1}{2}\mathbf{h}^{-1}(\widehat{\mathbf{C}} - \mathbf{h}) \tag{25}$$

---

[7] See Epstein and Elżanowski [4] for the discussion of this point.

is its integrable part, $\mathbf{D}^T$ denotes the transpose, and the coordinate representation of the Cauchy-Green tensor of the intermediate configuration $\widehat{\mathbf{C}} \equiv \widehat{\phi}^* \mathbf{h}$ is given by

$$\widehat{C}_{IJ} = h_{ij} \widehat{\phi}^i_{,I} \widehat{\phi}^j_{,J}. \tag{26}$$

In this framework the **material strain**

$$\mathbf{E}^m = \frac{1}{2} \ln(\mathbf{h}^{-1} \mathbf{D} \mathbf{h} \mathbf{D}^T) \cong \frac{1}{2} \mathbf{h}^{-1}(\mathbf{D} \mathbf{h} \mathbf{D}^T - \mathbf{h}) \tag{27}$$

may be viewed as a measure of inhomogeneity of a material.

4 The Stress Space

We are now ready to present the construction of the stress space of Stojanovic [15] and Kröner [7] modified to encompass inhomogeneous materials. First, let us assume that the linear isomorphism

$$\mathfrak{F} : T^*M \to TM, \tag{28}$$

relating the covariant and contravariant tensor fields on the body manifold M, is given[8]. We shall call the isomorphism $\mathfrak{F}$ the **residual stress function** and use it to pull back the kinematic objects, such as deformation or strain, from the tangent space TM to the cotangent bundle T^*M, establishing this way the **stress space**. In particular, let

$$\theta \equiv \mathfrak{F}^* \mathbf{h} \tag{29}$$

be the **residual stress metric** on T^*M corresponding to the intrinsic (flat) Euclidean metric $\mathbf{h}$ of the body manifold M. Although the metric $\mathbf{h}$ is flat, the stress metric θ is, in general, not flat, unless the isomorphism $\mathfrak{F}$ is holonomic. Denote by ω_θ the Levi-Civita connection of the metric θ and let $\mathbf{R}_\theta$ be its Riemannian curvature tensor. Finally, let $\mathbf{Rc}_\theta$ be the corresponding Ricci tensor and $\mathfrak{R}_\theta$ its scalar curvature. Following Kröner's lead, let us postulate that the **residual stress** measured at the intermediate (unloaded) configuration $\widehat{\phi}$ is represented by the Ricci tensor $\mathbf{Rc}_\theta$ of the Levi-Civita connection ω_θ. If, in addition, we assume that the isomorphism $\mathfrak{F}$ is such that the Levi-Civita connection ω_θ has constant scalar curvature $\mathfrak{R}_\theta$, then the first Bianchi identity ($\nabla_\theta \mathbf{R}_\theta = 0$, where ∇_θ denotes the covaraint derivative of the connection θ) implies, as often postulated in the literature (see Kröner [7], Minogawa [13], Stojanovic [15]), that the residual stresses are self-equilibrated, that is, that

$$\operatorname{div}_\theta \mathbf{Rc}_\theta = 0 \tag{30}$$

where $\operatorname{div}_\theta$ denotes the covariant divergence.

---

[8] This should be viewed as an additional constitutive postulate we will try to reconcile later with the previously made assumptions leading to the introduction of the uniformity tensor **D**.

Remark 2. The Einstein tensor

$$\mathbf{E}_\theta \equiv \mathbf{Rc}_\theta - \frac{\mathfrak{R}_\theta}{2}\theta, \tag{31}$$

rather than the Ricci tensor $\mathbf{Rc}_\theta$, is the geometric object which in dimension 3 is always the covariant divergence free, Besse [1]. However, in dimension 2, as the Einstein tensor is identically zero, the Ricci tensor becomes its natural substitute. It is not self-equilibrated but the vanishing of its covariant divergence is equivalent, as we mentioned earlier, to postulating that the scalar curvature $\mathfrak{R}_\theta$ is constant.

Having the residual stress defined by the Ricci tensor $\mathbf{Rc}_\theta$, we are now in the position to look again at the constitutive assumption (28) that there exists a linear transformation relating the material tangent and cotangent spaces. But first, viewing the material strain as the natural counter part of the residual stress, let us postulate that

$$\mathbf{E}^m = \frac{1}{2}\mathfrak{F}_*\mathbf{Rc}_\theta. \tag{32}$$

For this definition to be consistent with the earlier definition of the material strain tensor (27) the material metric $\mathbf{g}$, as given by the equation (22), must obey the following relation:

$$\mathbf{g} = \mathbf{h}\exp(\mathbf{E}^m) = \mathbf{h}\exp(\mathfrak{F}_*\mathbf{Rc}_\theta). \tag{33}$$

In other words, postulating the above relation (32), between the residual stress and the material strain, we de facto assume that given the intermediate configuration $\hat{\phi}$ and the uniformity tensor $\mathbf{D}$ there exists an isomorphism $\mathfrak{F}$ such that

$$\ln(\mathbf{D}\mathbf{D}^T\mathbf{h}^{-1}) = \mathfrak{F}_*\mathbf{Rc}_\theta \tag{34}$$

where $\theta = \mathfrak{F}^*\mathbf{h}$. Conversely, given the constitutive isomorphism $\mathfrak{F}$ the stress metric θ defines the stress space and the uniformity tensor $\mathbf{D}$ is given (up to an isometry of the metric $\mathbf{h}$) by the equation (34). The relation (32), between the Ricci (residual stress) tensor $\mathbf{Rc}$ and the material strain tensor $\mathbf{E}^m$, plays the role of the Hooke's law of the linear elasticity. When presented in coordinates, it takes the form

$$E^m{}_{KJ} = \mathfrak{F}^M_K \mathfrak{F}^N_J Rc_{MN} \tag{35}$$

where the tensor $\mathfrak{F}^M_K \mathfrak{F}^N_J$ may be viewed as an (inelastic) analog of the material modula. Following this line of thought, we may want to replace the constitutive isomorphism $\mathfrak{F}$ by a more general linear isomorphism $\varUpsilon$ between the bundles of covariant $(0,2)$-tensors on the tangent and cotangent spaces of the body manifold M. Indeed, given the isomorphism

$$\varUpsilon : S_2(T^*M) \rightarrow S_2(TM), \tag{36}$$

where $S_2(\cdot)$ denotes the bundle of covariant symmetric $(0,2)$-tensors, the stress metric $\theta = \varUpsilon^{-1}\mathbf{h}$ and the material strain

$$E^m{}_{KL} = \varUpsilon^{IJ}_{KL} Rc_{IJ}, \tag{37}$$

while the uniformity tensor **D** is such that

$$\ln(\mathbf{D}\mathbf{D}^T\mathbf{h}^{-1}) = \mathit{\Upsilon}\mathbf{Rc}_\theta. \tag{38}$$

Note that the existence of the isomorphism $\mathit{\Upsilon}$ implies the existence of the isomorphism $\mathfrak{F}$ as its base map.

5 Examples

We present here two simple examples of the stress space and the objects associated with it.

Example 1. Einstein metric
Consider a test case when the stress metric θ is the Einstein metric in dimension 3, that is, when

$$\mathbf{Rc}_\theta = \frac{\mathfrak{R}_\theta}{3}\theta. \tag{39}$$

In such a stress space the uniform material metric **g**, the uniformity tensor **D**, and the material strain $\mathbf{E}^m$ are given by:

$$\mathbf{g} = e^{\frac{\mathfrak{R}_\theta}{3}}\mathbf{h},$$
$$\mathbf{D} = e^{\frac{\mathfrak{R}_\theta}{6}}\mathbf{h}, \tag{40}$$
$$\mathbf{E}^m = e^{\frac{\mathfrak{R}_\theta}{6}}\mathbf{h}.$$

Example 2. Isotropic material
Consider the material isomorphism $\mathit{\Upsilon}$ such that the tensor

$$\mathit{\Upsilon}_{IJ}^{KL} = -\frac{\lambda}{2\mu(3\lambda+2\mu)}h_{IJ}h^{KL} + \frac{1}{2\mu}\delta_I^K\delta_J^L. \tag{41}$$

It can be interpreted as the inverse elasticity tensor of an isotropic material with the inhomogeneous Lame constants λ and μ, Landau and Lifshitz [8]. The corresponding residual stress metric θ is conformally equivalent to the metric **h**, that is, given a material point X

$$\theta(X) = \mathit{\Upsilon}^{-1}\mathbf{h}(X) = \frac{1}{3K(X)}\mathbf{h}, \tag{42}$$

where $K(X) = \mu(X) + \frac{2}{3}\lambda(X)$ is the inhomogeneous bulk module. The Levi-Civita connection of θ has the Christoffel symbols given by

$$\Gamma_{JL}^i = \delta_J^I N_{,L} + \delta_L^I N_{,J} - h^{IS}h_{JL}N_{,S} \tag{43}$$

where $N(X) \equiv -\frac{1}{2}\ln(3K(X))$. Hence, the Ricci tensor $\mathbf{Rc}_\theta$ is represented by a matrix with coordinates:

$$Rc_{KL} = N_{,KL} - 3N_{,K}N_{,L} + [||dN||^2 - \triangle N]h_{KL}, \tag{44}$$

Besse [1]. This, in turn, implies that the material strain (37) has the following representation:

$$E^m{}_{IJ} = -\frac{\lambda}{18K^2\mu}\mathfrak{R}_\theta h_{IJ} + \frac{1}{6K\mu}Rc_{IJ}. \tag{45}$$

One may view this relation as the residual stress analog of the (linear) Hooke's law for an isotropic material.

References

1. Besse, A.: Einstein Manifolds. Springer, Berlin (1987)
2. Bilby, B.A., Gardner, L.R.T., Stroh, A.N.: Continuous distribution of dislocations and the theory of plasticity. In: Proc. XIth ICTAM, vol. VIII, pp. 35–44. Presses de l'Universite de Bruxelles (1957)
3. Ciancio, V., Dolfin, M., Francaviglia, M., Preston, S.: Uniform Materials and the Multiplicative Decomposition of the Deformation Gradient in Finite Elastoplasticity. J. Non-Equilbrium Thermodyn. 33, 199–234 (2008)
4. Epstein, M., Elżanowski, M.: Material Inhomogeneitites and their Evolution. Springer, Berlin (2007)
5. Kobayashi, S., Nomizu, K.: Foundations of Differential Geometry. Wiley, New York (1963)
6. Kröner, E.: Kontinuumstheorie der Versetzungen und Eigenspannungen. Springer, Berlin (1958)
7. Kröner, E.: Stress Space and Strain Space in Continuum Mechanics. Phys. Stat. Sol. (B) 144, 39–44 (1987)
8. Landau, L.D., Lifshitz, E.M.: Elasticity Theory. Pergamon Press, Oxford (1986)
9. Lee, E.: Elastic-plastic deformation at finite strain. ASME Trans. J. Appl. Mech. 54, 1–6 (1969)
10. Lee, E., Agah-Tehrani, A.: The fusion of physical and continuum-mechanical concepts in the formulation of constitutive relations for elastic-plastic materials. In: Knops, R., Lacey, A. (eds.) Non-Classical Continuum Mechanics, pp. 244–259. Cambridge University Press, Cambridge (1987)
11. Marsden, J., Hughes, T.: Mathematical Foundations of Elasticity. Dover, New York (1983)
12. Maugin, G., Epstein, M.: Geometrical material structure of elastoplasticity. International J. of Plasticity 14, 90–115 (1998)
13. Minogawa, S.: On the stress functions in elastodynamics. Acta Mechanica 24, 209–217 (1976)
14. Schaefer, H.: Die Spannungsfunktionen des dreidimensionalen Kontinuums und des elastischen Körpers. Z. angew. Math. Mech. 33, 356–362 (1953)
15. Stojanovic, R.: Equilibrium conditions for internal stresses in non-euclidian continua and stress space. Int. J. Engng. Sci. 1, 323–327 (1963)
16. Wang, C.-C., Truesdell, C.: Introduction to Rational Elasticity. Noordhoff, Leyden (1973)

Coupled Nonlinear Thermoelastic Equations for an Orthotropic Beam with Thermal and Viscous Dissipation

M.B. Rubin

In honor of the 70th birthday of Professor Krzysztof Wilmanski

Abstract. Specific constitutive equations are presented for coupled nonlinear thermoelastic response of an orthotropic beam. The equations include thermal and viscous dissipation as well as temperature dependent specific heat, thermal expansion coefficients and elastic moduli.

1 Introduction

Nonlinear thermomechanical equations have been developed using the theory of a Cosserat curve by Green et al. [1, 2] and have been reconsidered in [5] using the thermodynamical formulation developed by Green and Naghdi [3, 4]. The objective of this paper is to present specific constitutive equations for an orthotropic beam which include temperature dependent specific heat, elastic moduli and thermal expansion coefficients. Attention is limited to a Helmholtz free energy function that is quadratic in Lagrangian strain which is limited to moderate strains but can be used for large rotations.

2 Basic Equations

Here, the equations are developed using the notation presented in [10]. Specifically, consider a beam which in its reference configuration has length L, and has a rectangular cross-section of uniform height H_1 and width H_2. Within the context of the

M.B. Rubin
Faculty of Mechanical Engineering, Technion - Israel Institute of Technology,
32000 Haifa, Israel
Tel.: (972-4) 829-3188
e-mail: mbrubin@tx.technion.ac.il

theory of a Cosserat rod, the kinematics of the reference configuration of a straight beam are specified by

$$\mathbf{X} = z^3 \mathbf{e}_3, \quad \mathbf{D}_\alpha = \mathbf{e}_\alpha, \quad \mathbf{D}_3 = \mathbf{X}_{,3} = \mathbf{e}_3, \quad D^{1/2} = \mathbf{D}_1 \times \mathbf{D}_2 \cdot \mathbf{D}_3 = 1, \qquad (1)$$

where $\mathbf{X}$ locates material points on the beam's centerline, z^3 is a convected axial coordinate, $\mathbf{D}_i$ $(i = 1,2,3)$ are director vectors with $\mathbf{D}_\alpha$ $(\alpha = 1,2)$ being identified with material line elements in the beam's cross-section, $\mathbf{D}_3$ is the tangent vector to the beam's centerline, a comma denotes partial differentiation with respect to z^3, and $\mathbf{e}_i$ are a fixed set of rectangular Cartesian base vectors. Throughout the text, Latin indices have the range $(i = 1,2,3)$, Greek indices have the range $(\alpha = 1,2)$ and the usual summation convention is implied for repeated indices.

The kinematics of the rod in its deformed present configuration are characterized by

$$\mathbf{x} = \mathbf{x}(z^3,t), \quad \mathbf{d}_\alpha = \mathbf{d}_\alpha(z^3,t), \quad \mathbf{d}_3 = \mathbf{x}_{,3},$$
$$d^{1/2} = \mathbf{d}_1 \times \mathbf{d}_2 \cdot \mathbf{d}_3 > 0, \quad d_{33}^{1/2} = |\mathbf{d}_3|, \qquad (2)$$

where $\mathbf{x}$ locates the beam's centerline, $\mathbf{d}_i$ are the present values of the directors $\mathbf{D}_i$ and t is time. Also, the velocity $\mathbf{v}$ and director velocities $\mathbf{w}_i$ are defined by

$$\mathbf{v} = \dot{\mathbf{x}}, \quad \mathbf{w}_i = \dot{\mathbf{d}}_i, \qquad (3)$$

where a superposed dot denotes material time differentiation holding z^3 fixed.

Within the context of the Cosserat theory the conservation of mass, balance of linear momentum and balances of director momentum are presented using the direct approach. The local forms of these equations can be written as

$$\dot{m} = 0, \quad m\dot{\mathbf{v}} = m\mathbf{b} + \mathbf{t}^3{}_{,3}, \quad my^{\alpha\beta}\dot{\mathbf{w}}_\beta = m\mathbf{b}^\alpha - \mathbf{t}^\alpha + \mathbf{m}^\alpha{}_{,3}, \qquad (4)$$

where m is the mass per unit reference length of the beam, $\mathbf{b}$ is the specific (per unit mass) assigned external force due to body forces and tractions on the beams lateral surfaces, $\mathbf{t}^3$ is the resultant force applied to the beam's end, $y^{\alpha\beta} = y^{\beta\alpha}$ is a constant symmetric positive definite matrix characterizing director inertia, $\mathbf{b}^\alpha$ are specific external applied director couples due to body force and tractions on the beams lateral surfaces, $\mathbf{t}^\alpha$ are intrinsic director couples and $\mathbf{m}^\alpha$ are director couples applied to the beam's end. Also, the kinetic quantities $\mathbf{t}^i$ and $\mathbf{m}^\alpha$ require constitutive equations.

For the simplest thermomechanical theory each material point is characterized by three temperature fields

$$\theta = \theta(z^3,t), \quad \phi_\alpha = \phi_\alpha(z^3,t), \qquad (5)$$

where θ is the average absolute temperature in the beam's cross-section and ϕ_α are average temperature gradients in the $\mathbf{d}_\alpha$ directions. These temperature fields are determined by three balances of entropy which in their local forms are proposed as

$$m\dot{\eta}^i = m(s^i + \xi^i) + \pi^i - p^i_{,3} \quad \text{with} \quad \pi^3 = 0. \tag{6}$$

In these equations, η^i are specific entropies, s^i are specific external rates of supply of entropy, ξ^i are specific internal rates of supply of entropy, π^i are intrinsic supplies of entropy and p^i are entropy fluxes. In addition, the values of the temperature fields (5) are assumed to satisfy the restriction

$$\theta - \frac{H_1}{2}|\phi_1| - \frac{H_2}{2}|\phi_2| > 0, \tag{7}$$

which ensures that the absolute temperature on the beam's lateral surfaces remains positive.

Also, the local forms of the balances of angular momentum and energy are given by

$$\frac{d}{dt}\left[\mathbf{x} \times m\mathbf{v} + \mathbf{d}_\alpha \times my^{\alpha\beta}\mathbf{w}_\beta\right] = \mathbf{x} \times m\mathbf{b}^3 + \mathbf{d}_\alpha \times m\mathbf{b}^\alpha + (\mathbf{x} \times \mathbf{t}^3 + \mathbf{d}_\alpha \times \mathbf{m}^\alpha)_{,3}, \tag{8a}$$

$$m\dot{\varepsilon} + \dot{\mathcal{K}} = \mathcal{W} + \mathcal{H}, \tag{8b}$$

where ε is the specific internal energy and the kinetic energy $\mathcal{K}$, external rate of work $\mathcal{W}$ and rate of heat $\mathcal{H}$ supplied are defined by

$$\mathcal{K} = \frac{1}{2}m(\mathbf{v} \cdot \mathbf{v} + y^{\alpha\beta}\mathbf{w}_\alpha \cdot \mathbf{w}_\beta),$$

$$\mathcal{W} = m\mathbf{b}^3 \cdot \mathbf{v} + m\mathbf{b}^\alpha \cdot \mathbf{w}_\alpha + (\mathbf{t}^3 \cdot \mathbf{v} + \mathbf{m}^\alpha \cdot \mathbf{w}_\alpha)_{,3}, \tag{9}$$

$$\mathcal{H} = m(\theta s^3 + \phi_\alpha s^\alpha) - (\theta p^3 + \phi_\alpha p^\alpha)_{,3}.$$

Next, using the balance laws (4) it can be shown that the balance of angular momentum requires symmetry of the second order tensor $\mathbf{T}$ which is defined by

$$\mathbf{T}^T = \mathbf{T} = d_{33}^{-1/2}(\mathbf{t}^i \otimes \mathbf{d}_i + \mathbf{m}^\alpha \otimes \mathbf{d}_{\alpha,3}). \tag{10}$$

Also, following the work in [8] it is convenient to introduce the rate of material dissipation $m\theta\xi'$ through the definition

$$m\theta\xi^3 + m\phi_\alpha\xi^\alpha = -(\pi^\alpha\phi_\alpha + p^3\theta_{,3} + p^\alpha\phi_{\alpha,3}) + m\theta\xi', \tag{11}$$

where the first term on the right-hand side of (11) characterizes the rate of dissipation due to thermal heat conduction. Then, with the help of the balance laws (4), (6) and the definition of the Helmholtz free energy ψ

$$\psi = \varepsilon - \theta\eta^3 - \phi_\alpha\eta^\alpha, \tag{12}$$

the balance of energy (8b) can be rewritten in the form

$$m\theta\xi' = \mathbf{t}^i \cdot \mathbf{w}_i + \mathbf{m}^\alpha \cdot \mathbf{w}_{\alpha,3} - m(\dot{\psi} + \eta^3\dot{\theta} + \eta^\alpha\dot{\phi}_\alpha). \tag{13}$$

In discussing the constitutive equations for the beam it is convenient to introduce the reciprocal vectors $\{\mathbf{D}^i, \mathbf{d}^i\}$ and the rate tensors $\{\mathbf{L}, \mathbf{D}\}$, such that

$$\mathbf{D}^i \cdot \mathbf{D}_j = \delta^i{}_j, \quad \mathbf{d}^i \cdot \mathbf{d}_j = \delta^i{}_j,$$

$$\mathbf{L} = \mathbf{w}_i \otimes \mathbf{d}^i = \mathbf{D} + \mathbf{W}, \quad \mathbf{D} = \frac{1}{2}\left(\mathbf{L} + \mathbf{L}^T\right) = \mathbf{D}^T, \tag{14}$$

where $\mathbf{D}$ is the rate of deformation tensor. Also, with the help of (1) and (14) the deformation tensors $\{\mathbf{F}, \mathbf{C}\}$ and the strains $\{\mathbf{E}, \boldsymbol{\beta}_\alpha\}$ are specified by

$$\mathbf{F} = \mathbf{d}_i \otimes \mathbf{D}^i, \quad \mathbf{C} = \mathbf{F}^T \mathbf{F}, \quad \mathbf{E} = \frac{1}{2}(\mathbf{C} - \mathbf{I}), \quad \boldsymbol{\beta}_\alpha = \mathbf{F}^{-1}\mathbf{d}_{\alpha,3}. \tag{15}$$

Then, using these definitions it follows that

$$\mathbf{w}_i = \mathbf{L}\mathbf{d}_i, \quad \dot{\mathbf{F}} = \mathbf{L}\mathbf{F}, \quad \dot{\mathbf{E}} = \mathbf{F}^T \mathbf{D} \mathbf{F}, \quad \mathbf{w}_{\alpha,3} = \mathbf{L}\mathbf{d}_{\alpha,3} + \mathbf{F}\dot{\boldsymbol{\beta}}_\alpha, \tag{16}$$

so that (13) can be rewritten in the form

$$m\theta\xi' = d_{33}^{1/2}\mathbf{T} \cdot \mathbf{D} + \mathbf{F}^T \mathbf{m}^\alpha \cdot \dot{\boldsymbol{\beta}}_\alpha - m(\dot{\psi} + \eta^3 \dot{\theta} + \eta^\alpha \dot{\phi}_\alpha). \tag{17}$$

Constitutive equations must be proposed for the response functions

$$\{\mathbf{T}, \mathbf{m}^\alpha, \psi, \eta^i, \pi^i, \xi^i, p^i\}. \tag{18}$$

Specifically, it is assumed that $\{\mathbf{T}, \mathbf{m}^\alpha\}$ separate into two parts

$$\mathbf{T} = \hat{\mathbf{T}} + \check{\mathbf{T}}, \quad \mathbf{m}^\alpha = \hat{\mathbf{m}}^\alpha + \check{\mathbf{m}}^\alpha, \tag{19}$$

where $\{\hat{\mathbf{T}}, \hat{\mathbf{m}}^\alpha, \psi, \eta_i\}$ characterize thermoelastic response with no material dissipation

$$d_{33}^{1/2}\hat{\mathbf{T}} \cdot \mathbf{D} + \mathbf{F}^T \hat{\mathbf{m}}^\alpha \cdot \dot{\boldsymbol{\beta}}_\alpha - m(\dot{\psi} + \eta^3 \dot{\theta} + \eta^\alpha \dot{\phi}_\alpha) = 0, \tag{20}$$

and $\{\check{\mathbf{T}}, \check{\mathbf{m}}^\alpha\}$ characterize viscous damping which must be dissipative. It then follows that the rate of material dissipation (17) requires

$$m\theta\xi' = d_{33}^{1/2}\check{\mathbf{T}} \cdot \mathbf{D} + \mathbf{F}^T \check{\mathbf{m}}^\alpha \cdot \dot{\boldsymbol{\beta}}_\alpha \geq 0. \tag{21}$$

For a thermoelastic rod the non-dissipative response functions have the forms

$$\hat{\mathbf{T}} = \hat{\mathbf{T}}(\mathbf{F}, \mathscr{V}), \quad \hat{\mathbf{m}}^\alpha = \hat{\mathbf{m}}^\alpha(\mathbf{F}, \mathscr{V}),$$

$$\psi = \hat{\psi}(\mathscr{V}), \quad \eta^i = \eta^i(\mathscr{V}), \quad \mathscr{V} = \{\mathbf{E}, \boldsymbol{\beta}_\alpha, \theta, \phi_\alpha\}. \tag{22}$$

Next, demanding that (20) is satisfied for all thermomechanical processes yields the constitutive restrictions

$$d_{33}^{1/2}\hat{\mathbf{T}} = m\mathbf{F}\frac{\partial\hat{\psi}}{\partial\mathbf{E}}\mathbf{F}^T, \quad \hat{\mathbf{m}}^\alpha = m\mathbf{F}^{-T}\frac{\partial\hat{\psi}}{\partial\boldsymbol{\beta}_\alpha}, \quad \eta^3 = -\frac{\partial\hat{\psi}}{\partial\theta}, \quad \eta^\alpha = -\frac{\partial\hat{\psi}}{\partial\phi_\alpha}. \tag{23}$$

Constitutive equations for the quantities $\{\pi^\alpha, \xi^i, p^i\}$ and details of the viscous terms $\{\check{\mathbf{T}}, \check{\mathbf{m}}^\alpha\}$ will be discussed later. However, once these quantities are known the intrinsic director couples $\mathbf{t}^i$ are determined by solving (10) to obtain

$$\mathbf{t}^i = \hat{\mathbf{t}}^i + \check{\mathbf{t}}^i, \quad \hat{\mathbf{t}}^i = (d_{33}^{1/2}\hat{\mathbf{T}} - \hat{\mathbf{m}}^\alpha \otimes \mathbf{d}_{\alpha,3})\mathbf{d}^i, \quad \check{\mathbf{t}}^i = (d_{33}^{1/2}\check{\mathbf{T}} - \check{\mathbf{m}}^\alpha \otimes \mathbf{d}_{\alpha,3})\mathbf{d}^i. \quad (24)$$

Also, it is noted that these equations are properly invariant under superposed rigid body motions.

3   Second Law of Thermodynamics

Within the context of the thermomechanical theory proposed by Green and Naghdi [3, 4], the constitutive equations (23) are obtained by ensuring that the reduced form of the balance of energy is satisfied for all thermomechanical processes. In addition, the constitutive equations must also satisfy a number of restrictions due to the second law of thermodynamics. For example, the restriction that heat flows from hot to cold requires

$$- (\pi^\alpha \phi_\alpha + p^3 \theta_{,3} + p^\alpha \phi_{\alpha,3}) > 0, \quad (25)$$

for nonzero values of $\{\phi_\alpha; \theta_{,3}; \phi_{\alpha,3}\}$.

Also, using the balances of entropy (6) and the definition (11) it follows that the rate of external supply of heat $\mathscr{H}$ in (9) can be rewritten in the form

$$\mathscr{H} = m(\theta\dot{\eta}^3 + \phi_\alpha \dot{\eta}^\alpha) - m\theta\xi', \quad (26)$$

Another statement of the second law requires non-negative material dissipation [6]

$$m\theta\xi' > 0. \quad (27)$$

It can be seen from (26) that a non-zero rate of material dissipation ($m\theta\xi' > 0$) causes a tendency for heat to be expelled from the beam.

4   Specific Thermoelastic Constitutive Equations

Here, constitutive equations are proposed for an orthotropic beam which are based on the work in [7-10]. In its reference configuration the beam has constant uniform mass density ρ_0^*. With the specification (1) the inertia properties of the beam are given in [10, section 5.26]

$$m = \rho_0^* H_1 H_2, \quad y^{11} = \frac{H_1^2}{\pi^2}, \quad y^{22} = \frac{H_2^2}{\pi^2}, \quad y^{12} = y^{21} = 0. \quad (28)$$

Moreover, the Helmholtz free energy is proposed in the generalized form

$$
m\psi = -m \left[\int_{\theta_0}^{\theta} \int_{\theta_0}^{\beta} \frac{C_V^*(\gamma)}{\gamma} d\gamma d\beta + \frac{C_V^*(\theta)}{\theta} \left(\frac{H_1^2 \phi_1^2}{2\pi^2} + \frac{H_2^2 \phi_2^2}{2\pi^2} \right) \right] \tag{29}
$$

$$
+ H_1 H_2 g(\theta) \left[\frac{1}{2} \mathbf{K}^* \cdot (\mathbf{E} \otimes \mathbf{E}) + \frac{1}{2} \mathbf{K}^{\alpha\beta} \cdot (\boldsymbol{\beta}_\alpha \otimes \boldsymbol{\beta}_\beta) - \mathbf{A}(\theta) \cdot \mathbf{E} - \phi_\beta \mathbf{a}^{\alpha\beta}(\theta) \cdot \boldsymbol{\beta}_\alpha \right],
$$

where $C_V^*(\theta)$ is the temperature dependent specific heat at zero strain and uniform temperature, θ_0 is the constant reference temperature, the stiffness tensors $\{\mathbf{K}^*, \mathbf{K}^{\alpha\beta}\}$ are constants and the tensors $\{\mathbf{A}, \mathbf{a}^{\alpha\beta}\}$ are functions of θ only to be specified. Also, the dependence of the elastic moduli on temperature is limited to a simplified form that is controlled by the scalar function $g(\theta) > 0$ with $g(\theta_0) = 1$. The fourth order tensor $\mathbf{K}^*$ has standard symmetries and the second order tensor $\mathbf{K}^{\alpha\beta}$ satisfies the restrictions

$$
\mathbf{K}^{\beta\alpha} = (\mathbf{K}^{\alpha\beta})^T. \tag{30}
$$

Now, using the constitutive restrictions (23) it follows that

$$
d_{33}^{1/2} \hat{\mathbf{T}} = H_1 H_2 g \mathbf{F} (\mathbf{K}^* \cdot \mathbf{E} - \mathbf{A}) \mathbf{F}^T,
$$

$$
\hat{\mathbf{m}}^\alpha = H_1 H_2 g \mathbf{F}^{-T} (\mathbf{K}^{\alpha\beta} \cdot \boldsymbol{\beta}_\beta - \phi_\beta \mathbf{a}^{\alpha\beta}),
$$

$$
m\eta^1 = H_1 H_2 \left[\left(\frac{\rho_0^* C_V^*}{\theta} \right) \frac{H_1^2 \phi_1}{\pi^2} + g \mathbf{a}^{\alpha 1} \cdot \boldsymbol{\beta}_\alpha \right],
$$

$$
m\eta^2 = H_1 H_2 \left[\left(\frac{\rho_0^* C_V^*}{\theta} \right) \frac{H_2^2 \phi_2}{\pi^2} + g \mathbf{a}^{\alpha 2} \cdot \boldsymbol{\beta}_\alpha \right], \tag{31}
$$

$$
m\eta^3 = m \left[\int_{\theta_0}^{\theta} \frac{C_V^*(\gamma)}{\gamma} d\gamma + \frac{1}{\theta} \left(\frac{dC_V^*}{d\theta} - \frac{C_V^*}{\theta} \right) \left(\frac{H_1^2 \phi_1^2}{2\pi^2} + \frac{H_2^2 \phi_2^2}{2\pi^2} \right) \right]
$$

$$
+ H_1 H_2 g \left[\frac{d\mathbf{A}}{d\theta} \cdot \mathbf{E} + \phi_\beta \frac{d\mathbf{a}^{\alpha\beta}}{d\theta} \cdot \boldsymbol{\beta}_\alpha \right]
$$

$$
- H_1 H_2 \frac{dg}{d\theta} \left[\frac{1}{2} \mathbf{K}^* \cdot (\mathbf{E} \otimes \mathbf{E}) + \frac{1}{2} \mathbf{K}^{\alpha\beta} \cdot (\boldsymbol{\beta}_\alpha \otimes \boldsymbol{\beta}_\beta) - \mathbf{A} \cdot \mathbf{E} - \phi_\beta \mathbf{a}^{\alpha\beta} \cdot \boldsymbol{\beta}_\alpha \right].
$$

Moreover, with the help of (12) the internal energy is given by

$$
m\varepsilon = m \left[\theta \int_{\theta_0}^{\theta} \frac{C_V^*(\gamma)}{\gamma} d\gamma - \int_{\theta_0}^{\theta} \int_{\theta_0}^{\beta} \frac{C_V^*(\gamma)}{\gamma} d\gamma d\beta + \frac{dC_V^*(\theta)}{d\theta} \left(\frac{H_1^2 \phi_1^2}{2\pi^2} + \frac{H_2^2 \phi_2^2}{2\pi^2} \right) \right]
$$

$$
+ H_1 H_2 \left(g - \theta \frac{dg}{d\theta} \right) \left[\frac{1}{2} \mathbf{K}^* \cdot (\mathbf{E} \otimes \mathbf{E}) + \frac{1}{2} \mathbf{K}^{\alpha\beta} \cdot (\boldsymbol{\beta}_\alpha \otimes \boldsymbol{\beta}_\beta) - \mathbf{A} \cdot \mathbf{E} \right] \tag{32}
$$

$$
+ H_1 H_2 \theta \left[g \frac{d\mathbf{A}}{d\theta} \cdot \mathbf{E} + \phi_\beta \left(\frac{dg}{d\theta} \mathbf{a}^{\alpha\beta} + g \frac{d\mathbf{a}^{\alpha\beta}}{d\theta} \right) \cdot \boldsymbol{\beta}_\alpha \right].
$$

In particular, for the case of zero strain and zero temperature gradient it follows that

$$\dot{\varepsilon} = C_V^*(\theta)\dot{\theta} \quad \text{for} \quad \mathbf{E} = 0, \quad \boldsymbol{\beta}_\alpha = 0 \quad \text{and} \quad \phi_\alpha = 0, \tag{33}$$

which justifies the interpretation of C_V^* as the temperature dependent specific heat at zero strain and uniform temperature.

Next, assuming that the principal directions $\mathbf{M}_i$ of orthotropy are aligned with the principal axes of the beam $\mathbf{M}_i = \mathbf{e}_i$ it can be shown [10, section 5.26] that

$$\begin{aligned}
\mathbf{K}^* \cdot \mathbf{E} = {} & (K_{1111}^* E_{11} + K_{1122}^* E_{22} + K_{1133}^* E_{33})(\mathbf{e}_1 \otimes \mathbf{e}_1) \\
& + (K_{1122}^* E_{11} + K_{2222}^* E_{22} + K_{2233}^* E_{33})(\mathbf{e}_2 \otimes \mathbf{e}_2) \\
& + (K_{1133}^* E_{11} + K_{2233}^* E_{22} + K_{3333}^* E_{33})(\mathbf{e}_3 \otimes \mathbf{e}_3) \\
& + (2K_{1212}^* E_{12})(\mathbf{e}_1 \otimes \mathbf{e}_2 + \mathbf{e}_2 \otimes \mathbf{e}_1) + (2K_{1313}^* E_{13})(\mathbf{e}_1 \otimes \mathbf{e}_3 + \mathbf{e}_3 \otimes \mathbf{e}_1) \\
& + (2K_{2323}^* E_{23})(\mathbf{e}_2 \otimes \mathbf{e}_3 + \mathbf{e}_3 \otimes \mathbf{e}_2), \tag{34}
\end{aligned}$$

$$\mathbf{K}^{1\beta} \cdot \boldsymbol{\beta}_\beta = K_{1212}\beta_{12}\mathbf{e}_2 + K_{3131}\beta_{13}\mathbf{e}_3,$$

$$\mathbf{K}^{2\beta} \cdot \boldsymbol{\beta}_\beta = K_{1212}\beta_{21}\mathbf{e}_1 + K_{3232}\beta_{23}\mathbf{e}_3,$$

where K_{ijmn}^* are the components of the stiffness tensor $\mathbf{K}^*$ of the three-dimensional orthotropic material relative to $\mathbf{e}_i$, $\{E_{ij}, \beta_{\alpha j}\}$ are the components of $\{\mathbf{E}, \boldsymbol{\beta}_\alpha\}$ with respect to $\mathbf{e}_i$

$$K_{ijmn}^* = \mathbf{K}^* \cdot (\mathbf{e}_i \otimes \mathbf{e}_j \otimes \mathbf{e}_m \otimes \mathbf{e}_n),$$

$$E_{ij} = \mathbf{E} \cdot (\mathbf{e}_i \otimes \mathbf{e}_j) = \frac{1}{2}(\mathbf{d}_i \cdot \mathbf{d}_j - \delta_{ij}), \quad \beta_{\alpha j} = \boldsymbol{\beta}_\alpha \cdot \mathbf{e}_j = \mathbf{d}_{\alpha,3} \cdot \mathbf{d}^j, \tag{35}$$

and where the constitutive constants $\{K_{1212}, K_{3131}, K_{3232}\}$ are specified by

$$K_{1212} = \frac{H_1 H_2}{6}(K_{1313}^* K_{2323}^*)^{1/2} b^*(\zeta),$$

$$K_{3131} = \frac{H_1^2}{12}\left(\frac{1}{C_{3333}^*}\right), \quad K_{3232} = \frac{H_2^2}{12}\left(\frac{1}{C_{3333}^*}\right),$$

$$C_{3333}^* = \frac{1}{C^*}(K_{1111}^* K_{2222}^* - K_{1122}^* K_{1122}^*),$$

$$C^* = K_{1111}^*(K_{2222}^* K_{3333}^* - K_{2233}^* K_{2233}^*) + K_{1122}^*(K_{1133}^* K_{2233}^* - K_{1122}^* K_{3333}^*) \tag{36}$$

$$+ K_{1133}^*(K_{1122}^* K_{2233}^* - K_{1133}^* K_{2222}^*),$$

$$b^*(\zeta) = \zeta\left[1 - \frac{192}{\pi^5}\zeta \sum_{n=1}^{\infty} \frac{1}{(2n-1)^5}\tanh\left\{\frac{\pi(2n-1)}{2\zeta}\right\}\right],$$

$$\zeta = \text{Min}\left(\frac{H_1}{H_2}, \frac{H_2}{H_1}\right).$$

Next, letting $\{\alpha_{11}^*, \alpha_{22}^*, \alpha_{33}^*\}$ be the orthotropic temperature dependent coefficients of linear thermal expansion it is convenient to introduce the functions $\{a_{11}, a_{22}, a_{33}\}$ by

$$a_{11}(\theta) = \int_{\theta_0}^{\theta} \alpha_{11}^*(\gamma)d\gamma, \quad a_{22}(\theta) = \int_{\theta_0}^{\theta} \alpha_{22}^*(\gamma)d\gamma, \quad a_{33}(\theta) = \int_{\theta_0}^{\theta} \alpha_{33}^*(\gamma)d\gamma. \quad (37)$$

Then, the tensors $\{\mathbf{A}, \mathbf{a}^{\alpha\beta}\}$ are defined by

$$\begin{aligned}
\mathbf{A} = &(K_{1111}^* a_{11} + K_{1122}^* a_{22} + K_{1133}^* a_{33})(\mathbf{e}_1 \otimes \mathbf{e}_1) \\
&+ (K_{1122}^* a_{11} + K_{2222}^* a_{22} + K_{2233}^* a_{33})(\mathbf{e}_2 \otimes \mathbf{e}_2) \\
&+ (K_{1133}^* a_{11} + K_{2233}^* a_{22} + K_{3333}^* a_{33})(\mathbf{e}_3 \otimes \mathbf{e}_3), \quad (38) \\
\mathbf{a}^{11} = &K_{3131} \alpha_{33}^* \mathbf{e}_3, \quad \mathbf{a}^{22} = K_{3232} \alpha_{33}^* \mathbf{e}_3, \quad \mathbf{a}^{12} = \mathbf{a}^{21} = 0,
\end{aligned}$$

and it can be seen from (31) and (34) that for free thermal expansion ($\hat{\mathbf{T}}$, $\hat{\mathbf{m}}^{\alpha} = 0$) the strains are given by

$$\begin{aligned}
E_{11} = a_{11}, \quad E_{22} = a_{22}, \quad E_{33} = a_{33}, \quad E_{12} = E_{13} = E_{23} = 0. \\
\beta_{13} = \phi_1 \alpha_{33}^*, \quad \beta_{12} = \beta_{21} = 0, \quad \beta_{23} = \phi_2 \alpha_{33}^*.
\end{aligned} \quad (39)$$

Next, motivated by the expressions in [7] for shells, the constitutive equations for the additional thermoelastic quantities of the beam are specified by

$$\pi^1 = -\frac{k_1^* H_1 H_2 \lambda_2 \lambda_3}{\theta \lambda_1} \phi_1, \quad \pi^2 = -\frac{k_2^* H_1 H_2 \lambda_1 \lambda_3}{\theta \lambda_2} \phi_2, \quad \pi^3 = 0,$$

$$p^1 = -\frac{k_3^* H_1^3 H_2 \lambda_1 \lambda_2}{12\theta \lambda_3} \phi_{1,3}, \quad p^2 = -\frac{k_3^* H_1 H_2^3 \lambda_1 \lambda_2}{12\theta \lambda_3} \phi_{2,3}, \quad (40)$$

$$p^3 = -\frac{k_3^* H_1 H_2 \lambda_1 \lambda_2}{\theta \lambda_3} \phi_{,3} \quad \xi^{\alpha} = 0,$$

where $k_i^*(\theta)$ are temperature dependent heat conduction coefficients in the principal directions of orthotropy and the stretches λ_i are defined by

$$\lambda_i = |\mathbf{d}_i|. \quad (41)$$

These equations can be motivated by considering weighted averages over the cross-section of the beam of the three-dimensional balance of entropy and confining attention to the simple homogeneous deformation where $\mathbf{d}_i$ deform but remain orthogonal. Also, use has been made of the approximation that division by the three-dimensional absolute temperature can be replaced by division by the average temperature θ. Furthermore, the values of λ_i are introduced to account for the fact that the entropy flux in the three-dimensional theory is measured per unit deformed area. Next, the restriction (25) due to the second law of thermodynamics will be satisfied provided that the heat conduction coefficients are positive

$$k_i^* > 0. \quad (42)$$

Also, using (11) and (40) it follows that

$$
m\theta\xi^3 = \frac{H_1 H_2}{\theta}\left[\frac{k_1^* \lambda_2 \lambda_3}{\lambda_1}\phi_1^2 + \frac{k_2^* \lambda_1 \lambda_3}{\lambda_2}\phi_2^2\right.
$$

$$
\left.+\frac{k_3^* \lambda_1 \lambda_2}{\lambda_3}\left\{(\theta_{,3})^3 + \frac{H_1^2}{12}(\phi_{1,3})^2 + \frac{H_2^2}{12}(\phi_{2,3})^2\right\}\right] + m\theta\xi'.
$$

(43)

For an isotropic material the constitutive constants and functions are given by

$$
K_{1111}^* = K_{2222}^* = K_{3333}^* = K^* + \frac{4}{3}\mu^* = \frac{2\mu^*(1-v^*)}{(1-2v^*)},
$$

$$
K_{1122}^* = K_{1133}^* = K_{2233}^* = K^* - \frac{2}{3}\mu^* = \frac{2\mu^* v^*}{(1-2v^*)},
$$

$$
K_{1212}^* = K_{1313}^* = K_{2323}^* = \mu^*, \quad E^* = 2\mu^*(1+v^*), \qquad (44)
$$

$$
K^* = \frac{E^*}{3(1-2v^*)}, \quad \frac{1}{C_{3333}^*} = E^*,
$$

$$
K_{1212} = \frac{H_1 H_2}{6}\mu^* b^*(\theta), \quad K_{3131} = \frac{E^* H_1^2}{12}, \quad K_{3232} = \frac{E^* H_2^2}{12},
$$

$$
\alpha_{11}^* = \alpha_{22}^* = \alpha_{33}^* = \alpha^*(\theta), \quad k_1^* = k_2^* = k_3^* = k^*(\theta),
$$

where $\{K^*, \mu^*, E^*, v^*\}$ are the small deformation bulk modulus, shear modulus, Young's modulus and Poisson's ratio, respectively, and α^* and k^* are functions of temperature only.

5 Specification of the Assigned Fields

The assigned fields $\{mb, mb^\alpha\}$ for a beam depend on the constant three-dimensional body force $\mathbf{b}^*$ and the values of the traction vector $\mathbf{t}^*$ on the beam's deformed lateral surfaces

$$
\mathbf{t}^* = \hat{\mathbf{t}}_1(z^2,z^3,t) \quad\text{on}\quad z^1 = \frac{H_1}{2}, \quad \mathbf{t}^* = \bar{\mathbf{t}}_1(z^2,z^3,t) \quad\text{on}\quad z^1 = -\frac{H_1}{2},
$$

$$
\mathbf{t}^* = \hat{\mathbf{t}}_2(z^1,z^3,t) \quad\text{on}\quad z^2 = \frac{H_2}{2}, \quad \mathbf{t}^* = \bar{\mathbf{t}}_2(z^1,z^3,t) \quad\text{on}\quad z^2 = -\frac{H_2}{2}.
$$

(45)

Specific expressions recorded in [10, section 5.26] are given by

$$
mb = mb^* + \int_{-H_2/2}^{H_2/2}\left[g^{1/2}\alpha\left(\frac{H_1}{2},z^2\right)\hat{\mathbf{t}}_1 + g^{1/2}\alpha\left(-\frac{H_1}{2},z^2\right)\bar{\mathbf{t}}_1\right]dz^2
$$

$$
+ \int_{-H_1/2}^{H_1/2}\left[g^{1/2}\alpha\left(z^1,\frac{H_2}{2}\right)\hat{\mathbf{t}}_2 + g^{1/2}\alpha\left(z^1,-\frac{H_2}{2}\right)\bar{\mathbf{t}}_2\right]dz^1,
$$

$$m\mathbf{b}^1 = \frac{H_1}{2} \int_{-H_2/2}^{H_2/2} \left[g^{1/2}\alpha\left(\frac{H_1}{2},z^2\right)\hat{\mathbf{t}}_1 - g^{1/2}\alpha\left(-\frac{H_1}{2},z^2\right)\bar{\mathbf{t}}_1 \right] dz^2$$

$$+ \int_{-H_1/2}^{H_1/2} \left[g^{1/2}\alpha\left(z^1,\frac{H_2}{2}\right)\hat{\mathbf{t}}_2 + g^{1/2}\alpha\left(z^1,-\frac{H_2}{2}\right)\bar{\mathbf{t}}_2 \right] z^1 dz^1, \qquad (46)$$

$$m\mathbf{b}^2 = \int_{-H_2/2}^{H_2/2} \left[g^{1/2}\alpha\left(\frac{H_1}{2},z^2\right)\hat{\mathbf{t}}_1 + g^{1/2}\alpha\left(-\frac{H_1}{2},z^2\right)\bar{\mathbf{t}}_1 \right] z^2 dz^2$$

$$+ \frac{H_2}{2} \int_{-H_1/2}^{H_1/2} \left[g^{1/2}\alpha\left(z^1,\frac{H_2}{2}\right)\hat{\mathbf{t}}_2 - g^{1/2}\alpha\left(z^1,-\frac{H_2}{2}\right)\bar{\mathbf{t}}_2 \right] dz^1,$$

where the quantities $g^{1/2}\alpha$ evaluated on the beam's lateral surfaces are specified by

$$\begin{aligned}
&\text{for } z^1 = \frac{H_1}{2} : &&g^{1/2}\alpha\left(\frac{H_1}{2},z^2\right)\mathbf{n}^* = \mathbf{d}_2 \times \left(\mathbf{x}+\frac{H_1}{2}\mathbf{d}_1+z^2\mathbf{d}_2\right)_{,3}, \\
&\text{for } z^1 = -\frac{H_1}{2} : &&g^{1/2}\alpha\left(-\frac{H_1}{2},z^2\right)\mathbf{n}^* = -\mathbf{d}_2 \times \left(\mathbf{x}-\frac{H_1}{2}\mathbf{d}_1+z^2\mathbf{d}_2\right)_{,3}, \\
&\text{for } z^2 = \frac{H_2}{2} : &&g^{1/2}\alpha\left(z^1,\frac{H_2}{2}\right)\mathbf{n}^* = \left(\mathbf{x}+z^1\mathbf{d}_1+\frac{H_2}{2}\mathbf{d}_2\right)_{,3} \times \mathbf{d}_1, \\
&\text{for } z^2 = -\frac{H_2}{2} : &&g^{1/2}\alpha\left(z^1,-\frac{H_2}{2}\right)\mathbf{n}^* = -\left(\mathbf{x}+z^1\mathbf{d}_1-\frac{H_2}{2}\mathbf{d}_2\right)_{,3} \times \mathbf{d}_1, \\
& && \mathbf{n}^*\cdot\mathbf{n}^* = 1,
\end{aligned} \qquad (47)$$

and the dependence of the scalar α on $\{z^3,t\}$ has not be shown explicitly.

Similarly, using the approximation of the three-dimensional temperature field θ^* in the form

$$\theta^*(z^i,t) = \theta(z^3,t) + z^\alpha \phi_\alpha(z^3,t), \qquad (48)$$

and specifying the outward normal components of the heat fluxes on the beam's lateral surfaces by

$$\begin{aligned}
&\theta^*\mathbf{p}^*\cdot\mathbf{n}^* = \hat{q}_1(z^2,z^3,t) \text{ on } z^1 = \frac{H_1}{2}, &&\theta^*\mathbf{p}^*\cdot\mathbf{n}^* = \bar{q}_1(z^2,z^3,t) \text{ on } z^1 = -\frac{H_1}{2}, \\
&\theta^*\mathbf{p}^*\cdot\mathbf{n}^* = \hat{q}_2(z^1,z^3,t) \text{ on } z^2 = \frac{H_2}{2}, &&\theta^*\mathbf{p}^*\cdot\mathbf{n}^* = \bar{q}_2(z^1,z^3,t) \text{ on } z^2 = -\frac{H_2}{2},
\end{aligned} \qquad (49)$$

the assigned thermal fields $\{ms^i\}$ depend on the three-dimensional external rate of heat supply r^* per unit mass and are given by

$$ms^1 = \int\limits_{-H_2/2}^{H_2/2}\int\limits_{-H_1/2}^{H_1/2}\left(\frac{m^*r^*}{\theta+z^\alpha\phi_\alpha}\right)z^1 dz^1 dz^2$$

$$-\frac{H_1}{2}\int\limits_{-H_2/2}^{H_2/2}\left[\frac{g^{1/2}\alpha\left(\frac{H_1}{2},z^2\right)}{\theta+\frac{H_1}{2}\phi_1+z^2\phi_2}\hat{q}_1 - \frac{g^{1/2}\alpha\left(-\frac{H_1}{2},z^2\right)}{\theta-\frac{H_1}{2}\phi_1+z^2\phi_2}\bar{q}_1\right]dz^2$$

$$-\int\limits_{-H_1/2}^{H_1/2}\left[\frac{g^{1/2}\alpha\left(z^1,\frac{H_2}{2}\right)}{\theta+z^1\phi_1+\frac{H_2}{2}\phi_2}\hat{q}_2 + \frac{g^{1/2}\alpha\left(z^1,-\frac{H_2}{2}\right)}{\theta+z^1\phi_1-\frac{H_2}{2}\phi_2}\bar{q}_2\right]z^1 dz^1,$$

$$ms^2 = \int\limits_{-H_2/2}^{H_2/2}\int\limits_{-H_1/2}^{H_1/2}\left(\frac{m^*r^*}{\theta+z^\alpha\phi_\alpha}\right)z^2 dz^1 dz^2$$

$$-\int\limits_{-H_2/2}^{H_2/2}\left[\frac{g^{1/2}\alpha\left(\frac{H_1}{2},z^2\right)}{\theta+\frac{H_1}{2}\phi_1+z^2\phi_2}\hat{q}_1 + \frac{g^{1/2}\alpha\left(-\frac{H_1}{2},z^2\right)}{\theta-\frac{H_1}{2}\phi_1+z^2\phi_2}\bar{q}_1\right]z^2 dz^2 \qquad (50)$$

$$-\frac{H_2}{2}\int\limits_{-H_1/2}^{H_1/2}\left[\frac{g^{1/2}\alpha\left(\theta^1,\frac{H_2}{2}\right)}{\theta+z^1\phi_1+\frac{H_2}{2}\phi_2}\hat{q}_2 - \frac{g^{1/2}\alpha\left(\theta^1,-\frac{H_2}{2}\right)}{\theta+z^1\phi_1-\frac{H_2}{2}\phi_2}\bar{q}_2\right]d\theta^1,$$

$$ms^3 = \int\limits_{-H_2/2}^{H_2/2}\int\limits_{-H_1/2}^{H_1/2}\left(\frac{m^*r^*}{\theta+z^\alpha\phi_\alpha}\right)dz^1 dz^2$$

$$-\int\limits_{-H_2/2}^{H_2/2}\left[\frac{g^{1/2}\alpha\left(\frac{H_1}{2},z^2\right)}{\theta+\frac{H_1}{2}\phi_1+z^2\phi_2}\hat{q}_1 + \frac{g^{1/2}\alpha\left(-\frac{H_1}{2},z^2\right)}{\theta-\frac{H_1}{2}\phi_1+z^2\phi_2}\bar{q}_1\right]dz^2$$

$$-\int\limits_{-H_1/2}^{H_1/2}\left[\frac{g^{1/2}\alpha\left(z^1,\frac{H_2}{2}\right)}{\theta+z^1\phi_1+\frac{H_2}{2}\phi_2}\hat{q}_2 + \frac{g^{1/2}\alpha\left(z^1,-\frac{H_2}{2}\right)}{\theta+z^1\phi_1-\frac{H_2}{2}\phi_2}\bar{q}_2\right]dz^1.$$

6 Specific Constitutive Equations for the Viscous Terms

Motivated by the expressions in [11], the viscous terms $\{\check{\mathbf{T}},\check{m}^\alpha\}$ are specified by the forms

$$d_{33}^{1/2}\check{\mathbf{T}} = H_1 H_2 g\left[C_1\dot{E}_{11}(\mathbf{d}_1\otimes\mathbf{d}_1)+C_2\dot{E}_{22}(\mathbf{d}_2\otimes\mathbf{d}_2)+C_3\dot{E}_{33}(\mathbf{d}_3\otimes\mathbf{d}_3)\right.$$
$$+2C_4\dot{E}_{12}(\mathbf{d}_1\otimes\mathbf{d}_2+\mathbf{d}_2\otimes\mathbf{d}_1)$$
$$\left.+2C_5\dot{E}_{13}(\mathbf{d}_1\otimes\mathbf{d}_3+\mathbf{d}_3\otimes\mathbf{d}_1)+2C_6\dot{E}_{23}(\mathbf{d}_2\otimes\mathbf{d}_3+\mathbf{d}_3\otimes\mathbf{d}_2 2)\right], \qquad (51)$$
$$\check{\mathbf{m}}^1 = H_1 H_2 g\left[C_7\dot{\beta}_{12}\mathbf{d}^2+C_8\dot{\beta}_{13}\mathbf{d}^3\right],$$
$$\check{\mathbf{m}}^2 = H_1 H_2 g\left[C_7\dot{\beta}_{21}\mathbf{d}^1+C_9\dot{\beta}_{23}\mathbf{d}^3\right],$$

where $C_i (i = 1, 2, \ldots, 9)$ are constant viscosity coefficients and the temperature dependent function $g(\theta)$ has been included for simplicity. Next, with the help of (16) and (35), the expression (21) can be rewritten in the form

$$m\theta\xi' = H_1 H_2 g \left[C_1 \left(\dot{E}_{11} \right)^2 + C_2 \left(\dot{E}_{22} \right)^2 + C_3 \left(\dot{E}_{33} \right)^2 \right.$$

$$\left. + 4C_4 \left(\dot{E}_{12} \right)^2 + 4C_5 \left(\dot{E}_{13} \right)^2 + 4C_6 \left(\dot{E}_{23} \right)^2 \right] \qquad (52)$$

$$+ H_1 H_2 g \left[C_7 \left\{ \left(\dot{\beta}_{12} \right)^2 + \left(\dot{\beta}_{21} \right)^2 \right\} + C_8 \left(\dot{\beta}_{13} \right)^2 + C_9 \left(\dot{\beta}_{23} \right)^2 \right] \geq 0,$$

which is satisfied provided that each of the constants C_i is non-negative

$$C_i \geq 0. \qquad (53)$$

Moreover, it can be shown that the constants $\{C_1 - C_6\}$ control viscosities to homogeneous deformations, C_7 controls viscosity to torsional deformation and $\{C_8, C_9\}$ control viscosities to bending deformations.

7 Initial and Boundary Conditions

The balance laws (4) and (6) represent three vector equations and three scalar equations to determine three kinematic vectors and three thermal quantities

$$\{\mathbf{x}, \mathbf{d}_\alpha\}, \quad \{\theta, \phi_\alpha\}. \qquad (54)$$

With the help of the constitutive equations it can be seen that these balance laws are second order in time for kinematic quantities and first order in time for the thermal quantities. This means that initial conditions must be specified for

$$\{\mathbf{x}, \mathbf{v}, \mathbf{d}_\alpha, \mathbf{w}_\alpha\}, \quad \{\theta, \phi_\alpha\} \quad \text{for} \quad t = 0. \qquad (55)$$

With regard to boundary conditions for a beam occupying the region $0 \leq z^3 \leq L$, it follows that the rate of work done $\mathscr{W}_c$ by the force $\mathbf{t}^3$ and director couples $\mathbf{m}^\alpha$ on the end $z^3 = L$ can be written in the form

$$\mathscr{W}_c = (\mathbf{t}^3 \cdot \mathbf{d}_i)(\mathbf{v} \cdot \mathbf{d}^i) + (\mathbf{m}^1 \cdot \mathbf{d}_2)(\mathbf{w}_1 \cdot \mathbf{d}^2) + (\mathbf{m}^1 \cdot \mathbf{d}_3)(\mathbf{w}_1 \cdot \mathbf{d}^3)$$

$$+ (\mathbf{m}^2 \cdot \mathbf{d}_1)(\mathbf{w}_2 \cdot \mathbf{d}^1) + (\mathbf{m}^2 \cdot \mathbf{d}_3)(\mathbf{w}_2 \cdot \mathbf{d}^3), \qquad (56)$$

where use has been made of (19), (29), (34), (38) and (51) to deduce that

$$\mathbf{m}^1 \cdot \mathbf{d}_1 = 0, \quad \mathbf{m}^2 \cdot \mathbf{d}_2 = 0, \qquad (57)$$

Thus, boundary conditions need to be specified at $z^3 = 0$ and $z^3 = L$ for the conjugate quantities

$$\{(\mathbf{t}^3 \cdot \mathbf{d}_1) \text{ or } (\mathbf{v} \cdot \mathbf{d}^1)\}, \quad \{(\mathbf{t}^3 \cdot \mathbf{d}_2) \text{ or } (\mathbf{v} \cdot \mathbf{d}^2)\}, \quad \{(\mathbf{t}^3 \cdot \mathbf{d}_3) \text{ or } (\mathbf{v} \cdot \mathbf{d}^3)\},$$
$$\{(\mathbf{m}^1 \cdot \mathbf{d}_2) \text{ or } (\mathbf{w}_1 \cdot \mathbf{d}^2)\}, \quad \{(\mathbf{m}^1 \cdot \mathbf{d}_3) \text{ or } (\mathbf{w}_1 \cdot \mathbf{d}^3)\}, \quad (58)$$
$$\{(\mathbf{m}^2 \cdot \mathbf{d}_1) \text{ or } (\mathbf{w}_2 \cdot \mathbf{d}^1)\}, \quad \{(\mathbf{m}^2 \cdot \mathbf{d}_3) \text{ or } (\mathbf{w}_2 \cdot \mathbf{d}^3)\}.$$

Also, the boundary conditions for the thermal quantities need to be specified at $z^3 = 0$ and $z^3 = L$ for the conjugate quantities

$$\{p_1 \text{ or } \phi_1\}, \quad \{p_2 \text{ or } \phi_2\}, \quad \{p_3 \text{ or } \theta\}. \quad (59)$$

In addition, it is noted that the mechanical moment $\mathbf{m}$ applied about the centroid of a cross-section with $z^3 = $ constant is determined by the expression

$$\mathbf{m} = \mathbf{d}_\alpha \times \mathbf{m}^\alpha. \quad (60)$$

Acknowledgements. This research was partially supported by MB Rubin's Gerard Swope Chair in Mechanics and by the fund for the promotion of research at the Technion.

References

1. Green, A.E., Naghdi, P.M., Wenner, M.L.: On the theory of rods, I. Derivations from the three-dimensional equations. Proc. Royal Society of London A 337, 451–483 (1974)
2. Green, A.E., Naghdi, P.M., Wenner, M.L.: On the theory of rods, II. Developments by direct approach. Proc. Royal Society of London A 337, 485–507 (1974)
3. Green, A.E., Naghdi, P.M.: On thermodynamics and the nature of the second law. Proc. Royal Soc. Lond. A 357, 253–270 (1977)
4. Green, A.E., Naghdi, P.M.: The Second Law Of Thermodynamics And Cyclic Processes. ASME J. Appl. Mech. 45, 487–492 (1978)
5. Green, A.E., Naghdi, P.M.: On thermal effects in the theory of rods. Int. J. Solids and Structures 15, 829–853 (1979)
6. Rubin, M.B.: An elastic-viscoplastic model for large deformation. Int. J. of Engng. Sci. 24, 1083–1095 (1986)
7. Rubin, M.B.: Heat conduction in plates and shells with emphasis on a conical shell. Int. J. of Solids Structures 22, 527–551 (1986)
8. Rubin, M.B.: Hyperbolic heat conduction and the second law of thermodynamics. Int. J. Engng. Sci. 30, 1665–1676 (1992)
9. Rubin, M.B.: Numerical solution of two- and three-dimensional thermomechanical problems using the theory of a Cosserat point. J. of Math. and Physics (ZAMP), Special Issue 46, S308–S334 (1995); In: Casey, J., Crochet, M.J. (eds.) Theoretical, Experimental, And Numerical Contributions To The Mechanics of Fluids And Solids. Brikhauser Verlag, Basel (1995)
10. Rubin, M.B.: Cosserat Theories: Shells, Rods and Points. In: Solid Mechanics and its Applications, vol. 79. Kluwer, The Netherlands (2000)
11. Yogev, O., Bucher, I., Rubin, M.B.: Dynamic Lateral Torsional Post-Buckling of a Beam-Mass System: Experiments. Journal of Sound and Vibrations 299, 1049–1073 (2007)

On the Mathematical Modelling of Functionally Graded Composites with a Determistinic Microstructure

Czesław Woźniak and Jowita Rychlewska

Abstract. The object of analysis is continuum modelling of composites with a deterministic space-varying (nonperiodic) material microstructure. The aim of the paper is to propose a new modelling method which makes it possible to describe response of a composite by means of PDEs with smooth (averaged) coefficients. In contrast to the known asymptotic homogenization theory the proposed approach leads to model equations describing phenomena depending on the microstructure size. The presented paper summarizes new results obtained in Poland during the last decade. Instead of limit passage with the microstructure size to zero the idea of the method is strictly related to the concept of tolerance relation and tolerance parameter as a certain upper bound for negligibles. That is why the proposed approach is referred to as tolerance modelling method.

1 Preface

The characteristic feature of the tolerance modelling is that in contrast to homogenization, [2], the proposed method is nonasymptotic, i.e. is not based on the limit passage with the microstructure size to zero. The crucial idea of this modelling method is strictly related to the concept of tolerance relation. This is a binary relation which is reflexive, symmetric but not transitive. This relation has a physical sense as a certain indiscernibility relation and for the first time was applied in [8].

Czesław Woźniak
Department of Structural Mechanics, Lodz University of Technology, al. Politechniki 6, 93-590 Łódź, Poland
e-mail: czeslaw.wozniak@p.lodz.pl

Jowita Rychlewska
Institute of Mathematics, Czestochowa University of Technology, ul. Dąbrowskiego 73, 42-200 Częstochowa, Poland
e-mail: rjowita@imi.pcz.pl

Fig. 1 A cross section of
functionally graded material
from the microscopic point
of view

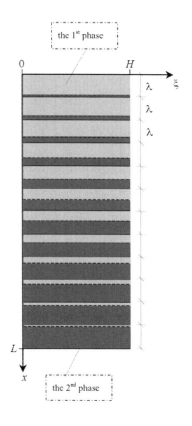

The tolerance relation in the linear normed space is uniquely determined by a sufficiently small real positive number which is called tolerance parameter. This number
was termed by Fichera as "an upper bound for negligibles", [1]. In the last decade the
aforementioned indiscernibility relation was applied by many authors to the mathematical non asymptotic modelling of periodic and functionally graded composite
materials with a deterministic microstructure, [3], [4]. An example of functionally
graded material is shown in Figure 1. General results of the tolerance modelling
method were summarised in monographs [5], [6], [7].

In this paper the main attention is focused on the analytical foundations of the
tolerance modelling for PDEs with functionally nonperiodic coefficients. The text
of this paper is strictly related to that presented in the first part of monograph [5].
Various applications to the special problems of the tolerance modelling method can
be find in the second part of the monograph [5].

2 Analytical Preliminaries

Let X be a linear normed space and let $\approx$ stand for a tolerance relation on X. This
relation for an arbitrary positive number δ will be defined by condition

$$\left(\forall (x_1,x_2) \in X^2\right) [x_1 \approx x_2 \Leftrightarrow \|x_1 - x_2\|_X \leq \delta] \tag{1}$$

Positive constant δ is said to be the tolerance parameter, [7].

Let Ω be a bounded domain in R^m. Points from domain Ω will be denoted by $x = (x_1, x_2, ..., x_m)$ or $z = (z_1, z_2, ..., z_m)$. The basic cell in R^m we define by setting $\square \equiv [-\lambda_1/2, \lambda_1/2] \times ... \times [-\lambda_m/2, \lambda_m/2]$ where $\lambda_1, ..., \lambda_m > 0$. By λ we denote diameter of $\square$, $\lambda \equiv diam\square$, and assume that $\lambda \ll L_\Omega$, where L_Ω is the smallest characteristic length dimension of domain Ω. Hence $\square(x) \equiv x + \square$ is a cell with centre at $x \in R^m$. Moreover we define $\Omega_x \equiv \Omega \cap \underset{z \in \square(x)}{\bigcup} \square(z)$, $x \in \overline{\Omega}$ as a cluster of 2^m cells having common sides. Family of cells $(\Omega, \square) = \{\square(x), x \in \overline{\Omega}\}$ will be referred to as the uniform cell distribution assigned to Ω.

The concept of tolerance periodic function is based on notions of tolerance parameter δ and cell distribution $(\Omega, \square)$. Let $\partial^k f$ be the $k - th$ gradient of function $f = f(x)$, $x \in \Omega$, $k = 0, 1, ..., \alpha$ for some $\alpha \geq 0$, where $\partial^0 f \equiv f$.

In order to present an unified approach to the tolerance modelling all subsequent considerations will be restricted to Sobolev spaces $H^k(\Omega)$ for $k = 0, 1, ..., \alpha$, where nonnegative α integer is assumed to be specified in every problem under consideration. By $H^0(\square)$ we denote the space of $\square$-periodic square-integrable functions defined in R^m. Moreover let $\widetilde{f}^{(k)}(\cdot, \cdot)$ be a function defined in $\overline{\Omega} \times R^m$, $k = 0, 1, ..., \alpha$. For the sake of simplicity the above denotations are assumed to hold both for scalar as well as vector functions.

Function $f \in H^\alpha(\Omega)$ will be called the tolerance periodic function (with respect to cell $\square$ and tolerance parameter δ), $f \in TP_\delta^\alpha(\Omega, \square)$, if for $k = 0, 1, ..., \alpha$ the following conditions hold

(i) $(\forall x \in \Omega) \left(\exists \widetilde{f}^{(k)}(x, \cdot) \in H^0(\square)\right) \left[\left\|\partial^k f |_{\Omega_x}(\cdot) - \widetilde{f}^{(k)}(x, \cdot)\right\|_{H^0(\Omega_x)} \leq \delta\right]$,

(ii) $\underset{\square(\cdot)}{\int} \widetilde{f}^{(k)}(\cdot, z) dz \in C^0(\overline{\Omega})$.

Function $\widetilde{f}^{(k)}(x, \cdot)$ will be referred to as the periodic approximation of $\partial^k f$ in $\square(x)$, $x \in \Omega$, $k = 0, 1, ..., \alpha$. It can be observed that for the given a priori cell $\square$ every constant function can be treated as an element of $H^0(\square)$.

It can be proved that the following properties of the tolerance periodic functions hold

(i) $TP_0^\alpha \equiv \underset{\delta > 0}{\bigcap} TP_\delta^\alpha = H^\alpha(\square)$,

(ii) $TP^\alpha \equiv \underset{\delta \geq 0}{\bigcup} TP_\delta^\alpha = H^\alpha(\Omega)$,

(iii) $(\forall \gamma \in R)(\forall f \in TP_\delta^\alpha) \left[\gamma f \in TP_{|\gamma|\delta}^\alpha\right]$,

(iv) $(\forall \delta_1, \delta_2 \geq 0) \left[TP_{\delta_1}^\alpha \oplus TP_{\delta_2}^\alpha \subset TP_{\delta_1 + \delta_2}^\alpha\right]$,

(v) $(\forall \delta \geq 0) \left[TP_\delta^\alpha \oplus TP_0^\alpha = TP_\delta^\alpha\right]$

where we have denoted $TP_\delta^\alpha \equiv TP_\delta^\alpha(\Omega, \square) \subset H^\alpha(\Omega)$, $\delta \geq 0$.

Function $v \in H^\alpha(\Omega)$ will be called the slowly varying function (with respect to the cell $\Box$ and tolerance parameter δ), $v \in SV_\delta^\alpha(\Omega,\Box)$, if

(i) $v \in TP_\delta^\alpha(\Omega,\Box)$,

(ii) $(\forall x \in \Omega)\left[\widetilde{v}^{(k)}(x,\cdot)\,|_{\Box(x)} = \partial^k v(x), k = 0,1,...,\alpha\right]$.

It means that periodic approximation $\widetilde{v}^{(k)}$ of $\partial^k v(\cdot)$ in $\Box(x)$ is a constant function for every $x \in \Omega$. It can be seen that

$$\left(v \in SV_\delta^\alpha(\Omega,\Box)\right) \Longrightarrow (\forall x \in \Omega)\left[\left\|\partial^k v(\cdot) - \partial^k v(x)\right\|_{H^0(\Box(x))} \le \delta, k = 0,1,...,\alpha\right].$$

Function $h \in H^\alpha(\Omega)$ will be called the highly oscillating function (with respect to the cell $\Box$ and tolerance parameter δ), $h \in HO_\delta^\alpha(\Omega,\Box)$, if

(i) $h \in TP_\delta^\alpha(\Omega,\Box)$,

(ii) $(\forall x \in \Omega)\left[\widetilde{h}^{(k)}(x,\cdot)\,|_{\Box(x)} = \partial^k \widetilde{h}(x,\cdot), k = 0,1,...,\alpha\right]$.

Moreover, for every $v \in SV_\delta^\alpha(\Omega,\Box)$ function $f \equiv hv \in TP_\delta^\alpha(\Omega,\Box)$ satisfies condition

(iii) $\widetilde{f}^{(k)}(x,\cdot)\,|_{\Box(x)} = v(x)\partial^k \widetilde{h}(x,\cdot)\,|_{\Box(x)}, k = 0,1,...,\alpha.$

If $\alpha = 0$ then we denote $\widetilde{f} \equiv \widetilde{f}^{(0)}$.

3   Averaging of Tolerance Periodic Functions

Let $f(\cdot) \in TP_\delta^\alpha(\Omega,\Box)$. By the averaging of tolerance periodic function $f \equiv \partial^0 f$ and its derivatives $\partial^k f$, $k = 1,2,...,\alpha$, we shall mean function $\langle \partial^k f \rangle(x)$, $x \in \overline{\Omega}$, defined by

$$\left\langle \partial^k f \right\rangle(x) \equiv \frac{1}{|\Box|} \int_{\Box(x)} \widetilde{f}^{(k)}(x,z)\,dz, \qquad k = 0,1,...,\alpha, \qquad x \in \overline{\Omega} \qquad (2)$$

where for the sake of simplicity we have denoted $\langle \partial^k f \rangle(x) \equiv \langle \partial^k f(\cdot) \rangle(x)$.

It follows that if $h \in HO_\delta^\alpha(\Omega,\Box)$ and $v \in SV_\delta^\alpha(\Omega,\Box)$ then

$$\left\langle \partial^k h \right\rangle(x) = \frac{1}{|\Box|} \int_{\Box(x)} \partial^k \widetilde{h}(x,z)\,dz$$

$$\left\langle \partial^k v \right\rangle(x) = \frac{1}{|\Box|} \int_{\Box(x)} \partial^k v(x)\,dz = \partial^k v(x)$$

$$\left\langle \partial^k(vh) \right\rangle(x) = \frac{1}{|\Box|} \int_{\Box(x)} \partial^k \widetilde{h}(x,z)\,dz v(x), \qquad x \in \Omega$$

for $k = 0,1,...,\alpha.$

It can be seen that if f is a periodic function then $\widetilde{f}$ is independent of x and $\langle f \rangle$ is constant.

Let $f\left(z, \partial^k g\left(z\right)\right)$, $k = 0, 1, ..., \alpha$ be a composite function defined in Ω such that $f\left(\cdot, \partial^k g\right) \in HO_\delta^0\left(\Omega, \Box\right)$, $g\left(\cdot\right) \in TP_\delta^\alpha\left(\Omega, \Box\right)$. The tolerance averaging of this function is defined by

$$\left\langle f\left(\cdot, \partial^k g\left(\cdot\right)\right)\right\rangle(x) \equiv \frac{1}{|\Box|} \int\limits_{\Box(x)} \widetilde{f}\left(x, z, \widetilde{g}^{(k)}\left(x, z\right)\right) dz, \qquad x \in \overline{\Omega} \qquad (3)$$

where $\widetilde{f} \equiv \widetilde{f}^{(0)}$.

Subsequently we shall focus the attention mainly on highly oscillating and slowly varying functions. The averaging of slowly varying functions is rather trivial, since, roughly speaking, slowly varying functions can be regarded as invariant under averaging. On the other hand, averaging of highly oscillating functions requires the knowledge of periodic approximation of these functions which are assumed to exist but are not given *a priori*. That is why the averaged values of highly oscillating functions will be calculated by using the procedure outlined below.

Let us assume that $f \in HO_\delta^\alpha\left(\Omega, \Box\right)$. We introduce the orthonormal vector basis $(e_1, ..., e_m)$ in R^m. By I we denote a set of all integers and define

$$\Lambda \equiv \{i_1 e_1 \lambda_1 + ... + i_m e_m \lambda_m, \qquad i_1, ..., i_m \in I\}$$

as a lattice in R^m. Let Ω_0 be a certain subset of points in $\Lambda \cap \Omega$. For an arbitrary function $f \in HO_\delta^\alpha\left(\Omega, \Box\right)$ and its derivatives $\partial^k f$, $k = 1, 2, ..., \alpha$ we define

$$\left\langle \partial^k f\right\rangle_0(x) \equiv \frac{1}{|\Box|} \int\limits_{\Box(x)} \partial^k f(z)\, dz, \qquad k = 0, 1, ..., \alpha, \qquad x \in \Omega_0$$

Moreover we define functions $\left\langle \partial^k f\right\rangle_E(x)$, $k = 0, 1, ..., \alpha$, $x \in \Omega$, as smooth extrapolations of functions $\left\langle \partial^k f\right\rangle_0(x)$, $k = 0, 1, ..., \alpha$, $x \in \Omega_0$, on Ω.

Let us conjecture that for every $f \in HO_\delta^\alpha\left(\Omega, \Box\right)$ there exist $\widetilde{f}^{(k)}(x, z)$, $x \in \Omega$, $z \in \Box(x)$ and a subset Ω_0 of Ω such that

(i) $\left\langle \partial^k f\right\rangle_E(x) = \left\langle \partial^k f\right\rangle(x)$, $x \in \Omega$,
(ii) $\left\langle \partial^k f\right\rangle(\cdot) \in SV_\delta^\alpha\left(\Omega, \Box\right)$.

Using this conjecture we can realize tolerance averaging procedure for function $f \in HO_\delta^\alpha\left(\Omega, \Box\right)$ without the specification of functions $\widetilde{f}(x, \cdot)$ for every $x \in \Omega$.

4   Tolerance Averaging of Integral Functionals

Let $\Omega \times \Xi$ be a bounded domain in R^n such that $\Omega \subset R^m$ and $\Xi \subset R^{n-m}$ for $n > m$. Points from domain Ω will be denoted by $x = (x_1, x_2, ..., x_m)$ or $z = (z_1, z_2, ..., z_m)$ and points from Ξ by $\xi = (\xi_1, \xi_2, ..., \xi_{n-m})$. If $n = m$ then Ξ and ξ drop out from considerations. Moreover, we introduce gradient operators of a smooth function f defined on $\Omega \times \Xi$ setting $\partial f \equiv grad_z f(z,\xi)$, $\overline{\nabla} f \equiv grad_\xi f(z,\xi)$ for $n > m$ such that $\nabla f \equiv \partial f + \overline{\nabla} f$. When $n = m$ then $\overline{\nabla}$ drop out from considerations and the total gradient of f is denoted by $\nabla f = \partial f$.

Subsequently we shall write $f(\cdot,\xi) \in TP_\delta^\alpha(\Omega,\square)$, provided that for every $x \in \Omega$ there is known the periodic approximation $\widetilde{f}^{(k)}(x,z.\xi)$ of $\nabla^k f(z,\xi)$, $z \in \square(x)$. If f is a composite function of the form $f = f(z,\nabla^k g(z,\xi))$, $k = 0,1,...,\alpha$, $z \in \Omega$, $\xi \in \Xi$ such that $f(\cdot,\nabla^k g) \in TP_\delta^0(\Omega,\square)$, $g(\cdot,\xi) \in TP_\delta^\alpha(\Omega,\square)$, $\xi \in \Xi$ then instead of (3) we obtain

$$\left\langle f\left(\cdot,\nabla^k g(\cdot,\xi)\right)\right\rangle(x) \equiv \frac{1}{|\square|} \int_{\square(x)} \widetilde{f}\left(x,z,\widetilde{g}^{(k)}(x,z,\xi)\right) dz, \quad x \in \overline{\Omega}, \quad \xi \in \Xi \quad (4)$$

Here and subsequently for the sake of simplicity we restrict ourselves to the case $\alpha = 1$. However, the obtained results can be easily generalized on the case in which α is an arbitrary positive integer. Let us assume that $w \in H^1(\Omega \times \Xi)$, $w = w(z,\xi)$, $z \in \Omega$, $\xi \in \Xi$. We shall deal with the integral functional

$$\mathscr{A}(w(\cdot)) = \iint_{\Omega\Xi} \mathscr{L}(z,\nabla w(z,\xi), w(z,\xi)) d\xi dz$$

where $\mathscr{L}(\cdot,\nabla w, w) \in HO_\delta^0(\Omega,\square)$. It implies that for every $x \in \overline{\Omega}$ there exists periodic approximation $\widetilde{\mathscr{L}} = \widetilde{\mathscr{L}}(x,z,\nabla w, w)$ of integrand $\mathscr{L}(z,\nabla w, w)$, $z \in \square(x)$.

Let $h = \{h^A(\cdot) \in HO_\delta^1(\Omega,\square), \quad A = 1,...,N\}$ be a system of N linear independent functions which is assumed to be postulated *a priori* in every modelling problem under consideration. We shall assume that for every $x \in \overline{\Omega}$ condition $\langle \rho h^A\rangle(x) = 0$ is satisfied for a certain given $\rho \in TP_\delta^0(\Omega,\square)$ positive function . In a special case $\langle h^A\rangle(x) = 0$, $A = 1,...,N$.

The fundamental assumption imposed on integrand $\mathscr{L}$ in the framework of the tolerance averaging approach is that field w in $\mathscr{L}$ will be assumed in the form

$$w(z,\xi) = w_h(z,\xi) \equiv u(z,\xi) + h^A(z) v_A(z,\xi), \quad A = 1,...,N \quad (5)$$

where summation over $A = 1,...,N$ holds and $u(\cdot,\xi)$, $v_A(\cdot,\xi) \in SV_\delta^1(\Omega,\square)$, $\overline{\nabla} u(\cdot,\xi)$, $\overline{\nabla} v_A(\cdot,\xi) \in SV_\delta^0(\Omega,\square)$ and $w(\cdot,\xi) \in TP_\delta^1(\Omega,\square)$ for every $\xi \in \Xi$. Formula (5) will be referred to as the micro-macro decomposition.

Functions u, v_A, $A = 1,...,N$ will play the role of the arguments of the averaged functional and are called averaged (macroscopic) variable and fluctuaction (microscopic) amplitudes, respectively. The postulated *a priori* functions $h^A(\cdot)$, $A = 1,...,N$ are referred to as fluctuation shape functions.

Let $\widetilde{h}^A(x,\cdot)$, $\partial \widetilde{h}^A(x,\cdot)$ stand for periodic approximations of $h^A(\cdot)$, $\partial h^A(\cdot)$ in $\square(x)$, $x \in \overline{\Omega}$, respectively. It can be observed that the periodic approximations of $w_h(\cdot,\xi)$ and $\nabla w_h(\cdot,\xi)$ in $\square(x)$, $x \in \overline{\Omega}$ have the form

$$\widetilde{w}_h^{(0)}(x,z,\xi) = u(x,\xi) + \widetilde{h}^A(x,z) v_A(x,\xi)$$

$$\widetilde{w}_h^{(1)}(x,z,\xi) = \nabla u(x,\xi) + \partial \widetilde{h}^A(x,z) v_A(x,\xi) + \widetilde{h}^A(x,z) \overline{\nabla} v_A(x,\xi) \qquad (6)$$

for every $x \in \overline{\Omega}$, almost every $z \in \square(x)$ and every $\xi \in \Xi$.

Setting $w = w_h$, we define $\mathscr{L}_h \equiv \mathscr{L}(z, \nabla w_h, w_h)$. At the same time we assume that $\mathscr{L}_h(\cdot, \nabla w_h, w_h) \in HO_\delta^0(\Omega, \square)$. Hence there exists periodic approximation of $\mathscr{L}_h(\cdot, \nabla w_h, w_h)$ in every $\square(x)$ denoted by $\widetilde{\mathscr{L}}_h^{(0)}(x, z, \nabla w_h, w_h)$, $z \in \square(x)$. Taking into account approximations (6) we obtain a periodic approximation of composite function $\mathscr{L}_h(z, \nabla w_h, w_h)$ in the form $\widetilde{\mathscr{L}}_h = \widetilde{\mathscr{L}}_h^{(0)}\left(x, z, \widetilde{w}_h^{(1)}(x,z,\xi), \widetilde{w}_h^{(0)}(x,z,\xi)\right)$.

The tolerance averaging of $\mathscr{L}_h$ under micro-macro decomposition (5) is given by

$$\langle \mathscr{L}_h \rangle \left(x, \nabla u, \overline{\nabla} v_A, u.v_A\right) \equiv \frac{1}{|\square|} \int\limits_{\square(x)} \widetilde{\mathscr{L}}_h \left(x, z, \widetilde{w}_h^{(1)}(x,z,\xi), \widetilde{w}_h^{(0)}(x,z,\xi)\right) dz \qquad (7)$$

where $\widetilde{w}_h^{(1)}$ and $\widetilde{w}_h^{(0)}$ are given by (6).

Functional

$$\mathscr{A}_h(u(\cdot), v_A(\cdot)) = \iint\limits_{\Omega \Xi} \langle \mathscr{L}_h \rangle \left(x, \nabla u, \overline{\nabla} v_A, u.v_A\right) d\xi dx \qquad (8)$$

where $\langle \mathscr{L}_h \rangle$ is given by (7) will be called the tolerance averaging of functional $\mathscr{A}(w(\cdot))$ under decomposition $w = w_h = u + h^A v_A$, $u(\cdot, \xi), v_A(\cdot, \xi) \in SV_\delta^1(\Omega, \square)$, $w_h(\cdot, \xi) \in TP_\delta^1(\Omega, \square)$, $\xi \in \Xi$.

5 Model Formulation

It is known that partial differential equations of mathematical physics obtained from the principle of stationary action in its classical form can involve exclusively the even highest derivatives with respect to all arguments of the unknown function. This drawback can be eliminated by introducing certain extension of the stationary action principle. In this section we restrict ourselves to lagrangians which depend only on the first derivative of unknown function. Hence functional $\mathscr{A}(w(\cdot))$ will be assumed in the form

$$\mathscr{A}(w(\cdot), p(\cdot)) = \iint\limits_{\Omega \Xi} \mathscr{L}(z, \nabla w(z, \xi), w(z, \xi), p(z, \xi)) d\xi dz \qquad (9)$$

where $\mathscr{L}(\cdot, \nabla w, w, p) \in HO_\delta^0(\Omega, \square)$, $w(\cdot, \xi) \in TP_\delta^1(\Omega, \square)$, $p(\cdot, \xi) \in TP_\delta^0(\Omega, \square)$, $\xi \in \Xi$. So far, ∇w, w are arguments of lagrangian $\mathscr{L}(\cdot)$ and $p(\cdot)$ plays the role of parameter. It means that the well known Euler-Lagrange equations

$$\nabla \cdot \frac{\partial \mathscr{L}}{\partial \nabla w} - \frac{\partial \mathscr{L}}{\partial w} = 0 \qquad (10)$$

are assumed to hold in $\Omega \times \Xi$. In general the derivatives $\partial\mathscr{L}/\partial\nabla w$ have to be continuous and divergence $\nabla \cdot (\partial\mathscr{L}/\partial\nabla w)$ in (10) has to be understood in the weak form.

Now, let us assume that function p is determined by independent equation

$$p(z,\xi) = r(z,\xi,\nabla w(z,\xi), w(z,\xi)) \tag{11}$$

where $r(\cdot,\xi,\nabla w, w) \in TP_\delta^0(\Omega,\square)$ in every problem under consideration is a known function.

Equation (11) will be called the constitutive equation for function p. The passage from (9) to Euler-Lagrange equation (10) in which p is given by constitutive equation (11) represent what will be called the extended principle of stationary action or the principle of stationary action extended by constitutive equation. This principle has to be understood as a certain formal analytical tool for deriving equations which cannot be obtained from the principle of stationary action in its well known form. Obviously, substituting the right-hand side of (11) into lagrangian $\mathscr{L}(x,\nabla w, w, p)$ in equation (10) we arrive at equations which are not implied by the principle of stationary action for functional (9).

As a simple example assume $\mathscr{L} = \mathscr{L}(x,\partial w, w, p)$ and $p = r\left(x, \overline{\nabla} w\right)$. In this case equation (10) is the second order PDE with respect to argument $x \in \Omega$ and the first order PDE with respect to argument $\xi \in \Xi$. It follows that the principle of extended stationary action makes it possible to obtain PDEs of the parabolic type.

So far, we restricted ourselves to the case in which $\mathscr{L}(\cdot)$, $r(\cdot)$ depend only on the first order derivatives of w. However, the above results can be easily generalized on the case in which $\mathscr{L}(\cdot)$, $r(\cdot)$, depend on higher order derivatives of w.

Subsequently action functional (9) will be assumed in the form

$$\mathscr{A}(w(\cdot), p(\cdot)) = \iint_{\Omega\Xi} (\mathscr{L}(z,\nabla w(z,\xi), w(z,\xi)) + p(z,\xi)w(z,\xi))\,d\xi dz \tag{12}$$

where $\mathscr{L}(\cdot,\nabla w, w) \in HO_\delta^0(\Omega,\square)$, $p(\cdot,\xi) \in HO_\delta^0(\Omega,\square)$. Instead of (10) we obtain

$$\nabla \cdot \frac{\partial\mathscr{L}}{\partial\nabla w} - \frac{\partial\mathscr{L}}{\partial w} = p \tag{13}$$

and the constitutive equation for $p(\cdot)$ retain the form (11).

Tolerance modelling procedure for equation (13) is realized in two steps. The first step is the tolerance averaging of action functional (12) by means of (8) under the micro-macro decomposition (5). It has to be emphasized that in this step function p is treated as a certain parameter. The second step is to apply the extended principle of stationary action. Using this procedure from (8) and (12) we obtain averaged functional

$$\mathscr{A}_h(u, v_A, p) = \iint_{\Omega\Xi} \left(\langle\mathscr{L}_h\rangle\left(x,\nabla u, u, \overline{\nabla} v_A, v_A\right) + \langle p\rangle(x)u + \langle ph^A\rangle(x)v_A\right)d\xi dx \tag{14}$$

where

$$\langle \mathscr{L}_h \rangle \left(x, \nabla u, u, \overline{\nabla} v_A, v_A \right) = \frac{1}{|\Box|} \int\limits_{\Box(x)} \mathscr{L} \left(x, z, \nabla u + \partial \tilde{h}^A v_A + \tilde{h}^A \overline{\nabla} v_A, u + \tilde{h}^A v_A \right) dz$$

$$\langle p \rangle (x) \equiv \frac{1}{|\Box|} \int\limits_{\Box(x)} \tilde{p}(x, z, \xi) \, dz$$

$$\langle p h^A \rangle (x) \equiv \frac{1}{|\Box|} \int\limits_{\Box(x)} \tilde{h}^A (x, z) \, \tilde{p}(x, z, \xi) \, dz \qquad (15)$$

for every $\xi \in \Xi$.

Applying the principle of stationary action to $\mathscr{A}_h$ we obtain the following system of equations

$$\nabla \cdot \frac{\partial \langle \mathscr{L}_h \rangle}{\partial \nabla u} - \frac{\partial \langle \mathscr{L}_h \rangle}{\partial u} = \langle p \rangle,$$

$$\overline{\nabla} \cdot \frac{\partial \langle \mathscr{L}_h \rangle}{\partial \overline{\nabla} v_A} - \frac{\partial \langle \mathscr{L}_h \rangle}{\partial v_A} = \langle p h^A \rangle, \quad A = 1, ..., N \qquad (16)$$

Bearing in mind the concept of the extended principle of stationary action we assume now that function p is determined by the constitutive equation (11). Under denotations

$$r_h \left(x, \nabla u, u, \overline{\nabla} v_A, v_A \right) \equiv \frac{1}{|\Box|} \int\limits_{\Box(x)} \tilde{r} \left(x, z, \xi, \nabla u + \partial \tilde{h}^A v_A + \tilde{h}^A \overline{\nabla} v_A, u + \tilde{h}^A v_A \right) dz$$

$$r_h^A \left(x, \nabla u, u, \overline{\nabla} v_A, v_A \right) \equiv \frac{1}{|\Box|} \int\limits_{\Box(x)} \tilde{h}^A \tilde{r} \left(x, z, \xi, \nabla u + \partial \tilde{h}^A v_A + \tilde{h}^A \overline{\nabla} v_A, u + \tilde{h}^A v_A \right) dz$$

the averaged constitutive equations take the form

$$\langle p \rangle (x) = r_h \left(x, \nabla u, u, \overline{\nabla} v_A, v_A \right)$$

$$\langle p h^A \rangle (x) = r_h^A \left(x, \nabla u, u, \overline{\nabla} v_A, v_A \right) \qquad (17)$$

Substituting the right-hand side of equations (17) into (16) we obtain the final system of equations for $u = u(x, \xi)$, $v_A = v_A(x, \xi)$, $A = 1, ..., N$ as the new basic unknowns. The above equations represent the tolerance model equations under the micro-macro decomposition (5).

It has to be emphasized that solutions to equations (16), (17) have the physical sense if the micro-macro decomposition (5) can be regarded as a certain reasonable approximation of function $w(\cdot)$ describing the behaviour of a microheterogeneous medium in the problem under consideration. Moreover, the value of tolerance parameter δ in definition of slowly varying functions $u(\cdot,\xi)$, $v_A(\cdot,\xi)$, $\xi \in \Xi$, have to be sufficiently small. This requirement can be verified only *aposteriori* i.e., after obtaining the form of functions $u(\cdot,\cdot)$, $v_A(\cdot,\cdot)$.

6 Example

In order to illustrate the general results let us assume lagrangian $\mathscr{L}$ in action functional (12) in the form

$$\mathscr{L} = \frac{1}{2}\nabla w \cdot A \cdot \nabla w + f w \tag{18}$$

where A is a positive definite symmetric matrix, $A = A^T$, of an order n and f is a scalar function. Moreover, function p in (12) will be assumed in the form

$$p = k \cdot \nabla w \tag{19}$$

where k is a vector from R^n. Hence arguments of lagrangian $\mathscr{L}$ are w and ∇w where w is a scalar function defined on $\Omega \times \Xi$, $w(\cdot) \in H^1(\Omega \times \Xi)$.

We recall that $x = (x_1,...,x_m) \in \Omega \subset R^m$ and $\xi = (\xi_1,...,\xi_{n-m}) \in \Xi \subset R^{n-m}$. Gradients in R^m and R^{n-m} are denoted by ∂ and $\overline{\nabla}$, respectively, provided that $n > m$. Gradients in R^n will be denoted by ∇. If $n = m$ then denotations ξ, Ξ, $\overline{\nabla}$ drop out from considerations.

It is assumed that A, k are highly-oscillating functional coefficients of $\mathscr{L}$, p, respectively, defined in $\square$, $A(\cdot),k(\cdot) \in HO^0_\delta(\Omega,\square) \subset L^\infty(\Omega)$. Moreover, $f(\cdot)$ is assumed to be an integrable tolerance periodic function defined in $\Omega \times \Xi$. In every problem under considerations functions $A(\cdot)$, $k(\cdot)$ and $f(\cdot,\xi)$, $\xi \in \Xi$ are assumed to be the known tolerance periodic functions. We recall the heuristic condition that diameter λ of basic cell $\square$ has to be treated as sufficiently small when compared to the smallest characteristic length dimension of Ω.

Using the extended principle of stationary action from (18) and (19) we obtain

$$\nabla \cdot (A \cdot \nabla w) - k \cdot \nabla w = f \tag{20}$$

Equation (20) in a general case has discontinuous highly oscillating functional coefficients. The aim of the modelling procedure is to "replace" this equation by equation with slowly varying smooth coefficients.

The tolerance modelling procedure begins with the micro-macro decomposition for the scalar field w

$$w(x,\xi) = u(x,\xi) + h^A(x) v_A(x,\xi), \quad x \in \Omega, \ \xi \in \Xi \tag{21}$$

where $u(\cdot,\xi)$, $v_A(\cdot,\xi)$ are slowly varying functions,

$$u(\cdot,\xi), v_A(\cdot,\xi) \in SV^1_\delta(\Omega,\square) \subset H^1(\Omega) \tag{22}$$

for every $\xi \in \Xi$. Moreover, $h^A(\cdot)$, $A = 1,...,N$ are the known highly oscillating linear independent continuous functions having the first order piecewise continuous derivatives.

We recall that for every tolerance periodic function $f(\cdot)$ defined in $\square$ the tolerance averaging $\langle f \rangle (\cdot)$ is defined by

$$\langle f \rangle (x) \equiv \frac{1}{|\square|} \int_{\square(x)} \widetilde{f}(x,z)\, dz \tag{23}$$

where $\widetilde{f}(x,\cdot)$ is a certain periodic approximation of $f(\cdot)$ in $\square(x)$. So that $\langle f \rangle (\cdot)$ is a slowly varying function.

The tolerance modelling leads from lagrangian $\mathscr{L}$ which is a quadratic function of ∇w to a new lagrangian which is a certain quadratic function of ∇u, v_A and $\overline{\nabla} v_A$. This quadratic function has to be positive definite and symmetric. It can be shown that the symmetry of the pertinent quadratic matrix implies condition

$$\left\langle \left(h^A \partial h^B - \partial h^A h^B \right) \cdot A \right\rangle = 0 \tag{24}$$

for every $A, B = 1,...,N$. Let us substitute the right hand side of micro-macro decomposition (21) into lagrangian (18) and formula (19). Hence, by using tolerance averaging formula (23), we obtain

$$\langle \mathscr{L}_h \rangle (x) = \frac{1}{2} \begin{bmatrix} \nabla u \\ v_A \\ \overline{\nabla} v_A \end{bmatrix}^T \begin{bmatrix} \langle A \rangle (x) & \langle A \partial h^B \rangle (x) & \langle A h^B \rangle (x) \\ \langle A \partial h^B \rangle (x) & \langle A \partial h^A \partial h^B \rangle (x) & 0 \\ \langle A h^B \rangle (x) & 0 & \langle h^B h^A A \rangle (x) \end{bmatrix} \begin{bmatrix} \nabla u \\ v_B \\ \overline{\nabla} v_B \end{bmatrix} + $$
$$+ \langle f \rangle (x) u + \langle f h^A \rangle (x) v_A \tag{25}$$

and

$$\langle p \rangle (x) = \langle k \rangle (x) \cdot \nabla u + \langle k h^A \rangle (x) \overline{\nabla} v_A + \langle k \partial h^A \rangle (x) v_A$$
$$\langle p h^A \rangle (x) = \langle k h^A \rangle (x) \cdot \nabla u + \langle k h^A h^B \rangle (x) \overline{\nabla} v_B + \langle k h^A \partial h^B \rangle (x) v_B \tag{26}$$

where

$$\langle f h^A \rangle (x) \equiv \frac{1}{|\square|} \int_{\square(x)} \widetilde{f}(x,z)\, \widetilde{h}^A(x,z)\, dz,$$

$$\langle k \rangle (x) \equiv \frac{1}{|\square|} \int_{\square(x)} \widetilde{k}(x,z)\, dz,$$

$$\langle k h^A \rangle (x) \equiv \frac{1}{|\square|} \int_{\square(x)} \widetilde{k}(x,z)\, \widetilde{h}^A(x,z)\, dz,$$

$$\langle k h^A h^B \rangle (x) \equiv \frac{1}{|\square|} \int_{\square(x)} \widetilde{k}(x,z)\, \widetilde{h}^A(x,z)\, \widetilde{h}^B(x,z)\, dz,$$

$$\left\langle k\partial h^A \right\rangle (x) \equiv \frac{1}{|\square|} \int\limits_{\square(x)} \widetilde{k}(x,z)\,\partial \widetilde{h}^A(x,z)\,dz,$$

$$\left\langle kh^A \partial h^B \right\rangle (x) \equiv \frac{1}{|\square|} \int\limits_{\square(x)} \widetilde{k}(x,z)\,\widetilde{h}^A(x,z)\,\partial \widetilde{h}^B(x,z)\,dz.$$

Moreover, we shall assume that the square matrix in formula (25) is positive definite.

Using the extended principle of stationary action we obtain

$$
\begin{aligned}
&\nabla \cdot \left(\langle A \rangle (x)\,\nabla u + \left\langle A \cdot \partial h^B \right\rangle (x)\,v_B + \left\langle A h^B \right\rangle (x)\,\overline{\nabla} v_B \right) - \\
&- \langle k \rangle (x) \cdot \nabla u - \left\langle kh^A \right\rangle (x)\,\overline{\nabla} v_A - \left\langle k \cdot \partial h^A \right\rangle (x)\,v_A = \langle f \rangle (x) \\
&\overline{\nabla} \cdot \left(\left\langle A h^A h^B \right\rangle (x)\,\overline{\nabla} v_B + \left\langle h^A A \right\rangle (x) \cdot \nabla u \right) - \left(\left\langle A \cdot \partial h^A \right\rangle (x) + \left\langle kh^A \right\rangle (x) \right) \cdot \nabla u - \\
&- \left\langle kh^A h^B \right\rangle (x)\,\overline{\nabla} v_B - \left\langle \partial h^A \cdot A \cdot \partial h^B \right\rangle (x)\,v_B - \left\langle h^A k \cdot \partial h^B \right\rangle (x)\,v_B = \left\langle fh^A \right\rangle (x)
\end{aligned}
\tag{27}
$$

The above equations together with micro-macro decomposition (21) represent the tolerance model for the problem under consideration.

References

1. Fichera, G.: Is the Fourier theory of heat propagation paradoxical? Rendiconti del Circolo Matematico di Palermo (1992)
2. Jikov, V., Kozlov, S., Oleinik, O.: Homogenization of differential operators and integral functionals. Springer, Berlin (1994)
3. Rychlewska, J., Woźniak, C.: Boundary layer phenomena in elastodynamics of functionally graded laminates. Arch. Mech. 58, 431–444 (2006)
4. Szymczyk, J., Woźniak, C.: Continuum modelling of laminates with a slowly graded microstructure. Arch. Mech. 58, 445–458 (2006)
5. Woźniak, C., et al. (eds.): Mathematical Modelling and Analysis in Continuum Mechanics of Microstructured Media. Publishing House of Silesian Technical University, Gliwice (in the course of publication)
6. Woźniak, C., Michalak, B., Jędrysiak, J. (eds.): Thermomechanics of microheterogeneous solids and structures. Tolerance averaging approach. Publishing House of Technical University of Lodz, Łódź (2008)
7. Woźniak, C., Wierzbicki, E.: Averaging Techniques in Thermomechanics of Composite Solids. Publishing House of Czestochowa University of Technology, Częstochowa (2000)
8. Zeeman, E.C.: The topology of the brain. In: Biology and Medicine. Medical Research Council (1965)

Part IV
MICRO- AND NANOSCALE MECHANICS

On the Derivation of Biological Tissue Models from Kinetic Models of Multicellular Growing Systems

N. Bellomo, A. Bellouquid, and E. De Angelis

Abstract. This paper deals with the derivation of macroscopic equations for a class of equations modelling complex multicellular systems delivered by the kinetic theory for active particles. The analysis is focused on growing cancer tissues. A critical analysis is proposed to enlighten the technical difficulties generated by dealing with living tissues and to focus the strategy to overcome them by new mathematical approaches.

1 Introduction

The derivation of mathematical models at the macroscopic scale of biological tissues is one of the challenging frontiers of applied mathematics and mechanics. An important application is the modelling of growing cancer tissues. The review paper [14] reports about various approaches, known in the literature, starting from the conceptual difficulties, which are generated by the fact that the system under consideration belongs to the living matter. Indeed, biological tissues are constituted by large systems of cells condensed into almost continuous masses constituted by chemical components and cells that are characterized by the ability of expressing a variety of biological functions.

N. Bellomo · E. De Angelis
Dipartimento di Matematica, Politecnico di Torino, Corso Duca degli Abruzzi 24,
10129 Torino, Italy
e-mail: {nicola.bellomo, elena.deangelis}@polito.it

A. Bellouquid
University Cadi Ayyad, Ecole Nationale des Sciences Appliquées, Safi, Maroc
e-mail: bellouquid@gmail.com

Referring to the above class of biological systems, let us select, among several ones, five key issues:

i) Cells show proliferative and/or destructive ability due to competition for survivance that induce mass variable phenomena;

ii) The material behavior rapidly evolves in time due to a Darwinian type selective adaptation to environmental conditions;

iii) The biological functions expressed by cells are heterogeneously distributed among them;

iv) All phenomena have a multiscale origin, namely the dynamics of cells is ruled by the dynamics at the molecular scale (genes), while the mechanics of tissues is determined by the underlying description at the cellular scale, and hence by the molecular scale.

v) Conservation law often lack in living systems, which have the ability of extracting energy from the outer environment for their own benefit.

Although our analysis is limited to the above five issues, while many others can be stated, it can be shown that the existing approaches fails in dealing with all of them. Nevertheless, it is a stimulating research perspective and this paper will give some hints based on a critical analysis focused on the derivation, by asymptotic methods, of macroscopic models from the underlying description delivered at the cellular scale.

Asymptotic methods applied to classical models of the mathematical kinetic theory lead to the so called parabolic and hyperbolic limits or equivalently low and high field limits. The parabolic (low field) limit of kinetic equations leads to a drift–diffusion type system (or reaction–diffusion system) in which the diffusion processes dominate the behavior of the solutions. The specialized literature offers a number of recent contributions concerning various limits for parabolic diffusive models of the mathematical kinetic theory of classical particles, [20]. On the other hand, in the hyperbolic (high field) limit the influence of the diffusion terms is of lower (or equal) order of magnitude in comparison with other convective or interaction terms and the aim is to derive hyperbolic macroscopic models [19].

The same methodological approach has been developed in the last decade to derive macroscopic equations from the underlying microscopic models for multicellular systems derived by methods of the generalized kinetic theory. Various contributions are available in the literature, among other [4], [6], [7], [15], [16], [18], [22], [24], while the book [12] and the survey [8] report about applications of the kinetic theory to model complex systems in biology.

Technically, asymptotic methods amount to expanding the distribution function in terms of a small dimensionless parameter related to the intermolecular distances (the space-scale dimensionless parameter) that is equivalent to the connections between the biological constants. The limit that we obtain is singular and the convergence properties can be proved under suitable technical assumptions. This paper specifically deals with the hyperbolic scaling, while bibliographical indications are given for the diffusion approximation.

The contents of this paper are as follows: Section 2 provides a description of the class of equations of the kinetic theory for active particles that describe multicellular systems, where cellular interactions modify the biological functions expressed by cells, and generate proliferative or destructive events. Moreover, the scaling problem is treated by writing the evolution equation in a suitable dimensionless form and extracting the parameters that characterize the model. Section 3 provides a review on the known results concerning the derivation of hyperbolic macroscopic models by methods of asymptotic analysis. Finally, Section 4 critically analyzes the approach presented in this present paper focusing on two specific stages of tumor growth: the early stage when mutations are predominant, while proliferation has not yet taken place and the late stage, when cells have reached their late stage and proliferative events characterize the dynamics. A critical analysis involves also the traditional approaches of continuum mechanics and lead to some research perspectives suitable to tackle the conceptual difficulties that have been outlined above.

2 The Mathematical Model and Scaling

Let us consider a physical system constituted by a large number of cells interacting in the environment of a vertebrate. The physical variable used to describe the state of each cell, called *microscopic state*, is denoted by the variable $\mathbf{w} = \{\mathbf{x}, \mathbf{v}, u\}$, where $\{\mathbf{x}, \mathbf{v}\} \in \Omega \times V \subset \mathbb{R}^n \times \mathbb{R}^n$ is the *mechanical microscopic state* and $u \in D_u \subseteq \mathbb{R}$ is the *biological microscopic state*. The statistical collective description of the system is encoded in the statistical distribution $f = f(t, \mathbf{x}, \mathbf{v}, u)$, which is called a *generalized distribution function*. Weighted moments provide, under suitable integrability properties, the calculation of macroscopic variables [12].

In detail, let us consider the following class of equations that model the evolution of $f = f(t, \mathbf{x}, \mathbf{v}, u)$:

$$\left(\partial_t + \mathbf{v} \cdot \nabla_{\mathbf{x}}\right) f = \nu L(f) + \eta\, G(f,f) + \mu\, I(f,f), \tag{1}$$

where:

- ν is the turning rate or turning frequency and hence $\tau = \frac{1}{\nu}$ is the mean run time.
- The linear transport term describes the dynamics of biological organisms modelled by a velocity-jump process,

$$L(f) = \int_V \left[T(\mathbf{v}^*, \mathbf{v})\, f(t, \mathbf{x}, \mathbf{v}^*, u) - T(\mathbf{v}, \mathbf{v}^*)\, f(t, \mathbf{x}, \mathbf{v}, u) \right] d\mathbf{v}^*, \tag{2}$$

where $T(\mathbf{v}, \mathbf{v}^*)$ is the probability kernel for the new velocity $\mathbf{v} \in V$ assuming that the previous velocity was $\mathbf{v}^*$. This corresponds to the assumption that cells choose any direction with bounded velocity. Specifically, the set of possible velocities is denoted by V, where $V \subset R^3$.

- η models the biological interaction rate, which is here assumed (for simplicity) constant.

• The operator G defined as

$$G(f,f) = \int_{\Lambda} w(\mathbf{x}, \mathbf{x}^*) B(u_* \to u | u_*, u^*) f(t, \mathbf{x}, \mathbf{v}, u_*) f(t, \mathbf{x}^*, \mathbf{v}, u^*) d\mathbf{x}^* du_* du^*$$
$$- f(t, \mathbf{x}, \mathbf{v}, u) \int_{\Gamma} w(\mathbf{x}, \mathbf{x}^*) f(t, \mathbf{x}^*, \mathbf{v}, u^*) d\mathbf{x}^* du^*,$$

where $\Lambda = D_u \times D_u \times \Omega$ and $\Gamma = D_u \times \Omega$, describes the gain minus the loss of cells in state u due to conservative encounters, namely those which modify the biological state without generating proliferation or destruction phenomena. The kernel B models the transition probability density of the candidate cell with state u_* into the state u of the test cell, after interaction with the field cell with state u^*, Ω is the interaction domain of the candidate and test cells, $w(\mathbf{x}, \mathbf{x}^*)$ is a normalized (with respect to space integration over Ω) weight function that accounts for the influence of the distance on the intensity of the interactions. The kernel B is a probability density not symmetrical with respect to u:

$$\int_{D_u} B(u_* \to u | u_*, u^*) du^* = 1, \forall u_*, u^*, \quad B(u_* \to u | u_*, u^*) \neq B(u \to u_* | u_*, u^*).$$

• The operator I models the proliferative and/or destructive actions and is defined as follows,

$$I(f,f) = f(t, \mathbf{x}, \mathbf{v}, u) \int_{\Gamma} w(\mathbf{x}, \mathbf{x}^*) p(u, u^*) f(t, \mathbf{x}^*, \mathbf{v}, u^*) d\mathbf{x}^* du^*.$$

where $w(\mathbf{x}, \mathbf{x}^*)$ has been defined above and $\mu\, p(u, u^*)$ is the net proliferative and/or destructive rate which may depend on the biological state of the interacting pairs.

Remark 1. The above equation describes the evolution in the space $\mathbf{x} \in \mathbb{R}^n$ and in the biological state $u \in D_u \subseteq \mathbb{R}$ of a large system of interacting cells. The distribution function $f(t, \mathbf{x}, \mathbf{v}, u)$ refers to the *test* cell, while interactions occur between pairs of a *test* and a *field* cell that generate proliferative or destructive outputs, and between a *candidate* and *field* cell with mutation of the state of the candidate cell into the state of the test cell, [13].

Remark 2. The assumption that the microscopic state u is a scalar variable can be technically related to the theory of modules by Hartwell [21], that corresponds to refer the collective behavior of the population to one biological function only. This interpretation is documented in paper [13].

Remark 3. The model includes nonlocal interactions that is important when dealing with biological systems: the dialogue between cells suggests to consider long range interactions as already documented in the paper [11]. The same issues are dealt with in the paper [25], where it is shown that modelling nonlocal interactions has a remarkable influence on the description of macroscopic biological phenomena.

Remark 4. The above modelling approach is based on the assumption that interactions occur, and are weighted, within the action domain Ω of the test cell. If Ω is small enough so that only binary localized encounters are relevant, the mathematical structures correspond to assuming that $w(\mathbf{x}, \mathbf{x}^*)$ is a delta function over $\mathbf{x}$. In this case the structure of the operators G and I is as follows:

$$G(f,f) = \int_{D_u^2} B(u_* \to u | u_*, u^*) f(t, \mathbf{x}, \mathbf{v}, u_*) f(t, \mathbf{x}, \mathbf{v}, u^*) du_* du^*$$

$$-f(t, \mathbf{x}, \mathbf{v}, u) \int_{D_u} f(t, \mathbf{x}, \mathbf{v}, u^*) du^*,$$

and

$$I(f,f) = f(t, \mathbf{x}, \mathbf{v}, u) \int_{D_u} p(u, u^*) f(t, \mathbf{x}, \mathbf{v}, u^*) du^*.$$

Remark 5. Simplified cases of Equation (1) can be obtained suppressing one or more of the three terms that characterize the model. For instance, the model in absence of proliferative and/or destructive events writes:

$$\left(\partial_t + \mathbf{v} \cdot \nabla_\mathbf{x}\right) f(t, \mathbf{x}, \mathbf{v}) = \nu L(f)(t, \mathbf{x}, \mathbf{v}) + \eta\, G(f, f)(t, \mathbf{x}, \mathbf{v}), \qquad (3)$$

this situation occurs in the early stage of tumor onset when cells have initiated to mutate, but have not yet in the stage of proliferation and competition with the immune system. Similarly, the following model:

$$\left(\partial_t + \mathbf{v} \cdot \nabla_\mathbf{x}\right) f = \nu L(f) + \mu I(f, f), \qquad (4)$$

correspond to the late stage, when cells have reached a steady distribution of their progression and only proliferative/destructive events take place.

Before developing the asymptotic analysis, we need to give some definitions, which will be used later.

Remark 6. The operator $G(f, g)$ can be written as follows:

$$G(f,g) = \frac{1}{2}\left[\int_{D_u}\int_{D_u} \varphi(u_*, u_{**}, u)(f(t, \mathbf{x}, \mathbf{v}, u_*) g(t, \mathbf{x}, \mathbf{v}, u_{**})\right.$$

$$+ g(t, \mathbf{x}, \mathbf{v}, u_*) f(t, \mathbf{x}, \mathbf{v}, u_{**})) du_* du_{**} - f(t, \mathbf{x}, \mathbf{v}, u) \int_{D_u} g(t, \mathbf{x}, \mathbf{v}, u_{**}) du_{**}$$

$$\left. - g(t, \mathbf{x}, \mathbf{v}, u) \int_{D_u} f(t, \mathbf{x}, \mathbf{v}, u_{**}) du_{**}\right]. \qquad (5)$$

Definition 1. *Let us consider the operator $H = L + G$. Local and global equilibrium solutions are defined as follows:*

1. A function $h(t, \mathbf{x}, \mathbf{v}, u)$ is said to be a local equilibrium *for the operator H if h is an equilibrium for L and for G: $L(h) = G(h, h) = 0$.*
2. A function h is said to be a global equilibrium *for the operator H if h is a local equilibrium independent of $t, \mathbf{x}$.*

Remark 7. The existence of at least one solution $k(u)$ for the operator G, such that $G(k,k) = 0$, can be proved, under suitable assumptions on the probability transition $B(u_* \to u|u_*, u^*)$, see paper [3]. Moreover, if $k(u)$ is an equilibrium for G and $M(\mathbf{v})$ is an equilibrium for L, then $h(\mathbf{v}, u) = M(\mathbf{v})k(u)$ is a global equilibrium for the operator H.

Bearing all above in mind, let us show how the class of equations (1), which acts as a fundamental paradigm for the derivation of various models of interest in biology, can be put in a dimensionless form to extract the key parameters to be used in the asymptotic analysis. Specifically, let us consider the *hyperbolic scaling* corresponding to:

$$t \to \varepsilon t, \quad x \to \varepsilon x \quad \Rightarrow \quad tv = \frac{1}{\varepsilon}, \tag{6}$$

and introduce the parameters

$$\eta = \varepsilon^{q-1}, \qquad \mu = \varepsilon^{\delta}, \qquad q \geq 1, \qquad \delta \geq 0,$$

where ε is a small parameter which will be allowed to tend to zero.

Therefore, the scaled non-dimensional model is as follows:

$$\left(\partial_t + \mathbf{v} \cdot \nabla_{\mathbf{x}}\right) f_{\varepsilon} = \frac{1}{\varepsilon}\left(L(f_{\varepsilon}) + \varepsilon^q G(f_{\varepsilon}, f_{\varepsilon}) + \varepsilon^{q+\delta} I(f_{\varepsilon}, f_{\varepsilon})\right). \tag{7}$$

3 Asymptotic Analysis

This section reports about the main results obtained in [7] and [10] concerning the derivation, in the hyperbolic case, of macroscopic equations delivered by the class of models reported in the preceding section. These results will be critically analyzed in the next section focussing on cancer tissues with reference to the stages of tumor growth corresponding to Remark 5.

Let us consider the asymptotic analysis for the class of models (7) for various different orders of magnitude of the parameters q and δ. Specifically we refer to [10], where the derivation needs the following assumptions:

Assumption 1. *There exists a global equilibrium $h(\mathbf{v}, u)$ for the operator $H = L + G$. stated in Definition 1.*

Assumption 2. *(Solvability conditions) The turning operator L satisfies the following solvability conditions:*

$$\int_V L(f)\,d\mathbf{v} = \int_V \mathbf{v}L(f)\,d\mathbf{v} = 0.$$

Assumption 3. *(Kernel of L) There exists a unique function $M_{\rho,U} \in L^1(V, (1 + |\mathbf{v}|)\,d\mathbf{v})$, for all $\rho \in [0, +\infty)$ and $U \in \mathbb{R}^n$, such that*

$$L(M_{\rho,U}) = 0, \quad \int_V M_{\rho,U}(\mathbf{v}) \, d\mathbf{v} = \rho, \quad \int_V \mathbf{v} M_{\rho,U}(\mathbf{v}) \, d\mathbf{v} = \rho U.$$

The hyperbolic approximation asymptotic limit can be obtained by developing the solution near a global equilibrium h by means of appropriate moments of f. Bearing all above in mind, let

$$f_\varepsilon = h(\mathbf{v}, u) + g_\varepsilon, \qquad \text{as} \qquad L(h) = G(h,h) = 0.$$

Therefore, the equilibrium equation, satisfied by g_ε, can be written as follows:

$$\partial_t g_\varepsilon + \mathbf{v} \cdot \nabla_{\mathbf{x}} g_\varepsilon = \frac{1}{\varepsilon} \Big(L g_\varepsilon + \varepsilon^q \left(2G(h, g_\varepsilon) + G(g_\varepsilon, g_\varepsilon) \right)$$
$$+ \varepsilon^{q+\delta} (I(h,h) + 2I(h, g_\varepsilon) + I(g_\varepsilon, g_\varepsilon)) \Big). \tag{8}$$

Suppose that as ε goes to zero, the solution g_ε converges at least in the distributional sense to a function g. Multiplying (8) by ε and letting ε go to zero yields $Lg = 0$.

Let us define the macroscopic variables ρ and U as follows

$$\rho = \int_V g(\mathbf{v}) \, d\mathbf{v}, \qquad \rho U = \int_V \mathbf{v} g(\mathbf{v}) \, d\mathbf{v}.$$

Therefore, by Assumption 3, the equilibrium distribution g is obtained as follows

$$g = M_{\rho,U}. \tag{9}$$

Integrating (8) over $\mathbf{v}$ and using Assumption (2), yields

$$\partial_t \langle g_\varepsilon \rangle + \langle \mathbf{v} \cdot \nabla_{\mathbf{x}} g_\varepsilon \rangle = \varepsilon^{q-1} \langle G(g_\varepsilon, g_\varepsilon) \rangle + 2 \langle G(h, g_\varepsilon) \rangle$$
$$+ \varepsilon^{\delta+q-1} \langle I(h,h) + 2I(h, g_\varepsilon) + I(g_\varepsilon, g_\varepsilon) \rangle. \tag{10}$$

Analogously, multiplying (8) by $\mathbf{v}$ and integrating with respect to $\mathbf{v}$, and using Assumption (2), yields

$$\partial_t \langle \mathbf{v} g_\varepsilon \rangle + \nabla_{\mathbf{x}} \cdot \langle \mathbf{v} \otimes \mathbf{v} g_\varepsilon \rangle = \varepsilon^{q-1} \langle \mathbf{v} G(g_\varepsilon, g_\varepsilon) + 2 \mathbf{v} G(h, g_\varepsilon) \rangle$$
$$+ \varepsilon^{\delta+q-1} \langle \mathbf{v} I(h,h) + \mathbf{v} I(h, g_\varepsilon) + \mathbf{v} I(g_\varepsilon, g_\varepsilon) \rangle. \tag{11}$$

Under a suitable hypothesis on the convergence of the moments, namely that for a certain (integral) operator Λ acting on g_ε, the sequence $\Lambda(g_\varepsilon)$ converges to $\Lambda(g)$ in the distributional sense, we get for $\varepsilon \to 0$

$$\partial_t \langle g_\varepsilon \rangle + \nabla_{\mathbf{x}} \cdot \langle \mathbf{v} g_\varepsilon \rangle \longrightarrow \partial_t \rho + \nabla_{\mathbf{x}} \cdot (\rho U),$$

$$\partial_t \langle \mathbf{v} g_\varepsilon \rangle + \nabla_{\mathbf{x}} \cdot \langle \mathbf{v} \otimes \mathbf{v} g_\varepsilon \rangle \longrightarrow \partial_t (\rho U) + \nabla_{\mathbf{x}} \cdot \left(\int_V \mathbf{v} \otimes \mathbf{v} M_{\rho,U} \, d\mathbf{v} \right).$$

Considering that the field U denotes the expected mean velocity of the particles, we can measure the statistical variation in velocity by the pressure tensor given by

$$P(t,\mathbf{x},u) = \int_V (\mathbf{v} - U) \otimes (\mathbf{v} - U) M_{\rho,U} \, d\mathbf{v}, \tag{12}$$

which is related to the second order moment:

$$\int_V \mathbf{v} \otimes \mathbf{v} M_{\rho,U} \, d\mathbf{v} = P + \rho U \otimes U.$$

The asymptotic limit of the right hand side of (10) and (11) depends on parameters q and δ. Therefore, the main result, under suitable assumptions on convergence of moments, can be finally stated as follows:

Theorem 1. *Let $g_\varepsilon(t,\mathbf{x},\mathbf{v},u)$ be a sequence of solutions to the scaled kinetic equation (8) such that g_ε converges, in the distributional sense, to a function g as ε goes to zero. Furthermore, assume that the moments*

$$\langle g_\varepsilon \rangle, \quad \langle \mathbf{v} \otimes \mathbf{v} g_\varepsilon \rangle, \quad \langle G(h,g_\varepsilon) \rangle, \quad \langle G(g_\varepsilon,g_\varepsilon) \rangle, \quad \langle I(h,g_\varepsilon) \rangle, \quad \langle I(g_\varepsilon,g_\varepsilon) \rangle,$$

$$\langle \mathbf{v} G(h,g_\varepsilon) \rangle, \quad \langle \mathbf{v} G(g_\varepsilon,g_\varepsilon) \rangle, \quad \langle \mathbf{v} I(h,g_\varepsilon) \rangle, \quad \langle \mathbf{v} I(g_\varepsilon,g_\varepsilon) \rangle,$$

converge in $D'(t,\mathbf{x},u)$ to the corresponding moments

$$\langle g \rangle, \quad \langle \mathbf{v} \otimes \mathbf{v} g \rangle, \quad \langle G(h,g) \rangle, \quad \langle G(g,g) \rangle, \quad \langle I(h,g) \rangle, \quad \langle I(g,g) \rangle,$$

$$\langle \mathbf{v} G(h,g) \rangle, \quad \langle \mathbf{v} G(g,g) \rangle, \quad \langle \mathbf{v} I(h,g) \rangle, \quad \langle \mathbf{v} I(g,g) \rangle,$$

and that all small terms vanish. Then, the asymptotic limit has the form (9) where $\rho(t,\mathbf{x},u)$ and ρU are weak solutions of the following equations:

Case 1. *$\delta \geq 0$, and $q > 1$: $(\rho,\rho U)$ is a weak solution of the following hyperbolic system without source term:*

$$\partial_t \rho + \nabla_\mathbf{x} \cdot (\rho U) = 0,$$
$$\tag{13}$$
$$\partial_t (\rho U) + \nabla_\mathbf{x} \cdot (\rho U \otimes U + P) = 0.$$

Case 2. *$\delta > 0$, and $q = 1$: $(\rho,\rho U)$ is a weak solution of the following hyperbolic system with source terms corresponding to conservative interactions:*

$$\partial_t \rho + \nabla_\mathbf{x} \cdot (\rho U) = \int_V G(M_{\rho,U}, M_{\rho,U}) \, d\mathbf{v} + 2 \int_V G(M_{\rho,U}, h) \, d\mathbf{v},$$

$$\partial_t (\rho U) + \nabla_\mathbf{x} \cdot (\rho U \otimes U + P) = \int_V \mathbf{v} G(M_{\rho,U}, M_{\rho,U}) \, d\mathbf{v} + 2 \int_V \mathbf{v} G(M_{\rho,U}, h) \, d\mathbf{v}.$$
$$\tag{14}$$

Case 3. $\delta = 0$, and $q = 1$: $(\rho, \rho U)$ is a weak solution of the following hyperbolic system with source terms corresponding to both conservative and proliferative interactions:

$$\partial_t \rho + \nabla_{\mathbf{x}} \cdot (\rho U) = \int_V G(M_{\rho,U}, M_{\rho,U}) \, d\mathbf{v} + 2 \int_V G(M_{\rho,U}, h) \, d\mathbf{v}$$

$$+ \int_V I(M_{\rho,U}, M_{\rho,U}) \, d\mathbf{v} + \int_V I(h, h) \, d\mathbf{v} + 2 \int_V I(M_{\rho,U}, h) \, d\mathbf{v},$$

$$\partial_t (\rho U) + \nabla_{\mathbf{x}} \cdot (\rho U \otimes U + P) = \int_V \mathbf{v} G(M_{\rho,U}, M_{\rho,U}) \, d\mathbf{v} + 2 \int_V \mathbf{v} G(M_{\rho,U}, h) \, d\mathbf{v}$$

$$+ \int_V \mathbf{v} I(M_{\rho,U}, M_{\rho,U}) \, d\mathbf{v} + \int_V \mathbf{v} I(h, h) \, d\mathbf{v} + 2 \int_V \mathbf{v} I(M_{\rho,U}, h) \, d\mathbf{v}.$$

$$(15)$$

Equations (14)-(15) are not closed. Detailed expressions of the source terms are obtained, as documented in [10] for relaxation models. Specifically, consider the case where the set for velocity is the sphere of radius $r > 0$, $V = r\mathbb{S}^{n-1}$. Moreover, let us take a kernel $T(\mathbf{v}, \mathbf{v}^*)$ in (2) of the form $T(\mathbf{v}, \mathbf{v}^*) = \lambda + \beta \mathbf{v} \cdot \mathbf{v}^*$, so that the operator $L(f)$ can be computed as follows:

$$L(f) = \int_V \left((\lambda + \beta \mathbf{v} \cdot \mathbf{v}^*) f(\mathbf{v}^*) - (\lambda + \beta \mathbf{v} \cdot \mathbf{v}^*) f(\mathbf{v}) \right) d\mathbf{v}^*$$

$$= \lambda \rho + \beta \rho \mathbf{v} \cdot U - \lambda |V| f(\mathbf{v}) = \lambda |V| \left(\frac{\rho}{|V|} \left(1 + \frac{\beta}{\lambda} \mathbf{v} \cdot U \right) - f(\mathbf{v}) \right). \quad (16)$$

Moreover, let $\beta r^2 = \lambda n$. Then $L(f)$ verifies Assumptions 2 and 3 for a function $M_{\rho,U}(\mathbf{v})$ given by

$$M_{\rho,U}(\mathbf{v}) = \frac{\rho}{|V|} \left(1 + \frac{\beta}{\lambda} \mathbf{v} \cdot U \right) = \frac{\rho}{|V|} \left(1 + \frac{n}{r^2} \mathbf{v} \cdot U \right), \quad (17)$$

where $L(f)$ is the relaxation operator

$$L(f) = \lambda |V| \left(M_{\rho,U}(\mathbf{v}) - f(\mathbf{v}) \right).$$

Moreover, the pressure tensor P associated with $M_{\rho,U}(\mathbf{v})$ is given by

$$P = \frac{r^2}{n} \rho \mathbb{I} - \rho U \otimes U.$$

Further technical calculations provide the explicit calculations of the source terms that appear in Eqs. (14) and (15). Analogous calculations have been developed in [7] without the assumption of existence of an equilibrium configuration. By these calculations, Eqs. (14) and (15) become:

$$\partial_t\rho + \nabla_x \cdot (\rho U) = \frac{1}{|V|}\left(G(\rho,\rho) + 2G(\rho,\langle h\rangle_v)\right.$$

$$\left. + \frac{n}{r^2}\left(G(\rho U,\rho U) + 2G(\rho U,\langle vh\rangle_v)\right)\right),$$

$$\partial_t(\rho U) + \frac{r^2}{n}\nabla_x\rho = \frac{2}{|V|}\left(G(\rho,\rho U) + G(\rho,\langle vh\rangle_v) + \frac{n}{r^2}G(\rho U,\langle v\otimes vh\rangle)\right),$$

(18)

if $\delta > 0$, and $q = 1$, and

$$\partial_t\rho + \nabla_x \cdot (\rho U) = \frac{1}{|V|}\left(G(\rho,\rho) + 2G(\rho,\langle h\rangle_v) + \frac{n}{r^2}\left(G(\rho U,\rho U)\right.\right.$$

$$\left.\left. + 2G(\rho U,\langle vh\rangle_v)\right)\right) + \frac{1}{|V|}\left(I(\rho,\rho) + 2I(\rho,\langle h\rangle_v) + \frac{n}{r^2}\left(I(\rho U,\rho U)\right.\right.$$

$$\left.\left. + 2I(\rho U,\langle vh\rangle_v)\right)\right) + \int_V I(h,h)\,dv,$$

$$\partial_t(\rho U) + \frac{r^2}{n}\nabla_x\rho = \frac{2}{|V|}\left[G(\rho,\rho U) + G(\rho,\langle vh\rangle_v) + I(\rho U,\rho U) + I(\rho,\langle vh\rangle_v)\right.$$

$$\left. + \frac{n}{r^2}\left(G(\rho U,\langle v\otimes vh\rangle_v) + I(\rho U,\langle v\otimes vh\rangle_v)\right)\right] + \int_V vI(h,h)\,dv,$$

(19)

if $\delta = 0$, and $q = 1$, where, for any vector function $F = F(t,x,v,u)$, the operators $I(F,F)$ and $G(F,F)$ are given by the following scalar quantities:

$$I(F,F) = F(t,x,v,u) \cdot \int_\Gamma w(x,x^*)\,p(u,u^*)\,F(t,x,v,u^*)\,dx^*\,du^*,$$

$$G(F,F) = \int_\Lambda w(x,x^*)\,B(u_* \to u|u_*,u^*)\,F(t,x,v,u_*) \cdot F(t,x^*,v,u^*)\,dx^*\,du_*\,du^*$$

$$ - F(t,x,v,u) \cdot \int_\Gamma w(x,x^*)\,F(t,x^*,v,u^*)\,dx^*\,du^*,$$

and for any scalar function $f = f(t,x,v,u)$ and any vector function $F = F(t,x,v,u)$, $I(f,F)$ and $G(f,F)$ are given by the following vector quantities:

$$I(f,F) = \frac{1}{2}\left(f(t,\mathbf{x},\mathbf{v},u) \int_\Gamma w(\mathbf{x},\mathbf{x}^*) \, p(u,u^*) F(t,\mathbf{x}^*,\mathbf{v},u^*) \, d\mathbf{x}^* du^* \right.$$

$$\left. + F(t,\mathbf{x},\mathbf{v},u) \int_\Gamma w(\mathbf{x},\mathbf{x}^*) \, p(u,u^*) f(t,\mathbf{x}^*,\mathbf{v},u^*) \, d\mathbf{x}^* \, du^* \right),$$

$$G(f,F) = \frac{1}{2} \int_\Lambda w(\mathbf{x},\mathbf{x}^*) \, B(u_* \to u|u_*,u^*) \left(f(t,\mathbf{x}^*,\mathbf{v},u^*) F(t,\mathbf{x},\mathbf{v},u_*) \right.$$

$$\left. + f(t,\mathbf{x},\mathbf{v},u_*) F(t,\mathbf{x}^*,\mathbf{v},u^*) \right) du_* du^* d\mathbf{x}^* - I(f,F),$$

which generalize in a natural way the definition of operators G and I given in section 2, so that we preserve the same name for these operators.

Remark 8. Equations (18) and (19) are closed nonlinear hyperbolic equations for the macroscopic variables ρ and ρU.

Although the contents of this paper essentially refers to the hyperbolic case, it is worth mentioning that the same analysis can be developed corresponding to the diffusion approximation. In this case the scaling uses the following parameters $\eta = \varepsilon^r$, $\mu = \varepsilon^q$, $r \geq 1$, and $\nu = \dfrac{1}{\varepsilon^p}$, with $p \geq 1$, where ε is dealt with as a small parameter to be let tend to zero. In addition, the diffusion scale time $\tau = \varepsilon t$ is used so that the following scaled equation is obtained:

$$\varepsilon \partial_t f_\varepsilon + \mathbf{v} \cdot \nabla_\mathbf{x} f_\varepsilon = \frac{1}{\varepsilon^p} L f_\varepsilon + \varepsilon^r G(f_\varepsilon, f_\varepsilon) + \varepsilon^q I(f_\varepsilon, f_\varepsilon). \qquad (20)$$

The literature in the field is documented in the papers [4], [5] and [6] , and therein cited bibliography. It is shown, also in the case of binary mixtures, that different regimes, e.g. linear and nonlinear diffusion and the onset of source terms, correspond to different values of the parameters that characterize the afore stated scaling.

4   Critical Analysis and Perspectives

The mathematical approach reviewed in Section 3 has shown how the continuum approximation delivered by hyperbolic equations can be obtained by asymptotic methods applied to the underlying description at the cellular level delivered by the kinetic theory for active particles. The need to derive macroscopic models, rather than using directly kinetic equations is motivated by the need of reducing complexity. On the other hand, if we look at the five key issues proposed in the first section, it can be verified that the traditional approach of continuum mechanics cannot take into account the complexity of living matter.

Actually, it has to be recognized that the seminal paper [23] has introduced a substantial improvement to the traditional approach of continuum mechanics by a deep analysis of the implications on the actual structure of conservation equations induced by mass variability due to growth phenomena. The research line opened in [23] has motivated a subsequent research activity of several authors who have applied this theory also to the case of cancer modelling as documented, among others, in [1], [2], [17]. On the other hand, continuum mechanics approach is based on conservation equations coupled with reaction diffusion models, which describe the evolution in time and space of locally averaged quantities related to the behaviour of cell populations. These equations need closure approximations derived on heuristic arguments and generally valid only in the proximity of equilibrium.

This issue can be made precise recalling, see [14], that a large variety of models at the macroscopic scale are derived by mass balance equations for the cellular components, extracellular matrix, and the extracellular fluid, coupled to a system of reaction-diffusion equations for the concentration of extracellular chemicals:

$$
\begin{cases}
\rho_j \left[\partial_t \phi_j + \nabla_{\mathbf{x}} \cdot (\phi_j \mathbf{v}_j) \right] = \Gamma_j(\rho, \phi, c), & j = 1, \ldots, L, \\
\partial_t c_i + \nabla_{\mathbf{x}} \cdot (c_i \mathbf{v}_\ell) = \nabla_{\mathbf{x}} \cdot (Q_i(\rho, \phi, c) \nabla c_i) + \Lambda_i(\rho, \phi, c), & i = 1, \ldots, M,
\end{cases}
\tag{21}
$$

where $\phi_j = \phi_j(t, \mathbf{x})$ denotes the concentration of each component, e.g. cells, matrix, or fluid, and $c_i = c_i(t, \mathbf{x})$ denotes the concentrations of the chemicals and nutrients. In Eq. $(21)_1$, ρ_j are the mass densities of cellular components and $\mathbf{v}_j$ is the mass velocity of the jth component, and $\mathbf{v}_\ell$ is the velocity of the liquid. Moreover, $\Gamma_j(\rho, \phi, c)(t, \mathbf{x})$ is a source term for the particular component, which might include, for example, production and death terms. In Eq. $(21)_2$, $Q_i(\rho, \phi, c)(t, \mathbf{x})$ is the diffusion coefficient of the ith chemical factor, and $\mathbf{v}_i$ is the velocity of the chemical component. Finally, $\Lambda_i(\rho, \phi, c)(t, \mathbf{x})$ is the source term for the particular nutrient or chemical, which similarly might include, for example, production and uptake by the cells.

Therefore, the component equations are coupled to the chemical equations via the source terms, $\Lambda_i(\rho, \phi, c)$ and $\Gamma_j(\rho, \phi, c)$. For example, blood vessels might be the source of oxygen, which is consumed by the tumour cells and in turn alters the proliferation or death rate of the tumour cells. A vast literature is reviewed in [14] to report about different ways of closing the above system.

It can be remarked that heterogeneity is lost in the averaging process, while models do not take into account the dynamics that occurs at the lower scale. Possibly, this aspect can be taken into account by coupling models at the lower scale to those at the higher scale. However, this idea still appears to be implemented at an effective level. Finally reaction diffusion equations offer a mathematical structure postulated a priori rather than generated by cellular dynamics.

Although the approach reviewed in the preceding section does respond to some of the five conceptual issues posed in Section 1, further developments are necessary to deal effectively with all of them. Specifically, the dynamics of cells should be linked to the dynamics at the molecular scale, and consequently the mechanics of tissues is effectively linked to the underlying description at the cellular scale and takes into account their Darwinian evolution. Bearing all in mind some perspectives are indicated in conclusion of this paper.

Some hints are already given in [9], where it is suggested to consider a system constituted by two interacting populations of active particles, genes and cells respectively, corresponding to two different scales, namely the molecular (genetic) scale and the cellular scale. The overall state of the system, at the higher and lower scales, respectively, is defined by the probability density distributions:

$$f = f(t,u) : \quad [0,T] \times D_u \to \mathbb{R}_+, \qquad \varphi = \varphi(t,v) : \quad [0,T] \times D_v \to \mathbb{R}_+, \quad (22)$$

over the microscopic states $u \in D_u$ and $v \in D_v$.

The interaction scheme from the lower to the higher scale can be represented as follows:

$$\left\{ \partial_t \varphi = N_1[\varphi, \varphi] + N_2[\varphi, \psi] \right\} \quad \to \quad \left\{ \partial_t f = M_g[f, \varphi] \right\}, \qquad (23)$$

where ψ represents the state of the outer environment.

The above formal scheme, that can be seen in a vector form, corresponds to the following dynamics:

• The evolution of the system at the lower scale is determined by the interaction between genes among themselves and with the outer environment that is supposed known and acting at the same scale.

• The evolution of the system at the higher scale is determined by the interaction between active particles, of the population, among themselves and with particles of the lower system that is obtained by solution of the evolution equation for such a system.

The over or lower-expression of genes can be measured, by an appropriate weighted distance between φ and the natural expression, say φ_0. The key aspect of the modelling, which is at present an open problem, consists linking the afore mentioned distance to the parameters of the model at the cellular scale δ and q. Consequently, the equations at the macroscopic scale change of type with rules modelled at the lower scale.

Acknowledgements. Partially supported by the European Union FP7 Health Research Grant number FP7-HEALTH-F4-2008-202047.

References

1. Ambrosi, D., Mollica, F.: On the mechanics of a growing tumour. Internat. J. Engrg. Sci. 40, 1297–1316 (2002)
2. Ambrosi, D., Preziosi, L.: On the closure of mass balance models for tumour growth. Math. Models Methods Appl. Sci. 12, 737–754 (2002)
3. Arlotti, L., Bellomo, N.: Solution of a new class of nonlinear kinetic models of population dynamics. Appl. Math. Lett. 9, 65–70 (1996)

4. Bellomo, N., Bellouquid, A.: On the onset of nonlinearity for diffusion models of binary mixtures of biological materials by asymptotic analysis. Internat. J. Non-Linear Mech. 41, 281–293 (2006)
5. Bellomo, N., Bellouquid, A.: On the derivation of macroscopic tissue equations from hybrid models of the kinetic theory of multicellular growing systems—The effect of global equilibrium. Nonlinear Anal. Hybrid Syst. 3, 215–224 (2009)
6. Bellomo, N., Bellouquid, A., Herrero, M.A.: From microscopic to macroscopic description of multicellular systems and biological growing tissues. Comput. Math. Appl. 53, 647–663 (2007)
7. Bellomo, N., Bellouquid, A., Nieto, J., Soler, J.: Multicellular growing systems: Hyperbolic limits towards macroscopic description. Math. Models Methods Appl. Sci. 17, 1675–1692 (2007)
8. Bellomo, N., Delitala, M.: From the mathematical kinetic, and stochastic game theory to modelling mutations, onset, progression and immune competition of cancer cells. Physics of Life Reviews 5, 183–206 (2008)
9. Bellomo, N., Delitala, M.: On the coupling of higher and lower scales using the mathematical kinetic theory of active particles. Appl. Math. Lett. 22, 646–650 (2009)
10. Bellouquid, A., De Angelis, E.: From Kinetic Models of Multicellular Growing Systems to Macroscopic Biological Tissue Models (submitted)
11. Bellouquid, A., Delitala, M.: Mathematical models and tools of kinetic theory towards modelling complex biological systems. Math. Models Methods Appl. Sci. 15, 1639–1666 (2005)
12. Bellouquid, A., Delitala, M.: Mathematical Modeling of Complex Biological Systems. In: A Kinetic Theory Approach. Birkäuser, Boston (2006)
13. Bellomo, N., Forni, G.: Complex multicellular systems and immune competition: New paradigms looking for a mathematical theory. Current Topics in Developmental Biology 81, 485–502 (2008)
14. Bellomo, N., Li, N.K., Maini, P.K.: On the foundations of cancer modelling: selected topics, speculations, and perspectives. Math. Models Methods Appl. Sci. 18, 593–646 (2008)
15. Chalub, F.A., Markovich, P., Perthame, B., Schmeiser, C.: Kinetic models for chemotaxis and their drift–diffusion limits. Monatsh. Math. 142, 123–141 (2004)
16. Chalub, F.A., Dolak-Struss, Y., Markowich, P., Oeltz, D., Schmeiser, C., Sorefs, A.: Model hierarchies for cell aggregation by chemotaxis. Math. Models Methods Appl. Sci. 16, 1173–1198 (2006)
17. Chaplain, M.A.J., Graziano, L., Preziosi, L.: Mathematical modelling of the loss of tissue compression responsiveness and its role in solid tumour development. IMA J. Mathematics Applied to Medicine and Biology 23, 197–229 (2003)
18. Filbet, F., Laurençot, P., Perthame, B.: Derivation of hyperbolic models for chemosensitive movement. J. Math. Biol. 50, 189–207 (2005)
19. Goudon, T., Nieto, J., Soler, J., Poupaud, F.: Multidimensional high-field limit of the electrostatic Vlasov-Poisson-Fokker-Planck system. J. Differential Equations 213, 418–442 (2005)
20. Goudon, T., Sánchez, O., Soler, J., Bonilla, L.L.: Low-field limit for a nonlinear discrete drift-diffusion model arising in semiconductor superlattices theory. SIAM J. Appl. Math. 64, 1526–1549 (2004)
21. Hartwell, H.L., Hopfield, J.J., Leibner, S., Murray, A.W.: From molecular to modular cell biology. Nature 402, c47–c52 (1999)

22. Hillen, T., Othmer, H.: The diffusion limit of transport equations derived from velocity–jump processes. SIAM J. Appl. Math. 61, 751–775 (2000)
23. Humphrey, J.D., Rajagopal, K.R.: A constrained mixture model for growth and remodeling of soft tissues. Math. Models Methods Appl. Sci. 12, 407–430 (2002)
24. Othmer, H.G., Hillen, T.: The diffusion limit of transport equations II: chemotaxis equations. SIAM J. Appl. Math. 62, 1222–1250 (2002)
25. Szymańska, Z., Morale Rodrigo, C., Lachowicz, M., Chaplain, M.A.J.: Mathematical modelling of cancer invasion of tissue: the role and effect of nonlocal interactions. Math. Models Methods Appl. Sci. 19, 257–281 (2009)

Instabilites in Arch Shaped MEMS

K. Das and R.C. Batra

Abstract. Arch shaped microelectromechanical systems (MEMS) have been used as mechanical memories, micro-relays, micro-valves, optical switches, and digital micro-mirrors. A bi-stable structure is characterized by a multivalued load deflection curve. Here, the symmetry breaking, the snap-through instability, and the pull-in instability of a sinusoidal shaped MEMS under static and dynamic electric loads have been studied. The electric load is a nonlinear function of the a priori unknown deformed shape of the arch, and is thus a follower type load. The nonlinear partial differential equation governing transient deformations of the arch is solved numerically using the Galerkin method and the resulting ordinary differential equations are integrated by using the Livermore solver for ordinary differential equations. For the static problem, the displacement control and the pseudo-arc length continuation methods are used to obtain the bifurcation curve of the MEMS displacement versus a load parameter. The displacement control method fails to compute asymmetric deformations of the MEMS, which are found by the pseudo-arc-length continuation method. Two distinct mechanisms of the snap-through instability for the dynamic problem are demonstrated. It is found that critical loads and geometric parameters for instabilities of an arch under an electric load with and without the consideration of mechanical inertia effects are quite different.

1 Introduction

An electrically actuated microelectromechanical system (MEMS) consists of a deformable electrode made of a conductive material suspended over a rigid conductive electrode with a dielectric medium, generally air, between them (Fig. 1). An electric potential difference applied between the two electrodes induces the Coulomb pressure on the electrodes, which deflects the deformable electrode towards the rigid

K. Das · R.C. Batra
Department of Engineering Science and Mechanics,
Virginia Polytechnic Institute and State University, Blacksburg, VA 24061 USA
e-mail: {kdas,rbatra}@vt.edu

one. The elastic restoring force induced in the deformed electrode restricts its motion. Electric charges redistribute on the deformable electrode's surface and the gap between it and the rigid electrode decreases, which in turn increases the Coulomb pressure and deforms the deformable electrode more until the Coulomb force balances the elastic restoring force. MEMS of dimensions in the range of a few to a hundred micrometers are used as radio frequency (RF) switches, varactors and inductors [24], accelerometers [25], pressure sensors, controllers for micro-mirrors [28], micro-pumps [3], and bio-MEMS.

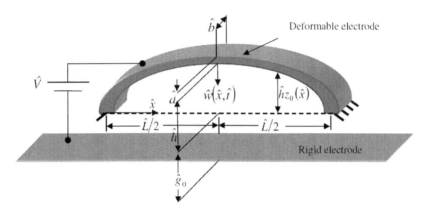

Fig. 1 Schematic sketch of the problem studied.

For electrically actuated MEMS, the applied electric potential has an upper limit, beyond which the corresponding Coulomb force is not balanced by the elastic restoring force, resulting in collapse of the deformable electrode on the rigid one. This phenomenon, called the pull-in instability, was observed experimentally by Taylor [27] and Nathanson et al. [19]. The corresponding values of the potential difference and the peak displacement of the deformable electrode are called the pull-in voltage and the pull-in displacement, respectively. Collectively, the two are called the pull-in parameters.

Accurate estimates of pull-in parameters are crucial for designing electrically actuated MEMS. In switching applications [21], the pull-in instability is necessary for the switch to operate. However, for micro-mirrors [12] and micro-resonators [18] the pull-in instability restricts the range of operational displacement of the device.

In an arch shaped deformable electrode of Fig. 1, in addition to the pull-in instability, the snap-through instability can occur under the Coulomb pressure; these two instabilities have been studied in [15, 13] with a one degree of freedom model. Figure 2(a) shows a bifurcation diagram between the non-dimensional peak displacement $w = \hat{w}/\hat{g}_0$ and the nondimensional electric potential difference parameter $\beta = \hat{\varepsilon}_0 \hat{b} \hat{L}^4 \hat{V}^2 / 2\hat{E}\hat{I}\hat{g}_0^3$. Here $\hat{\varepsilon}_0$ is the vacuum permittivity, $\hat{b}$ the arch width, $\hat{L}$ the arch length, $\hat{g}_0$ the initial gap, $\hat{V}$ the electric potential difference between the

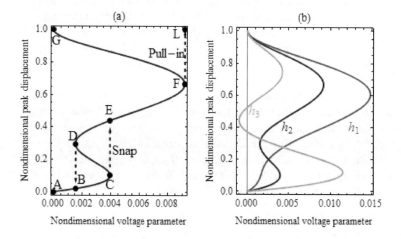

Fig. 2 (a) Bifurcation diagram for snap-through and pull-in instabilities of an arch shaped MEMS (from [13]), (b) bifurcation diagrams for different arch elevations ($h_1 < h_2 < h_3$) (from [13]).

two electrodes, $\hat{E}$ Young's modulus of the arch material, and $\hat{I} = \hat{b}\hat{d}^3/12$ the second moment of the cross-section of the arch about a horizontal axis passing through the centroid of the cross-section. The bifurcation diagram has two stable branches AC and DF. Initially, with increase in β, w increases gradually from point A to point C, the arch maintains its initial curved shape, and the resultant elastic restoring force balances the Coulomb force. At point C, the deflection increases suddenly, and the arch is inverted to a new equilibrium position corresponding to point E. This sudden jump in the deflection is called the snap-through instability, and the corresponding voltage the snap-through voltage. From point E to point F, the elastic restoring force induced in the arch again balances the Coulomb force. However, just after point F the deformable electrode collapses on to the rigid one, and the pull-in happens. Configurations of the arch corresponding to points on parts CD and FG of the curve are unstable and experimentally non-observable. During the loading process, the arch follows the path ACEFL, and it follows the path FDBA during unloading. Depending upon the initial elevation $\hat{h}$, either a snap-through instability or a pull-in instability or both instabilities occur. Advantages of the snap-through instability have been exploited in actuators [29, 30, 26, 22], microvalves [10], and transducers [17]. The snap-through instability of an arch shaped MEMS under slowly applied electric loads has been observed experimentally and studied with reduced order models in [31, 15, 16, 13]. Zhang et al. [31] and Krylov et al. [16] have studied static problems involving the snap-through and the pull-in instabilities in circular and bell shaped arches, respectively. Depending on the arch rise $\hat{h}$, the arch thickness $\hat{d}$, load types (step or ramp), and the gap $\hat{g}_0$ between the electrodes the arch shape and the load type, following three scenarios arise: either only the pull-in instability occurs (red

curve in Fig. 2b), or the arch undergoes the snap-through and then the pull-in instability (blue curve in Fig. 2b), or the snap-through and the pull-in happen simultaneously (green curve in Fig. 2b). In each case, the pull-in instability occurs.

The pull-in instability in a MEMS under a transient electric load has been analyzed in [20, 23, 9, 6, 2, 14, 5]. Identifying the pull-in instability is easy because of
the physical phenomenon of the deformable electrode touching the rigid one. The
"dynamic snap-through" generally means a large increase in response resulting from
a small increase in a load parameter [11]. Conditions for the dynamic snap-through
to occur depend on the geometric parameters of the arch and on the type of loads.
Here we investigate these instabilities in a sinusoidal shaped MEMS. Results for an
arch shaped MEMS, and values of the load parameter β for which the snap-through
instability will occur under a step electric potential difference are given in [7]. We
also study effects of damping on the instability parameters.

2 Mathematical Model

The governing equation for a shallow micro-arch under an electrostatic load in terms
of non-dimensional variables is [13]

$$\ddot{w} + c\dot{w} + w^{IV} - \alpha \left(h z_0'' - w'' \right) \int\limits_0^1 \left(2h z_0' - w' - \left(w' \right)^2 \right) dx = \frac{\beta}{(1 + h z_0 - w)^2}, \quad (1)$$

$$x \in (0,1).$$

The boundary and initial conditions for a fixed-fixed arch initially at rest are

$$w(0,t) = w(1,t) = w'(0,t) = w'(1,t) = w(x,0) = \dot{w}(x,0) = 0 \quad (2)$$

In Eq. (1) and (2), a super-imposed dot and a prime denote derivative with respect
to time t and the space coordinate x, respectively, $\hat{\varepsilon}_0 = 8.854 \times 10^{-12} \mathrm{Fm}^{-1}$ the
vacuum permittivity, $\hat{\rho}$ the mass density, $x = \hat{x}/\hat{L}$, $w = \hat{w}/\hat{g}_0$, the transverse dis-
placement, $d = \hat{d}/\hat{g}_0$, $b = \hat{b}/\hat{g}_0$, $h = \hat{h}/\hat{g}_0$, $\alpha = \hat{g}_0^2 \hat{b} \hat{d}/2\hat{I}$ the stretching ratio, $c = \hat{c}\hat{L}^2/\sqrt{\hat{\rho}\hat{b}\hat{d}\hat{E}\hat{I}}$, $\hat{c}$ the damping coefficient, $z_0(x)$ the initial shape of the arch, and
$t = \hat{t}\sqrt{(\hat{E}\hat{I})/(\hat{\rho}\hat{b}\hat{d}\hat{L}^4)}$, $\hat{t}$ the dimensional time. As discussed in [5] the damping
provided by deformations of the air between the two electrodes can also be approx-
imated by the term $c\dot{w}$. We note that Eq. (1) with $h = 0$ is the well known governing
equation for the microbeam based MEMS [1, 4].

We solve Eq. (1) using the Galerkin method by approximating the transverse
displacement w by the series

$$w(x,t) \approx \sum_{i=1}^n q_i(t)\phi_i(x) \quad (3)$$

where $q_i(t)$ are the generalized coordinates and $\phi_i(x)$ eigenmodes of an undamped fixed-fixed straight beam. Numerical experiments have shown that $n = 6$ in Eq. (3) gives converged solutions of Eq. (1).

3   Results for a Sinusoidal Arch MEMS

Figure 3 depicts loci of the maximum deflection (dot-dashed curve) produced by different step potentials and static bifurcation curves (solid curves and a dashed red curve) obtained using the PALC algorithm for two different values of the arch height for a sinusoidal arch, $z_0 = \sin(\pi x)$ MEMS. The dashed portion of the bifurcation curve for $h = 0.5$ represents unstable deformations asymmetric about $x = 0.5$. The response of a sinusoidal arch is qualitatively similar to that of a bell-shaped arch [8]. For $h = 0.35$, the direct snap-through due to the step electric potential difference occurs when the locus of the maximum deflection intersects at point A the unstable branch of the bifurcation curve. Similarly, the pull-in instability occurs when the locus of the maximum deflection intersects again at point B the unstable branch of the bifurcation curve. However, for $h = 0.5$, the arch experiences the indirect snap-through when the locus of the maximum deflection intersects at point I the unstable branch of the bifurcation curve for asymmetric solutions and the snap-through and the pull-in instabilities ensue simultaneously.

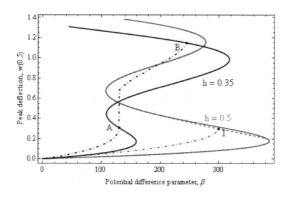

Fig. 3 Static bifurcation diagrams ($w(0.5)$ vs. β) of the arch for two different values of the arch height h. Solid curves are results from the PALC algorithm. Dot-dashed solid curves are the loci of the maximum displacement $\max_{0 \le t \le 2}(w(0.5,t))$ under the step load.

4   Effect of Viscous Damping

Referring to Fig. 1, we consider a bell shaped silicon arch with $\hat{L} = 1$ mm, $\hat{b} = 30\ \mu$m, $\hat{d} = 2.4\ \mu$m, $\hat{g}_0 = 10.1\ \mu$m, $\hat{h} = 3.0\ \mu$m, and its bottom-surface described by $z_0 = \sin^2(\pi\hat{x}/\hat{L})$. For viscous damping coefficient $c = 1.0$, and different values of the step potential difference β, Figure 4 evinces time histories of the peak displacement at the mid-span of the arch with $h = 0.3$ and $a = 106$. For each value of

β, the amplitude of oscillations of the arch decreases with time, and the peak deflection of the arch approaches that of the statically deformed arch. Figure 5 exhibits the static bifurcation curve (black curve) of the arch ($h = 0.3$, $a = 106$) , and for dynamic problems and different values of c loci of the maximum displacement (colored curves) under a step electric potential difference. As c increases, the response of the dynamic problem approaches that of the static problem earlier in time. However, the enhanced damping does not suppress the two instabilities, and the peak $w(0.5,t)$ vs. β curves for different values of c are qualitatively similar to each other. Table 1 lists the snap-through and the pull-in parameters for different values of c.

Fig. 4 Time histories of the downward displacement of the mid-span of the fixed-fixed bell shaped arch ($h = 0.297$, $a = 106$) for different applied step voltages. Dashed blue and dashed green curves show the statically deformed position after and before the snap-through instability.

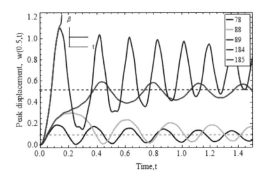

Fig. 5 The static bifurcation diagram ($w(0.5)$ vs. β, black curve) of the arch ($h = 0.3$, $a = 106$). For different values of the damping coefficients, dotted curves are the loci of the maximum displacement $\max_{0 \le t \le 2}(w(0.5,t))$ under a step electric potential difference.

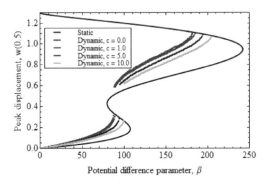

Table 1 Critical values of β for different values of the damping coefficient c.

Critical values of β	$c = 0$	$c = 1.0$	$c = 5.0$	$c = 10.0$	Static
Snap-through	87	88	94	99	106
Pull-in	181	184	194	204	240

5 Conclusions

We have investigated the snap-through and the pull-in instabilities in an electrically actuated micro-arch modeled as an undamped Euler-Bernoulli beam incorporating the nonlinear mid-plane stretching. Two distinct mechanisms, namely the 'direct' and the 'indirect', snap-through instabilities are found. It is found that the PALC algorithm can compute multiple branches in the bifurcation curve, which correspond to symmetric and asymmetric deformations of the arch. The DIPIE algorithm fails to compute asymmetric solutions.

This work contributes to our understanding of the nonlinear behavior and instabilities of a micro-arch under electrical loads, and enables studying bi-stable MEMS.

Other conclusions are summarized below:

1. An undamped arch under a step electric load may experience either direct or indirect snap-through instability.
2. For relatively small arch heights, the static problem has solutions with deformations symmetric about the mid-span of the arch and the direct snap-through happens when the locus of the maximum deflection of the dynamic problem intersects the unstable branch of the bifurcation curve for the static problem.
3. For relatively large arch heights, the static problem has solutions with deformations symmetric and asymmetric about the mid-span of the arch. The asymmetric solution has lower total potential energy than the corresponding symmetric solution.
4. The indirect snap-through happens when the locus of the maximum deflection of the dynamic problem intersects the unstable branch of the bifurcation curve of the asymmetric deformations for the static problem.

Acknowledgements

This work was supported by the Office of Naval Research grant N00014-06-1-0567 to Virginia Polytechnic Institute and State University with Dr Y. D. S. Rajapakse as the program manager. Views expressed in the paper are those of authors, and neither of the funding agency nor of VPI&SU.

References

1. Abdel-Rahman, E.M., Younis, M.I., Nayfeh, A.H.: Characterization of the mechanical behavior of an electrically actuated microbeam. Journal of Micromechanics and Microengineering 12, 759–766 (2002)
2. Ananthasuresh, G.K., Gupta, R.K., Senturia, S.D.: An approach to macromodeling of MEMS for nonlinear dynamic simulation. In: Proceedings of the ASME International Conference of Mechanical Engineering Congress and Exposition (MEMS), Atlalta, GA, pp. 401–407 (1996)
3. Bassous, E., Taub, H.H., Kuhn, L.: Ink jet printing nozzle arrays etched in silicon. Applied Physics Letters 31, 135–137 (1977)

4. Batra, R.C., Porfiri, M., Spinello, D.: Electromechanical Model of Electrically Actuated Narrow Microbeams. Journal of Microelectromechanical Systems 15, 1175–1189 (2006)
5. Chao, P.C.-P., Chiu, C.W., Liu, T.-H.: DC dynamic pull-in predictions for a generalized clamped-clamped micro-beam based on a continuous model and bifurcation analysis. Journal of Micromechanics and Microengineering 18, 0960–1317 (2008)
6. Chu, P.B., Nelson, P.R., Tachiki, M.L., Pister, K.S.J.: Dynamics of polysilicon parallel-plate electrostatic actuators. Sensors and Actuators A: Physical 52, 216–220 (1996)
7. Das, K., Batra, R.C.: Pull-in and snap-through instabilities in transient deformations of microelectromechanical systems. Journal of Micromechanics and Microengineering 19, 035008 (2009)
8. Das, K., Batra, R.C.: Symmetry breaking, snap-through, and pull-in instabilities under dynamic loading of microelectromechanical shallow arches. Smart Materials and Structures 18, 115008 (2009)
9. Flores, G., Mercado, G.A., Pelesko, J.A.: Dynamics and touchdown in electrostatic MEMS. In: Proceedings of International Conference on MEMS, NANO and Smart Systems, 2003, pp. 182–187 (2003)
10. Goll, C., Bacher, W., Buestgens, B., Maas, D., Menz, W., Schomburg, W.K.: Microvalves with bistable buckled polymer diaphragms. Journal of Micromechanics and Microengineering, 77–79 (1996)
11. Humphreys, J.S.: On dynamic snap buckling of shallow arches. AIAA Journal 4, 878–886 (1966)
12. Hung, E.S., Senturia, S.D.: Extending the travel range of analog-tuned electrostatic actuators. Journal of Microelectromechanical Systems 8, 497–505 (1999)
13. Krylov, S., Bojan, R.I., David, S., Shimon, S., Harold, C.: The pull-in behavior of electrostatically actuated bistable microstructures. Journal of Micromechanics and Microengineering, 055026 (2008)
14. Krylov, S., Maimon, R.: Pull-in Dynamics of an Elastic Beam Actuated by Continuously Distributed Electrostatic Force. Journal of Vibration and Acoustics 126, 332–342 (2004)
15. Krylov, S., Serentensky, S., Schreiber, D.: Pull-in behavior of electrostatically actuated multistable microstructures. In: ASME 2007 International Design Engineering Technical Conference & Computers and Information in Engineering Conference, Las Vegas, Nevada, USA (2007)
16. Krylov, S., Seretensky, S., Schreiber, D.: Pull-in behavior and multistability of a curved microbeam actuated by a distributed electrostatic force. In: Seretensky, S. (ed.) IEEE 21st International Conference on Micro Electro Mechanical Systems, 2008. MEMS 2008, pp. 499–502 (2008)
17. Kugel, V.D., Xu, B., Zhang, Q.M., Cross, L.E.: Bimorph-based piezoelectric air acoustic transducer: model. Sensors and Actuators A: Physical 69, 234–242 (1998)
18. Legtenberg, R., Tilmans, H.A.C.: Electrostatically driven vacuum-encapsulated polysilicon resonators Part I. Design and fabrication Sensors and Actuators A: Physical 45, 57–66 (1994)
19. Nathanson, H.C., Newell, W.E., Wickstrom, R.A., Davis Jr., J.R.: The resonant gate transistor. IEEE Transactions on Electron Devices 14, 117–133 (1967)
20. Nayfeh, A., Younis, M., Abdel-Rahman, E.: Dynamic pull-in phenomenon. MEMS resonators Nonlinear Dynamics 48, 153–163 (2007)
21. Nguyen, C.T.C., Katehi, L.P.B., Rebeiz, G.M.: Micromachined devices for wireless communications. Proceedings of the IEEE 86, 1756–1768 (1998)
22. Park, S., Hah, D.: Pre-shaped buckled-beam actuators: Theory and experiments. Sensors and Actuators A: Physical 148, 186–192 (2008)

23. Postma, H.W.C., Kozinsky, I., Husain, A., Roukes, M.L.: Dynamic range of nanotube- and nanowire-based electromechanical systems. Applied Physics Letters 86, 223105 (2005)
24. Rebeiz, G.M.: RF MEMS theory, design, and technology. John Wiley & Sons, Inc., Hoboken (2003)
25. Roylance, L.M., Angell, J.B.: A batch-fabricated silicon accelerometer. IEEE Transactions on Electron Devices 26, 1911–1917 (1979)
26. Saif, M.T.A.: On a tunable bistable MEMS-theory and experiment. Journal of Microelectromechanical Systems 9, 157–170 (2000)
27. Taylor, G.: The Coalescence of Closely Spaced Drops when they are at Different Electric Potentials. Proceedings of the Royal Society of London. Series A, Mathematical and Physical Sciences 306, 423–434 (1968)
28. Van Kessel, P.F., Hornbeck, L.J., Meier, R.E., Douglass, M.R.: A MEMS-based projection display. Proceedings of the IEEE 86, 1687–1704 (1998)
29. Vangbo, M.: An analytical analysis of a compressed bistable buckled beam. Sensors and Actuators A: Physical 69, 212–216 (1998)
30. Vangbo, M., Bcklund, Y.: A lateral symmetrically bistable buckled beam. Journal of Micromechanics and Microengineering 8, 29–32 (1998)
31. Zhang, Y., Wang, Y., Li, Z., Huang, Y., Li, D.: Journal of Snap-Through and Pull-In Instabilities of an Arch-Shaped Beam Under an Electrostatic Loading Microelectromechanical Systems 16, 684–693 (2007)

Towards Poroelasticity of Fractal Materials

M. Ostoja-Starzewski

Abstract. This study proposes a model of poroelasticity grasping the fractal geometry of the pore space occupied by the fluid phase as well as the fractal geometry of the solid (matrix) phase. The dimensional regularization approach employed is based on product measures which account for an arbitrary anisotropic structure. This, in turn, leads to a re-interpretation of spatial gradients (of both fluid velocity and displacement fields) appearing in the balance and constitutive equations; the latter are adapted from the classical poroelasticity. In effect, an initial-boundary value problem of a fractal medium can be mapped into one of a homogenized, non-fractal medium, and, should all the fractal dimensions become integer, all the equations reduce back to those of classical poroelasticty. Overall, the proposed methodology broadens the applicability of continuum mechanics/physics and sets the stage for poromechanics of fractal materials.

1 Background

In the Foreword to [1] James R. Rice identified the lack of *clear scale separation as "the Achilles heel of current formulation of the continuum mechanics of fluid-saturated rock masses."* From the standpoint of multiscale mechanics of materials [2], this fatal weakness in spite of overall strength is two-fold: (r) spatial randomness of heterogeneous rock structure, and (f) possibly fractal character of that structure. In fact, one may consider four cases: (r) only, (f) only, (r) and (f) jointly, no (r) and no (f). Thinking of rocks, or indeed any permeable materials, the latter case is hard to imagine, unless one assumes typical wavelengths of dependent fields to be much larger than the microstructural scales. Yet it is this case upon which most poromechanics theories are set up.

M. Ostoja-Starzewski
Department of Mechanical Science & Engineering University of Illinois at
Urbana-Champaign, 1206 W. Green Street, Urbana, IL, 61801-2906, USA
e-mail: martinos@illinois.edu

Now, in the case of (r) only, one may use techniques of scale-dependent homogenization [2], where hierarchies of mesoscale bounds allow one to determine the scaling trend to Representative Volume Element (RVE), so that the RVE size, and indeed its effective properties, are found. Below the RVE one has to deal with a Statistical Volume Element (SVE), which plays a key role in setting up continuum random field approximations and stochastic finite element models. Recalling the results on RVE and SVE pertaining to various types of linear and nonlinear elastic/dissipative materials obtained so far, one of these applies to the Stokes flow in the pores of a random skeleton, which, upon scale-dependent homogenization, yields the Darcy law [3]. That approach may be extended to poroelasticity models where the matrix itself is deformable. It would then yield RVE sizes for the scale where the Darcy law holds and possibly a different RVE size for the poroelasticity equation itself. Drawing on the well-known analogy to thermoelasticity of random materials (albeit knowing that even in the deterministic, homogeneous case it has to be taken with a grain of salt [4]), we would expect yet another RVE size for the coupling term between elastic and flow effects.

This paper is concerned with the case (f) only, which implies working on a scale where the underlying fractal structure has been homogenized and replaced by an effective deterministic medium; we do not consider the spatial material randomness of that structure. An approach of that type has been advanced since the work relying on dimensional regularization by Tarasov [5-7]. He developed continuum-type equations of conservation of mass, momenta, and energy for fractal porous media, and, on that basis studied several fluid mechanics and wave motion problems. In principle, one can then map a mechanics problem of a fractal [which is described by its mass (D) and surface (d) fractal dimensions plus the spatial resolution (R)] onto a problem in the Euclidean space in which this fractal is embedded, while having to deal with coefficients explicitly involving D, d and R. As it turns out, D is also the order of fractional integrals employed to state global balance laws. This approach's great promise stems from the fact that much of the framework of continuum mechanics/physics may be generalized and partial differential equations (with derivatives of integer order) may still be employed [8,9]. Whereas the original formulation of Tarasov was based on the Riesz measure – and thus more suited to isotropic media – the model proposed more recently is based on a product measure [10,11]. That measure grasps the anisotropy of fractal geometry (i.e. different fractal dimensions in different directions) on mesoscale, which is more suited to solid materials. As a an added benefit − or, depending on one's taste, a disadvantage − this potentially leads to asymmetry of the Cauchy stress [12], and hence to a framework of micropolar mechanics of fractal materials. In the dimensional regularization, while all the derived relations depend explicitly on D, d and R, upon setting $D = 3$ and $d = 2$, they reduce to conventional forms of governing equations for continuous media with Euclidean geometries. Prior research has already involved an extension to continuum thermomechanics and fracture mechanics, a generalization of extremum and variational principles, turbulent flows in fractal porous media and thermoelasticity [13-16]. We follow this approach to mechanics of fractal media, while adopting the formulation of poroelasticity dating back to Biot such as presented in [17].

2   Fractal Structures and Product Measures

We begin with a body B in the Euclidean 3-space made of two phases: solid (i.e., matrix) W_s and fluid W_f, which occupies the pores; i.e. $B = W_s \cup W_f$. In general, the solid is a fractal, and the fluid is a fractal. Thus, W_s is characterized by a volume fractal dimension D_s and a surface fractal dimension d_s, while W_f is characterized by its volume fractal dimension D_f and its surface fractal dimension d_f. As dictated by the fractal geometry [18,19], $D_s + D_f$ is not necessarily equal to 3, and $d_s + d_f$ is not necessarily equal to 2.

Following [5-14], the way to think of the fractal structure of the matrix is to consider a pore occupied by the fluid of mass m_f obeying a power law

$$m(R) = kR^{D_f} \quad D_f < 3, \tag{1}$$

Here R is the length scale of measurement (or resolution) and k is a proportionally constant. Certainly, (1) can be applied to a pre-fractal, i.e. a fractal-type, physical object with lower and upper cut-offs. Next, we use a fractional integral to represent the mass in a 3D region

$$m(W) = \int_W \rho(\mathbf{r}) dV_D = \int_W \rho(\mathbf{r}) c_3(D,R) dV_3, \tag{2}$$

where the first and second equalities, respectively, involve fractional (Riesz-type) integrals and conventional integrals, while the coefficient $c_3(D,R)$ provides a transformation between the two.

In order to deal with generally anisotropic rather than isotropic media, we replace (1) by a more general power law relation with respect to each coordinate

$$m_f(R) \sim x_1^{\alpha_1} x_2^{\alpha_2} x_3^{\alpha_3},$$

whereby the mass distribution is specified via a product measure

$$m_f(x_1, x_2, x_3) = \int \int \int_W \rho(x_1, x_2, x_3) d\mu(x_1) d\mu(x_2) d\mu(x_3). \tag{3}$$

This more general approach to anisotropy is motivated by the work of Carpinteri and co-workers [20,21] who considered a porous concrete microstructure modeled by a Sierpiński carpet in the cross-section and a Cantor set in the longitudinal coordinate. Here the length measurement in each coordinate is provided by

$$d\mu(x_k) = c_1^{(k)}(\alpha_k, x_k) dx_k, \quad k = 1, 2, 3. \tag{4}$$

Then, the total fractal dimension D_f of mass m_f is $\alpha_1 + \alpha_2 + \alpha_3$, while

$$c_3 = c_1^{(1)} c_1^{(2)} c_1^{(3)} = \prod_{i=1}^{3} c_1^{(i)}. \tag{5}$$

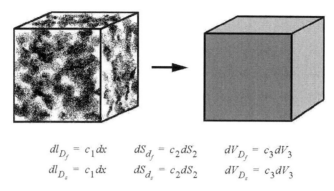

$$dl_{D_f} = c_1 dx \qquad dS_{d_f} = c_2 dS_2 \qquad dV_{D_f} = c_3 dV_3$$
$$dl_{D_s} = c_1 dx \qquad dS_{d_s} = c_2 dS_2 \qquad dV_{D_s} = c_3 dV_3$$

Fig. 1 Homogenization of a material with fractal geometries of fluid in pore space (black) and solid matrix (white). In general, the structure depdens on the direction (anisotropic dependence is present).

For clarity of notation, we do not put $_f$ subscripts on all the quantities. All that is stated in this paragraph applies to the solid body W_s, and then we would have m_s, D_s, and so forth, with all the quantities carrying $_s$ subscripts.

Next, continuing with the fluid phase for the sake of explicit discussion and focusing on the surface coefficient (c_2), we consider a cubic volume element, whose each surface element is specified by the normal vector (along axes x_i; see Fig. 1). Therefore, $c_2^{(k)}$ associated with the surface element $S_d^{(k)}$ is

$$c_2^{(k)} = c_1^{(i)} c_1^{(j)} = c_3/c_1^{(k)}, \; i \neq j \text{ and } i,j \neq k. \tag{6}$$

To express the $c_1^{(k)}$ coefficients, we adopt a modified Riemann-Liouville fractional integral of Jumarie [22,23], so that

$$c_1^{(k)} = \alpha_k (l_k - x_k)^{\alpha_k - 1}, \; k = 1,2,3, \tag{7}$$

where l_k is the total length (integral interval) along x_k. Note that, for point mass $\rho(x_1,x_2,x_3) = m_0 \delta(x_1)\delta(x_2)\delta(x_3)$, when $D_f \to 3$ ($\alpha_1, \alpha_2, \alpha_3 \to 1$), then $m(W_f) = \alpha_1 \alpha_2 \alpha_3 l^{D_f-3} m_0 \to m_0$, and so the conventional concept of point mass is recovered [24]. However, the Riesz fractional integral has a non-smooth transition of mass with respect to its fractal dimension ($m(W_f) = 0$ except when $D_f = 3$, $m(W_f) = m_0 0^{D_f-3} = m_0$).

At this point we note the fractional Gauss theorem [11], formulated within the framework of product measures discussed above,

$$\int_{\partial W_f} f_k n_k dS_{d_f} = \int_{W_f} \left(f_k c_2^{(k)} \right)_{,k} c_3^{-1} dV_{D_f}$$
$$= \int_{W_f} f_{k,k} \, c_2^{(k)} \, c_3^{-1} dV_{D_f} = \int_{W_f} \frac{f_{k,k}}{c_1^{(k)}} dV_{D_f}. \tag{8}$$

It is important to observe three properties of the (fractal gradient) operator $\nabla_k^{D_f} = (\cdot)_{,k}/c_1^{(k)}$ under product measures:

(1) It is the "inverse" operator of fractional integrals.

(2) The rule of "term-by-term" differentiation is satisfied, i.e. $\nabla_k^{D_f}(AB) = \nabla_k^{D_f}(A)B + A\nabla_k^{D_f}(B)$.

(3) Its operation on any constant is zero, which indeed is a desired property not possessed by the usual fractional derivative of Riemann-Liouville [25]. While we note that properties (2) and (3) do not hold in Tarasov's Riesz measure formulations, the fractional generalization of Reynold's transport theorem is

$$\frac{d}{dt}\int_{W_f} P dV_{D_f} = \int_{W_f}\left[\frac{\partial}{\partial t}P + (Pv_k)_{,k}\right]dV_{D_f}, \tag{9}$$

implying that the fractal material time derivative is the same as the conventional material time derivative (dP/dt):

$$\left(\frac{d}{dt}\right)_{D_f}P = \frac{d}{dt}P = \frac{\partial}{\partial t}P + P_{,k}\,v_k. \tag{10}$$

From a homogenization standpoint this allows an interpretation of the fractal (intrinsically discontinuous) medium as a continuum with a 'fractal metric' embedded in the equivalent homogenized continuum model, that is

$$dl_{D_f} = c_1 dx, \quad dS_{d_f} = c_2 dS_2, \quad dV_{D_f} = c_3 dV_3. \tag{11}$$

Here $dl_{D_f}, dS_{d_f}, dV_{D_f}$ represent the line, surface, and volume elements in the fractal fluid body, while dx, dS_2, dV_3, respectively, denote those in the homogenized model, see Fig. 1. The coefficients c_1, c_2, c_3 provide relations between both pictures. Standard image analysis techniques (such as the "box method" or the "sausage method") allow a quantitative calibration of these coefficients for every direction and every cross-sectional plane.

3 Governing Relations

Continuing within the framework of continuum mechanics of fractal media developed in [10,11], we will be replacing all the spatial derivatives $\partial/\partial x_k$ in the fluid phase by fractional derivatives $\nabla_k^{D_f}$ (and $\partial/\partial x_k$ by $\nabla_k^{D_f}$ in the solid phase) introduced above. This is motivated, on account of (11), by

$$\nabla_k^{D_f} = \frac{1}{c_1^{(k)}}\frac{\partial}{\partial x_k} = \frac{\partial}{\partial l_{D_f}^{(k)}} \quad \text{in } W_f. \tag{12}$$

By analogy, for small gradients in the solid phase we have a relation between the strain tensor and the displacement

$$\varepsilon_{ij} = \frac{1}{2}\left(\nabla_j^{D_s} u_i + \nabla_i^{D_s} u_j\right) = \frac{1}{2}\left(\frac{1}{c_1^{(j)}} u_{i,j} + \frac{1}{c_1^{(i)}} u_{j,i}\right) \quad \text{in } W_s. \tag{13}$$

We note that this approach renders the mechanical approach equivalent to that based on energy principles.

Assuming low fluid velocities everywhere, in view of (12), the Darcy law is expressed as

$$q_i = -\rho\,\kappa_{ij}\nabla_j^{D_f} p, \tag{14}$$

where q_i is the flux, ρ is the mass density, and κ_{ij} is the permeability tensor.

With the understanding that linear relations hold, the three constitutive equations are

$$\sigma_{ij} = L_{ijkl}\varepsilon_{kl} - M_{ij}p \equiv \frac{1}{2}L_{ijkl}\left(\nabla_l^{D_s} u_k + \nabla_k^{D_s} u_l\right) - M_{ij}p, \tag{15}$$

$$m - m_0 = R_{ij}\varepsilon_{ij} + Qp \equiv \frac{1}{2}R_{ij}\left(\nabla_l^{D_s} u_k + \nabla_k^{D_s} u_l\right) + Qp, \tag{16}$$

where σ_{ij} denotes the Cauchy stress tensor, p is the fluid mass content per unit volume, $L_{ijkl}(=L_{klij})$ is the stiffness tensor for drained conditions, $m = \rho v$ is the fluid mass per unit volume (v being the apparent volume fraction of voids), and p is the pore fluid pressure. The scalar Q and two tensors $M_{ij}(=M_{ji})$, and $R_{ij}(=R_{ji})$ are the constitutive quantities. On account of the Maxwell relation, $R_{ij} = \rho M_{ij}$ holds just as in the non-fractal case. The stiffness tensor for undrained conditions is $L_{ijkl}^u = L_{ijkl} + Q^{-1}\rho M_{ij}M_{kl}$, so that (15) may be replaced by $\sigma_{ij} = L_{ijkl}^u \varepsilon_{kl}$.

The modified balance equations of fractal media relevant in the present context include the fractional equation of continuity

$$\dot{\rho} = -\rho\nabla_k^{D_f} v_k, \tag{17}$$

and the fractional equation of balance of density of momentum

$$\rho\dot{v}_k = \rho f_k + \nabla_k^{D_s} \sigma_{kl}. \tag{18}$$

The first of these leads to

$$\frac{\partial m}{\partial t} = -\nabla_i^{D_f} q_i, \tag{19}$$

while the second one, specialized to quasi-statics and in the absence of body forces, yields

$$\nabla_k^{D_s} \sigma_{kl} = 0. \tag{20}$$

Focusing, for simplicity, on an isotropic porous material, we have

$$M_{ij} = \zeta\delta_{ij}, \quad \kappa_{ij} = \kappa\delta_{ij},$$

$$L_{ijkl} = G(\delta_{ik}\delta_{jl} + \delta_{il}\delta_{jk}) + (K - 2G/3)\delta_{ij}\delta_{kl}, \tag{21}$$

with

$$\zeta = 1 - \frac{K}{K_s'}, \tag{22}$$

where K_s' is a bulk modulus that can be roughly identified with bulk modulus of the solid making up the matrix. In that case, by combining the constitutive equations (15-16) with the balance equations (18-19), one finds

$$\frac{1}{c}\frac{\partial m}{\partial t} = \nabla_i^{D_s}\nabla_i^{D_s}m, \tag{23}$$

and

$$(K+G/3)\nabla_k^D\nabla_j^D u_k + G\nabla_k^D\nabla_k^D u_j - \zeta\nabla_j^D p = 0. \tag{24}$$

In the above

$$c = \frac{\kappa(K+4G/3)(K_u - K)}{\zeta^2(K_u + 4G/3)}, \tag{25}$$

just as in the classical poroelasticity theory.

4 Conclusion

As discussed in the Introduction, from the standpoint of multiscale mechanics of materials, the origin of the lack of separation of scales in geomaterials is two-fold: (r) spatial randomness of heterogeneous rock structure, and (f) possibly fractal character of that structure. The present study has made a step in the direction of (f) only. The approach taken builds on the dimensional regularization pioneered by Tarasov, but employs product measures to account for an arbitrary anisotropic structure. Given the fact that the fluid phase occupies the pore space, while the solid phase is its geometric complement, there are two sets of fractal dimensions, product measures and fractal coefficients — one for each phase. Our previous studies suggest the key balance equations as well as the interpretation of the gradients (of both fluid velocity and displacement fields) appearing in the constitutive equations. The latter are adapted from the classical poroelasticity.

With reference to Fig. 1, an initial-boundary value problem of a fractal medium can now be mapped into one of a homogenized, non-fractal medium. Given a specific material, the four fractal dimensions describing the pore phase (D_f, d_f) and the solid phase (D_s, d_s) can be determined by standard image analysis techniques. In general, should all the fractal dimensions become integer, all the equations reduce back to those of conventional poroelasticty. Overall, the proposed methodology broadens the applicability of continuum mechanics/physics and sets the stage for poromechanics of fractal materials.

References

1. Guéguen, Y., Boutéca, M.: Mechanics of Fluid-Saturated Rocks. Elsevier, Amsterdam (2004)
2. Ostoja-Starzewski, M.: Microstructural Randomness and Scaling in Mechanics of Materials. CRC Press, Boca Raton (2008)

3. Du, X., Ostoja-Starzewski, M.: On the size of representative volume element for Darcy law in random media. Proc. R. Soc. A 462, 2949–2963 (2006)
4. Rice, J.R., Cleary, M.P.: Some basic stress-diffusion solutions for fluid-saturated elastic porous media with compressible constituents. Review of Geophysics and Space Physics 14, 227–241 (1976)
5. Tarasov, V.E.: Continuous medium model for fractal media. Phys. Lett. A 336, 167–174 (2005)
6. Tarasov, V.E.: Fractional hydrodynamic equations for fractal media. Ann. Phys. 318(2), 286–307 (2005)
7. Tarasov, V.E.: Wave equation for fractal solid string. Mod. Phys. Lett. B 19, 721–728 (2005)
8. Ostoja-Starzewski, M.: Towards thermomechanics of fractal media. ZAMP 58(6), 1085–1096 (2007)
9. Ostoja-Starzewski, M.: Towards thermoelasticity of fractal media. J. Thermal Stresses 30(9-10), 889–896 (2007)
10. Ostoja-Starzewski, M., Li, J.: Fractal materials, beams, and fracture mechanics. ZAMP 60(6), 1194–1205 (2009)
11. Li, J., Ostoja-Starzewski, M.: Fractal solids, product measures and fractional wave equations. Proc. R. Soc. A 465, 2521–2536 (2009)
12. Li, J., Ostoja-Starzewski, M.: Fractal solids, product measures and continuum mechanics. In: Maugin, G.A. (ed.) Proc. EUROMECH Colloquium, vol. 510 (in press, 2009)
13. Ostoja-Starzewski, M.: Continuum mechanics models of fractal porous media: Integral relations and extremum principles. J. Mech. Mater. Struct. 4(5), 912 (2009)
14. Ostoja-Starzewski, M.: Extremum and variational principles for elastic and inelastic media with fractal geometries. Acta Mech. 205, 161–170 (2009)
15. Ostoja-Starzewski, M.: On turbulence in fractal porous media. ZAMP 59(6), 1111–1117
16. Ignaczak, Ostoja-Starzewski, M.: Thermoelasticity with Finite Wave Speeds. Oxford University Press, Oxford (2009)
17. Rudnicki, J.W.: Effect of pore fluid diffusion on deformation and failure of rock. In: Bažant, Z.P. (ed.) Mechanics of Geomaterials, pp. 315–347 (1985)
18. Mandelbrot, B.B.: The Fractal Geometry of Nature. W.H. Freeman & Co., New York (1982)
19. Feder, J.: Fractals (Physics of Solids and Liquids). Springer, Heidelberg (2007)
20. Carpinteri, A., Chiaia, B., Cornetti, P.: A disordered microstructure material model based on fractal geometry and fractional calculus. ZAMM 84, 128–135 (2004)
21. Carpinteri, A., Pugno, N.: Are scaling laws on strength of solids related to mechanics or to geometry? Nature Materials 4, 421–423 (2005)
22. Jumarie, G.: On the representation of fractional Brownian motion as an integral with respect to $(dt)^a$. Appl. Math. Lett. 18, 739–748 (2005)
23. Jumarie, G.: Table of some basic fractional calculus formulae derived from a modified Riemann–Liouville derivative for non-differentiable functions. Appl. Math. Lett. 22(3), 378–385 (2008)
24. Temam, R., Miranville, A.: Mathematical Modeling in Continuum Mechanics. Cambridge University Press, Cambridge (2005)
25. Oldham, K.B., Spanier, J.: The Fractional Calculus. Academic Press, London (1974)

The Maxwell Problem (Mathematical Aspects)

Evgeniy V. Radkevich

Abstract. We study the large-time behavior of global smooth solutions to the Cauchy problem for hyperbolic regularization of conservation laws. An attracting manifold of special smooth global solutions is determined by the Chapman projection onto the phase space of consolidated variables. For small initial data we construct the Chapman projection and describe its properties in the case of the Cauchy problem for moment approximations of kinetic equations. The existence conditions for the Chapman projection are expressed in terms of the solvability of the Riccati matrix equations with parameter.

1 Introduction

1.1 The State Equation. Closure

This paper is devoted to mathamatical aspects of the Maxwell problem [2] on the derivation of the Navier-Stokes equation from kinetics. Following [1] we study the behavior of solutions to the Cauchy problem for hyperbolic regularizations of conservation laws or (in another terminology) for systems of conservation laws with relaxation. Consider m conservation laws (1) and $N - m$ conservation laws with relaxation (2):

$$\partial_t u_i + \operatorname{div}_x f^i(u,v) = 0, \quad i = 1,\ldots,m, \tag{1}$$

$$\partial_t v_k + \operatorname{div}_x g^k(u,v) + b^k(u)v = 0, \quad k = m+1,\ldots,N., \tag{2}$$

then we have m conservative variables $u(x,t) : \mathbb{R}^d \times \mathbb{R}_+ \to \mathbb{R}^m$ and $N - m$ so-called nonequilibrium variables $v(x,t) : \mathbb{R}^d \times \mathbb{R}_+ \to \mathbb{R}^{N-m}$, where $x \in \mathbb{R}^d$, b is the relaxation $(N - m) \times (N - m)$-matrix,

Evgeniy V. Radkevich
Moscow State University, Department Mech.-Math., Vorobievy Gory,
119899, Moscow, Russia
e-mail: evrad07@gmail.com

$$f^i(u,v) \in \mathbb{R}^d, \; i = 1,\ldots,m; \;\; g^k(u,v) \in \mathbb{R}^d, \; k = 1,\ldots,N-m,$$

are currents. The leading part of system (1) is nonstrictly hyperbolic in the sense of the following definition.

Definition 1.1. *A system is nonstrictly hyperbolic if the characteristic matrix*

$$\tau E + \xi \cdot \begin{pmatrix} f_u(u,v) & f_v(u,v) \\ g_u(u,v) & g_v(u,v) \end{pmatrix} \tag{3}$$

has only real (possibly multiple) roots $\tau = \tau_j(\xi,u,v), \; j = 1,\ldots,N.$

The condition in Definition 1.1 is satisfied if the system (1) is symmetrizable. Examples of such systems are the following: moment approximations of kinetic equations and the Dirac-Schwinger extension of the Maxwell equations [3]. Hyperbolic regularizations of conservation laws (or systems of conservation laws with relaxation) were considered by many authors. First of all, this concerns the study of the relaxation phenomenon, in particular, the stability and singular limit as the relaxation time tends to zero (cf., for example, [4]-[7]). The so-called "intermediate attractor" for (1-2) was studied in connection with the Maxwell problem (cf. [8,9]). To derive equations of hydrodynamics from the kinetic gas theory, it is important to find a simple functional dependence of the transport coefficients on the interaction potential and thereby to simplify the analysis of the equations under consideration. We are interested in the Chapman conjecture [1], [9], on the existence problem of the state equation

$$v = \mathcal{Q}u, \tag{4}$$

(the so-called Chapman state equation or the Chapman projection) expressing the nonequilibrium variables in terms of the conservative variables (the projection into the phase space of conservative variables), where $\mathcal{Q}$ is an operator with respect to space variables x. This equation completes the system of conservation laws

$$\partial_t w + \partial_x f(w, \mathcal{Q}u(w)) = 0. \tag{5}$$

so that the solutions w to the Cauchy problem of the corresponding closure (5) define the set of the special solutions $U_{ChEns} = \{u = w, \; v = \mathcal{Q}w\}$ to the Cauchy problem for the system (1-2) and form an invariant attracting manifold $\mathcal{M}_{ChEns}$, called an intermediate attractor. In other words, for any solution $U = (u,v)$ to the Cauchy problem for the system (1-2) with the initial data $U|_{t=0} = (u^0,v^0)$ it is possible to choose initial data $w_0 = \mathcal{T}(u_0,v_0)$ for the closure (5) in such a way that some norm of the difference $U - U_{ChEns}$ between U and the special solution $U_{ChEns} = (w, \mathcal{Q}w)$ tends to zero as $t \to \infty$. Moreover, if, in the phase space of conservative variables,

$$w \to 0, \; \text{for} \; t \to \infty,$$

then U_H tends to zero faster than U_{ChEns}. We can say that in this case the influence of nonequilibrium variables is inessential (we have the separation of dynamics) [9].

Now we can define the approximation of the state equation and the corresponding closure (so-called Navier-Stokes approximation). Due to physical considerations [9] we assume that derivatives of nonequilibrium variables are small, then we find the following relation

$$v = -b^{-1} \operatorname{div}_x g(u,0) \tag{6}$$

and the corresponding closure

$$\partial_t u + \partial_x f(u, -b^{-1} \operatorname{div}_x g(u,0)) = 0 \tag{7}$$

(so-called Navier-Stokes approximation to (1-2)), where $\det b(u) \neq 0$. For the thirteen-moment Grad system to the Boltzmann kinetic equation the Navier-Stokes approximation (7) is exactly equivalent to the Navier-Stokes equations.

Considering conservation laws with stiff relaxation

$$\partial_t u + \operatorname{div}_x f(u,v) = 0, \quad \partial_t v + \operatorname{div}_x g(u,v) + \frac{1}{\varepsilon} b(u)v = 0, \tag{8}$$

we find that the Navier-Stokes approximation

$$v = -\varepsilon b^{-1} \operatorname{div}_x g(u,0), \quad \partial_t u + \operatorname{div}_x f(u, -\varepsilon b^{-1} \operatorname{div}_x g(u,0)) = 0, \tag{9}$$

is the first approximation to the so-called local equilibrium approach (see [1])

$$\partial_t u + \operatorname{div}_x f(u,0) = 0.$$

2   Linear Analysis. Reduction to a Quadratic Matrix Equation

2.1   *Reduction to a Quadratic Matrix Equation*

We consider the Cauchy problem for the first order linear hyperbolic system with constant coefficients and with relaxation [1]

$$\partial_t u + \mathscr{A}_j \partial_{x_j} u + Bu = 0, \tag{10}$$

where $x \in \mathbb{R}^n$, $u \in \mathbb{R}^N$, $\mathscr{A}_j$ and B are constant matrices. For system (10) the Chapman conjecture of the state equation existence [1, 9] asserts that

$$u = \Pi u_c = (u_c, \Pi_{21} u_c),$$

where $u_c = (u_1, .., u_m, 0, .., 0)^T$ and Π is a zero order pseudodifferential matrix operator. Suppose, following [1], that the matrix of operator Π corresponding to the Chapman-Enskog projection into m equations of the system (10) has the form

$$\Pi = \begin{pmatrix} \Pi_{11} & \Pi_{12} \\ \Pi_{21} & \Pi_{22} \end{pmatrix},$$

where $\Pi_{11} = E_m$ is the unit matrix of order m and $\Pi_{22} = 0_{N-m}$ is the zero square matrix of order $N - m$. We denote by $\Lambda(\xi)$ the resolvent matrix $\sum_{j=1}^{n} \mathscr{A} i \xi_j + B$ and represent it in block form:

$$\Lambda = \begin{pmatrix} \Lambda_{11} & \Lambda_{12} \\ \Lambda_{21} & \Lambda_{22} \end{pmatrix}.$$

Since Π is a projection,

$$\Pi \partial_t u_c + \mathscr{A} \Pi \partial_x u_c + B \Pi u_c = 0. \tag{11}$$

Since $\Pi^2 = \Pi$,

$$\Pi \partial_t u_c + \Pi \mathscr{A} \Pi \partial_x u_c + \Pi B \Pi u_c = 0. \tag{12}$$

Subtracting (12) from (11), we find

$$(E - \Pi)(\mathscr{A} \partial_x + B) \Pi u_c = 0.$$

We denote the Fourier image of Π with respect to x by P. After the Fourier transform with respect to x, the last equality takes the form $(E - P)\Lambda P v_c = 0$, i.e. $\Lambda P v_c \in Ker(E - P)$. We note that for $\forall v \in Ker(E - P)$ admitting the representation $v^T = (v_m^T, v_{N-m}^T)$, with $v_k \in \mathbb{C}^k$ the following equality holds: $v_{N-m} = P_{21} v_m$. Hence we find the system of equations for P_{21} which completely determines the projection Π:

$$P_{21}(\Lambda_{11} + \Lambda_{12} P_{21}) = \Lambda_{21} + \Lambda_{22} P_{21}.$$

After transformations this equation takes the form

$$P_{21} \Lambda_{12}(\xi) P_{21} - \Lambda_{22}(\xi) P_{21} + P_{21} \Lambda_{11}(\xi) - \Lambda_{21}(\xi) = 0, \tag{13}$$

i.e., we obtain a Riccati type matrix equation. This object is nontrivial object. For example, we will consider two special 2×2 cases (13):

$$X^2 = 0, \quad X^2 = \begin{pmatrix} 0 & 1 \\ 0 & 0 \end{pmatrix}$$

In the first case, there are infinitely many of such matrices and they form a two-dimensional cone in $C^4 (\det X = 0, \ \mathbf{tr} X = 0)$. There are no solutions to the second equation, since a matrix has only the zero eigenvalue if the squared matrix possesses this property, i.e. X is nilpotent and the squared nilpotent matrix of second order vanishes.

Lemma 2.1. For any $\gamma \in \mathbb{R}$ the set of solutions to the matrix equation (13) with a matrix Λ coincides with the set of solutions to the same equation (13) with the matrix $\Lambda + \gamma E$.

Proof. Indeed, with $\Lambda + \gamma E$ we associate the matrix equation

$$P_{21} \Lambda_{12} P_{21} - (\Lambda_{22} + \gamma E) P_{21} + P_{21}(\Lambda_{11} + \gamma E) - \Lambda_{21} = 0,$$

where the left-hand side differs from the left-hand side of (13) by $-\gamma E P_{21} + P_{21}\gamma E = 0$. It is obvious that the sets of solutions to these equations coincide.

Thus, to study the matrix equation (13), we can assume without loss of generality that $det(\Lambda) \neq 0$.

2.2 Solutions to the Quadratic Matrix Equation in the Case $|\Lambda| \neq 0$

This section is devoted to the solvability condition for the matrix equation.

Proposition 2.1. *Assume that* $|\Lambda| \neq 0$ *and*

$$P = \begin{pmatrix} P_{11} & P_{12} \\ P_{21} & P_{22} \end{pmatrix}, \tag{14}$$

where P_{11} *is the unit matrix of order m,* P_{22} *is the zero square matrix of order* $N - m$, *and* P_{12} *is the zero matrix. Then the quadratic matrix equation (13) is solvable if and only if there exists a matrix P of the form (14) such that P is a solution to the quadratic matrix equation*

$$(E - P)\Lambda P = 0. \tag{15}$$

Proof. We first assume that the matrix equation (13) is solvable. Taking P of the form (14) and representing the product $(E - P)\Lambda P$ in block form, we see that P is a solution to the matrix equation (15). Conversely, let P of the form (14) be a solution to the matrix equation (15). Representing $M = (E - P)\Lambda P$ via blocks of the same size as the blocks of P, we see that the blocks M_{11}, M_{12}, and M_{22} are zero. The equation for M_{21} coincides with (13) up to a sign, i.e., the matrix equation (13) is solvable.

As a consequence follows

Theorem 2.1. *Let a matrix* Π_{21} *be a solution to*

$$\Pi_{21}\Lambda_{12}\Pi_{21} - \Lambda_{22}\Pi_{21} + \Pi_{21}\Lambda_{11} - \Lambda_{21} = 0 \tag{16}$$

and $X = \Lambda\,\Pi$, *where* $\Pi = \begin{pmatrix} \Pi_{11} & \Pi_{12} \\ \Pi_{21} & \Pi_{22} \end{pmatrix}$ *is a quadratic matrix of order N,* Π_{11} *is the identity matrix of order m, and* Π_{12}, Π_{22} *are zero matrices. Then X is a solution to the quadratic matrix equation*

$$X^2 - \Lambda X = 0. \tag{17}$$

The matrix equation (17) is simpler than the general matrix equation (16) and it is not difficult to describe (17) completely. Solutions of the matrix equation (13) correspond to a part of the set of solutions of equation (17) alone. Thus, we have to define the selection rule.

Theorem 2.2. *Let* $|\Lambda| \neq 0$. *Then the quadratic matrix equation (13) is solvable if and only if there are two solutions* X_1 *and* X_2 *to the quadratic matrix equation*

$$X^2 - \Lambda X = 0, \tag{18}$$

such that

1. $X_1 e_j = 0$ *for all* $j > m$.
2. $e_j^T X_2 = e_j^T \Lambda$ *for all* $j \leq m$.
3. $\Lambda X_2 = X_1 \Lambda$.

Proof. Assume that the matrix equation (13) is solvable. Then the matrix equation (15) is also solvable. We note that a matrix P belongs to the above class if and only if $P e_j = 0 \; \forall j > m$ and $e_j^T P = e_j^T \; \forall j \leq m$. Multiplying the matrix equation (15) by Λ from the left and changing variables according to $X_1 = \Lambda P$, we see that the matrix X_1 is a solution to (18) and satisfies condition 1). Similarly, multiplying (15) by Λ from the right and changing variables according to $X_2 = P\Lambda$, we find that the matrix X_2 is a solution to (18) and satisfies condition 2). Since $X_1 = \Lambda P$ and $X_2 = P\Lambda$ condition 3) is also valid.

Assume that there exist two solutions X_1 and X_2 to the matrix equation (18) satisfying conditions 1)-3). We set $P = \Lambda^{-1} X_1 = X_2 \Lambda^{-1}$. Due to satisfaction of conditions 1) and 2) the matrix P has the required form. Substituting $X_1 = \Lambda P$ into (18) and multiplying by Λ^{-1} from the left, we find that P is a solution to (15).

Theorem 2.3. *Let* $|\Lambda| \neq 0$. *Then the quadratic matrix equation (13) is solvable if and only if there is a solution* X_1 *to the quadratic matrix equation (18) such that*
1. $X_1 e_j = 0$ *for all* $j > m$.
2. $e_j^T \Lambda^{-1} X_1 = e_j^T$ *for all* $j \leq m$.

Proof. We set $X_2 = \Lambda^{-1} X_1 \Lambda$. It is obvious that X_2 is a solution to the matrix equation (18). Furthermore, keeping condition 2) of Theorem 2.3, X_2 satisfies condition 2) of Theorem 2.2. Since $X_2 = \Lambda^{-1} X_1 \Lambda$ we have $\Lambda X_2 = X_1 \Lambda$, i.e., condition 3) of Theorem 2.2 is also satisfied.

For details of the proofs of the following results see [14, 17].

Lemma 2.2. *Suppose that* $\det(\Lambda) \neq 0$, X *is a solution to the matrix equation (18), and vectors* $h_1, \ldots, h_N$ *form the Jordan basis for* X. *Then there exists* $K \geq 0$ *such that* $h_1, \ldots, h_K$ *belong to the Jordan basis for* Λ *(moreover, if* $X h_j = \lambda h_j + h_{j-1}$, *then* $\Lambda h_j = \lambda h_j + h_{j-1}$*) and* $h_{K+1}, \ldots, h_N$ *are the eigenvectors corresponding to the eigenvalue* 0.

Lemma 2.3. *Let* $\det(\Lambda) \neq 0$. *For* $K \geq 0$ *we denote by* X *a matrix with the Jordan basis* $h_1, \ldots, h_N$, *where the vectors* $h_1, \ldots, h_K$ *form the Jordan basis for* Λ *(listed in such a way that if* $X h_j = \lambda h_j + h_{j-1}$, *then* $\Lambda h_j = \lambda h_j + h_{j-1}$*) and* $h_{K+1}, \ldots, h_N$

are the eigenvectors corresponding to the eigenvalue 0. Then X is a solution to the matrix equation (18).

Keep in mind the geometrical formulation of the necessary and sufficient conditions of the solvability of the quadratic matrix equation (16).

Theorem 2.4. *Let $|\Lambda| \neq 0$, and let vectors $v_1, \ldots, v_m$ satisfy the following conditions:*

1. $V = \mathbf{Lin}\{v_j\}_1^m$ is an eigenspace of the matrix Λ, i.e. $\Lambda V = V$.
2. $v_1, .., v_m, e_{m+1}, .., e_N$ form a basis.

Then the quadratic matrix equation (13) is solvable. The inverse assertion is also true.

2.3 Explicit Formula

Now, we discuss a possible explicit formula for solutions to the Riccati matrix equation.

Theorem 2.5. *Suppose that vectors $v_1, \ldots, v_m$ form a basis for a linear Λ-invariant subspace V and $v_1, \ldots, v_m, e_{m+1}, \ldots, e_n$ is a basis for R^n. We regard these vectors as columns of a matrix $\begin{pmatrix} C_{11} \\ C_{21} \end{pmatrix}$. Then the solution to the matrix equation (13), associated with these vectors, listed in the above order, is represented in the form*

$$P_{21} = C_{21} C_{11}^{-1}. \tag{19}$$

Proof. Since we can assume that $det(\Lambda) \neq 0$, for the solution to the matrix equation (13) we have

$$\begin{pmatrix} E & 0 \\ P_{21} & 0 \end{pmatrix} = \Lambda^{-1} \begin{pmatrix} C_{11} & 0 \\ C_{21} & E \end{pmatrix} \begin{pmatrix} J_1 & 0 \\ 0 & 0 \end{pmatrix} \begin{pmatrix} C_{11}^{-1} & 0 \\ -C_{21}C_{11}^{-1} & E \end{pmatrix},$$

where J_1 is a block from the Jordan form of the matrix Λ corresponding to the space V. Hence

$$\begin{pmatrix} E & 0 \\ P_{21} & 0 \end{pmatrix} = \Lambda^{-1} \begin{pmatrix} C_{11}J_1 C_{11}^{-1} & 0 \\ C_{21}J_1 C_{11}^{-1} & 0 \end{pmatrix}.$$

Multiplying both sides of the last equality from the left by Λ, we find

$$\begin{pmatrix} \Lambda_{11} + \Lambda_{12}P_{21} & 0 \\ \Lambda_{21} + \Lambda_{22}P_{21} & 0 \end{pmatrix} = \begin{pmatrix} C_{11}J_1 C_{11}^{-1} & 0 \\ C_{21}J_1 C_{11}^{-1} & 0 \end{pmatrix},$$

which implies $P_{21}C_{11}J_1 C_{11}^{-1} - C_{21}J_1 C_{11}^{-1} = 0$ in view of (13). Since Λ is invertible, the matrix J_1 is also invertible. Hence we can multiply the last equality by $C_{11}J_1^{-1}$ from the right. Then $P_{21}C_{11} = C_{21}$, which implies (19).

2.4   The Number of Solutions

Corollary 2.1. *With every m-dimensional eigenspace V of the matrix Λ at most one solution to the matrix equation (13) is associated.*

Proof. Indeed, neither V does provide any solution to (13) (if $\mathbf{Lin}\{v_1,\dots,v_m, e_{m+1},\dots,e_n\} = R^n$, where $V = \mathbf{Lin}\{v_1,\dots,v_m\}$) nor V can be associated with a solution to (13) by formula (19). In the second case, we show that the solution is independent of the choice of the basis for the space V. Let $w_1,\dots,w_m$ be another basis for V. We write the vectors $v_1,\dots,v_m$ as columns of a matrix W^0 and the vectors $w_1,\dots,w_m$ as columns of a matrix W^1. Since these bases generate the same linear space V, there exists a nonsingular matrix K such that $W^1 = W^0 K$ or, in the block form,

$$\begin{pmatrix} W_1^1 \\ W_2^1 \end{pmatrix} = \begin{pmatrix} W_1^0 \\ W_2^0 \end{pmatrix} K,$$

which implies $W_j^1 = W_j^0 K$, $j = 1,2$. Hence the solution of the form (19) corresponding to the basis for W^1 can be written as

$$P_{21,W} = W_2^1 (W_1^1)^{-1} = W_2^0 K K^{-1} (W_1^0)^{-1} = P_{21,V}.$$

Thus, the solutions defined by the bases $v_1,\dots,v_m$ and $w_1,\dots,w_m$ coincide.

The following results are only an information (for details see [19, 22])

Theorem 2.6. *Let the matrix equation (13) have infinite number of solutions. Then there exists $\lambda \in \mathbb{C}$ such that $dim(Ker(\Lambda - \lambda E)) \geq 2$.*

The set of solutions to the matrix equation (13) is infinite if and only if there exist eigenspaces V and W of the matrix Λ satisfying the following conditions:

1. *V defines the solution to equation (13).*
2. *W is an eigenspace of the matrix Λ corresponding to an eigenvalue λ.*
3. *W contains two incollinear eigenvectors.*
4. *$V \cap W \neq \{0\}$.*
5. *$W \setminus V \neq \emptyset$.*

2.5   The Lyapunov Equation. Separation of Dynamics

The Lyapunov matrix equation

$$-M_{11}Q_{12} + Q_{12}M_{22} - M_{12} = 0 \tag{20}$$

is a special case of the quadratic matrix equation (13) with vanishing quadratic term. The following assertion is proved in [20].

Theorem 2.7. *Suppose that $det(M_{11}) \neq 0$ and $det(M_{22}) \neq 0$. Assume that the matrix*

$$M = \begin{pmatrix} M_{11} & M_{12} \\ 0 & M_{22} \end{pmatrix}$$

has no eigenvalues λ such that, in the block form, the corresponding eigenvector has the form $v_0 = \begin{pmatrix} v_{0,1} \\ 0 \end{pmatrix}$ and the corresponding associated eigenvector has the form $v_1 = \begin{pmatrix} v_{1,1} \\ v_{1,2} \end{pmatrix}$, where $v_{1,2} \neq 0$. Then there exists a solution Q_{12} to the Lyapunov matrix equation (20) with the matrices M_{11}, M_{12}, and M_{22}.

We construct the canonical form of (10).

Lemma 2.4. *Suppose that a matrix S is invertible and can be written in the block form with blocks S_{ij}, $i,j = 1,2$, where S_{11} and S_{22} are square matrices. Assume also that $FS = SF = E$, where the matrix F can be represented by blocks of the same size. In this case, if $S_{11} = E$, then the matrix F_{22} is invertible.*

Proof. Assume the contrary. Since $FS = E$, we have

$$F_{21} + F_{22}S_{21} = 0. \tag{21}$$

Since F_{22} is noninvertible, there is a row $h \neq 0$ such that $hF_{22} = 0$. Using (21), we find $hF_{21} = 0$. But, in this case, the last rows of the matrix F are linearly dependent: there is a row v such that $v \neq 0$ and $vF = 0$. Thus, the matrix F is noninvertible. On the other hand, the matrix F is the inverse of S. We arrive at a contradiction.

Theorem 2.8. *Suppose that Λ is divided into blocks $\Lambda_{ij}, i,j,= 1,2$. Then the quadratic matrix equation (13) is solvable if and only if there exists a matrix S satisfying the following conditions:*

1. S is invertible,
2. $S_{11} = E$.
3. $(S^{-1}\Lambda S)_{21} = 0$.

Proof. Assume that there exists a matrix S satisfying conditions 1)-3). For $F = S^{-1}$ we have

$$F_{21} + F_{22}S_{21} = 0, \quad F_{21}(\Lambda_{11} + \Lambda_{12}S_{21}) + F_{22}(\Lambda_{21} + \Lambda_{22}S_{21}) = 0.$$

Expressing F_{21} by use of the first equation and substituting the result into the second equation, we find

$$F_{22}(-S_{21}(\Lambda_{11} + \Lambda_{12}S_{21})) + F_{22}(\Lambda_{21} + \Lambda_{22}S_{21}) = 0.$$

We note that the matrix S satisfies the assumptions of Lemma 2.4. Hence

$$(-S_{21}(\Lambda_{11} + \Lambda_{12}S_{21})) + (\Lambda_{21} + \Lambda_{22}S_{21}) = 0,$$

i.e., the matrix S_{21} satisfies the quadratic matrix equation (13). Assume that the quadratic matrix equation (13) is solvable. We set $S_{11} = E$, $S_{12} = 0$, $S_{21} = P_{21}$, $S_{22} = E$. It is easy to verify that the inverse matrix exists: $S^{-1} = 2E - S$. We see that the matrix S satisfies conditions 1) and 2). Computing $(S^{-1}\Lambda S)_{21}$, we find

$$(S^{-1}\Lambda S)_{21} = F_{21}(\Lambda_{11}+\Lambda_{12}S_{21})+F_{22}(\Lambda_{21}+\Lambda_{22}S_{21}) =$$

$$= (-P_{21})(\Lambda_{11}+\Lambda_{12}P_{21})+(\Lambda_{21}+\Lambda_{22}P_{21}) = 0,$$

since P_{21} is a solution to the quadratic matrix equation (13). Thus, the matrix S also satisfies condition 3). The theorem is proved. Thus, the existence of a Chapman-Enskog projection is equivalent to the possibility to represent the original system in the block form such that $(S^{-1}\Lambda S)_{21} = 0$, which allows us to separate dynamics. The following theorem (cf. the proof in [16]) provides us with conditions under which a matrix can be reduced to the block-diagonal form.

Theorem 2.9. *Assume that a matrix Λ is invertible and $v_1,\ldots,v_m$ is a basis for its eigenspace V such that* $\mathbf{Lin}\{v_1,\ldots,v_m,e_{m+1},\ldots,e_N\} = R^N$. *We also assume that V cannot be extended to an $m+1$-dimensional eigenspace of the matrix Λ by extending the basis $v_1,\ldots,v_m$ with an associated eigenvector of Λ. Then there exist matrices P_{21} and Q_{12} such that*

$$\begin{pmatrix} E & -Q_{12} \\ 0 & E \end{pmatrix}\begin{pmatrix} E & 0 \\ -P_{21} & E \end{pmatrix}\Lambda\begin{pmatrix} E & 0 \\ P_{21} & E \end{pmatrix}\begin{pmatrix} E & Q_{12} \\ 0 & E \end{pmatrix} = \begin{pmatrix} M_{11} & 0 \\ 0 & M_{22} \end{pmatrix}.$$

Now, we consider the representation of the solution as the sum of three terms and introduce the notion of the L_2-well-posedness in the sense of Chapman-Enskog.

Suppose that a matrix Λ satisfies the assumptions of Theorem 2.9. We change the variables according to $U = S^{-1}u$. Then a solution to the Cauchy problem (10) with the initial data

$$U|_{t=0} = \begin{pmatrix} u_0 \\ v_0 \end{pmatrix}$$

can be written in terms of the Fourier images as follows:

$$U = e^{-Mt}\begin{pmatrix} u_0 \\ v_0 \end{pmatrix},$$

where $M = S^{-1}\Lambda S$. By Theorem 2.9, the matrix M takes the form

$$M = \begin{pmatrix} E & Q_{12} \\ 0 & E \end{pmatrix}\begin{pmatrix} M_{11} & 0 \\ 0 & M_{22} \end{pmatrix}\begin{pmatrix} E & -Q_{12} \\ 0 & E \end{pmatrix},$$

which implies

$$U = \begin{pmatrix} E & Q_{12} \\ 0 & E \end{pmatrix}\exp\left(-\begin{pmatrix} M_{11} & 0 \\ 0 & M_{22} \end{pmatrix}t\right)\begin{pmatrix} E & -Q_{12} \\ 0 & E \end{pmatrix}\begin{pmatrix} u_0 \\ v_0 \end{pmatrix} = U_{Ch}+U_{Cor}+U_H,$$

where each of the terms is a solution to the system (10) with some initial data:

$$U_{Ch} = e^{-Mt}\begin{pmatrix} u_0 \\ 0 \end{pmatrix}, \quad U_{Cor} = e^{-Mt}\begin{pmatrix} -Q_{12}v_0 \\ 0 \end{pmatrix}, \quad U_H = \begin{pmatrix} Q_{12}e^{-M_{22}t}v_0 \\ e^{-M_{22}t}v_0 \end{pmatrix}.$$

The first term U_{Ch} corresponds to the projection onto the phase space of consolidated variables, the second term U_{Cor} is a corrector describing the influence of the initial data relative to nonequilibrium variables, and the third term U_H is a remainder.

Definition 2.1. *We say that a projection P satisfies the Chapman L_2-well-posedness condition for a class of initial data $\mathscr{H} = \{(\mathscr{U}_0, \mathscr{V}_0)\}$ if for any initial data $(\mathscr{U}_0, \mathscr{V}_0) \in \mathscr{H}$ there is a constant $T_0 > 0$ such that for all $t > T_0$*

$$\frac{||U_H||(t)}{||U_{Ch}||(t)} \le Ke^{-\delta t}, \ t > T_0, \tag{22}$$

where K and $\delta > 0$ are constants.

2.6 Crack Condition and the Existence of an Attracting Manifold

We find conditions that guarantee the validity of the estimate

$$||U_H|| = o(||U_{Ch}||), \ t \to \infty,$$

where $||f||$ denotes the norm of f in the space L_2. For this purpose, we prove several technical auxiliary assertions (see [21]):

Lemma 2.5. *Suppose that a matrix Λ polynomially depends on ξ and there exists $k_0 > 0$ such that for all $\xi : |\xi| > k_0$, all the eigenvalues $\lambda(\xi)$ of Λ are algebraically simple and $|\lambda(\xi)| \le C_1(1 + |\xi|)^{d_1}$, where C_1 and d_1 are constants. Let v be an eigenvector of Λ. Then for $|\xi| > k_0$:*

$$\frac{\max\{|e_i^T v|\}}{\min\{|e_i^T v| \ne 0\}} \le C_2(1 + |\xi|)^{d_2}, \tag{23}$$

where C_2 and d_2 are constants.

Lemma 2.6. *Suppose that a matrix Λ is defined for all $\xi \in \mathbb{R}$ and satisfies the assumptions of Theorem 2.9 for all $\xi \in \Xi$, where $\Xi = \mathbb{R} \backslash \Xi_-$ and the set Ξ_- is finite. Then P_{21} and Q_{12} are defined on Ξ. Assume that the matrices $P_{21}(\xi)$ and $Q_{12}(\xi)$ can be defined by continuity on the set Ξ_-. We also assume that the matrix Λ polynomially depends on ξ and there is $k_0 > 0$ such that for all $\xi : |\xi| > k_0$, all the eigenvalues of the matrix Λ are algebraically simple and satisfy the following estimate: $|\lambda(\xi)| \le C_1(1 + |\xi|)^{d_1}$, where C_1 and d_1 are constants. Then there is $d \in \mathbb{N}$ such that for all $\xi \in \mathbb{R}$*

$$|P_{21}| \le K_1(1 + |\xi|)^d, \ |Q_{12}| \le K_2(1 + |\xi|)^{3d},$$

where K_1 and K_2 are constants and $|A|$ is the matrix norm of A in L_∞.

Notation 2.1. *The minimal number $d \in \mathbb{N}$ satisfying the assumptions of Lemma 2.6 is denoted by d_Λ.*

We also need a two-sided estimate for $|e^{-Mt}v|$, where $|.|$ denotes the $L_\infty(\mathbb{R})$. For the sake of brevity, we introduce the following notation.

Notation 2.2. *Suppose that a square matrix M continuously depends on the parameter ξ. Let λ_j, $j = 1,\ldots,s$, be eigenvalues of M. We denote by d_j the maximal size of the Jordan cell corresponding to the eigenvalue λ_j . Let the eigenvalues λ_j be listed in ascending order of the real part. Let $l(M)$ and $L(M)$ denote the minimal and maximal eigenvalues respectively, i.e.*

$$l(M) = \operatorname{Re}\lambda_1 \le \operatorname{Re}\lambda_2 \le \ldots \le \operatorname{Re}\lambda_s = L(M).$$

We set $d(M) = d_1$.

We will use one more technical lemma (see [19]:

Lemma 2.7. *Let a square matrix M continuously depend on the parameter ξ. Then for any $\varepsilon > 0$ there is $T_0 > 0$ such that for all $t > T_0$ the following estimate holds:*

$$e^{-L(M)t}|v| \le |e^{-Mt}v| \le \frac{1+\varepsilon}{(d(M)-1)!}|M|^{d(M)-1}e^{-l(M)t}t^{d(M)-1}|v|, \qquad (24)$$

where $|A|$ denotes the matrix norm of A in $L_\infty(\mathbb{R})$.

Notation 2.3. *Let $\Gamma(\xi)$ be a finite set of continuous functions $\gamma_1(\xi),\ldots,\gamma_s(\xi)$ of the parameter ξ. Introduce the notation $l(\xi,\Gamma(\xi)) = \inf_s\{\operatorname{Re}\gamma_s(\xi) \mid \gamma_s(\xi) \in \Gamma(\xi)\}$, $l_0(\Gamma) = \inf_\xi l(\xi,\Gamma(\xi))$, $L(\xi,\Gamma(\xi)) = \sup_s\{\operatorname{Re}\gamma_s(\xi) \mid \gamma_s(\xi) \in \Gamma(\xi)\}$, $L_0(\Gamma) = \sup_\xi L(\xi,\Gamma(\xi)))$.*

Condition 2.1. *The pair of sets $\Gamma_1(\xi)$ and $\Gamma_2(\xi)$ satisfies the strong crack condition if*

$$\exists \gamma > 0 : \ l_0(\Gamma_2) - L_0(\Gamma_1) \ge \gamma. \qquad (25)$$

Now, we formulate the conditions for the existence of an attracting manifold.

Theorem 2.10. *Let the matrix Λ in the problem (10) satisfy the assumptions of Lemma 2.6. Suppose that Γ_1 is the set of all those eigenvalues of Λ that determine the separation of dynamics for the eigenspace V, and Γ_2 is the set of all the remaining eigenvalues of Λ. Assume that Γ_1 and Γ_2 satisfy the strong crack condition. Let the Fourier images of initial data $(\mathcal{U}_0,\mathcal{V}_0)$ belong to the set*

$$\mathscr{H} = \{(\mathcal{U}_0,\mathcal{V}_0) : \ ||\mathcal{U}_0|| \ne 0, \ (1+|\xi|)^{3d_\Lambda}|M_{22}|^{d(M_{22})-1}\mathcal{V}_0 \in L_2(\mathbb{R})\}.$$

Then the projection P corresponding to the separation of dynamics satisfies the Chapman-Enskog L_2-well-posedness condition (Definition 2.1) for the class of initial data $\mathscr{H}$ with constants K and δ such that

(i) K depends on $||\mathcal{U}_0||$, $||\mathcal{V}_0||$,
(ii) δ depends on δ and some properties of the matrix M.

Proof. Indeed,

$$\|U_H(t)\| = \left(\int_{\mathbb{R}} \left| \left(\begin{array}{c} Q_{12} e^{-M_{22}t} \mathcal{V}_0 \\ e^{-M_{22}t} \mathcal{V}_0 \end{array} \right) \right|^2 d\xi \right)^{\frac{1}{2}} \le \left(\int_{\mathbb{R}} |1 + |Q_{12}|^2| |e^{-M_{22}t} \mathcal{V}_0|^2 d\xi \right)^{\frac{1}{2}}.$$

Using Lemmas 2.6 and 2.7, we find

$$\|U_H(t)\|^2 \le \int_{\mathbb{R}} h_1 |M_{22}|^{2d(M_{22})-2} e^{-2l(M_{22})t} t^{2d(M_{22})-2} |\mathcal{V}_0|^2 d\xi,$$

$$h_1 = (1 + K_2^2 (1 + |\xi|)^{10d_\Lambda}) \left(\frac{1 + \varepsilon}{(d(M_{22}) - 1)!} \right)^2.$$

From (25) it follows that

$$e^{-l(M_{22})t} \le e^{-l_0(\Gamma_2)t} \le e^{-\eta} e^{-L_0(\Gamma_1)t}; \quad e^{-L(M_{11})t} \ge e^{-L_0(\Gamma_1)t}.$$

By Lemma 2.3,

$$\|U_{Ch}(t)\| \ge \left(\int_{\mathbb{R}} e^{-2L(M_{11})t} |\mathcal{U}_0|^2 d\xi \right)^{\frac{1}{2}}.$$

Combining the last four inequalities, we find

$$\left(\frac{\|U_H\|(t)}{\|U_{Ch}\|(t)} \right)^2 \le \frac{e^{-2\eta} t^{2d(M_{22})-2} \int_{\mathbb{R}} e^{-2L_0(\Gamma_1)t} h_1 |\mathcal{V}_0|^2 d\xi}{\int_{\mathbb{R}} e^{-2L_0(\Gamma_1)t} |\mathcal{U}_0|^2 d\xi},$$

which implies the required estimate (22) because $L_0(\Gamma_1)$ is independent of ξ.

3 Nonlinear Analysis. Chapman Projection

3.1 Statement of the Problem and Auxiliaries

We consider the nonlinear system of equations

$$\partial_t u + \sum_{j=1}^{n} \mathscr{A}_j \partial_{x_j} u + Bu = f(u), \tag{26}$$

with the initial condition $u|_{t=0} = \phi$, where u is an N-dimensional vector, A_j and B are constant matrices, $n \le 3$, and $f(u)$ is a vector-valued polynomial, i.e. $f(u) = \sum_{j=1}^{N} \left(\sum_{\sigma \in \Theta_j} K(j,\sigma) u^\sigma \right) e_j$, where $\sigma \in (\mathbb{N} \cup \{0\})^N$, $u^\sigma = \prod_{j=1}^{N} u_j^{\sigma_j}$. We set

$$\alpha = \min\{|\sigma| : \sigma \in \cup_{j=1}^{N} \Theta_j\}, \quad \alpha + \beta = \max\{|\sigma| : \sigma \in \cup_{j=1}^{N} \Theta_j\}.$$

Assume that $f(u)$ contains no terms of zero or first order, i.e. $\alpha \ge 2$. We denote by $\|\cdot\|$ the norm in L_2 with respect to the variable x. Let $|u| = \sqrt{u^T u}$ and let $|u|_0$ denote the norm of u in C. Following [10], we denote by ∂ the vector consisting of all first order derivatives and by ∂_x the vector consisting of first order derivatives with respect to the spatial variables, i.e. $\partial = (\partial_x, \partial_t)$. For the sake of brevity, we write ∂_j instead of ∂_{x_j} .

We begin with the following auxiliary assertion generalizing Lemma 2.7.

Lemma 3.1. *Let M be a square matrix. Then there are constants $C_M \in \mathbb{R}$ and $d_M \in \mathbb{Z}$ such that for any vector v and a number $t \geq 0$*

$$|e^{-Mt}v| \leq C_M(1+t^{d_M})e^{-l(M)t}|v|. \tag{27}$$

The following assertion concerns estimates for the norms of $f(u)$:

Lemma 3.2. *For a vector-valued function $u(x,t) \in C([0,T),H^2) \cap C^1([0,T),H^1)$ with $T > 0$ and $s \in \{1,2\}$, $j \in \{1,2,3\}$ the following estimates hold:*

$$||f(u)|| \leq C_{0,0}|u|_0^{\alpha-1}(1+|u|_0^{\beta})||u||, \tag{28}$$

$$||\partial_j^s f(u)|| \leq C_{s,0}|u|_0^{\alpha-1}(1+|u|_0^{\beta})||\partial_j^s u||. \tag{29}$$

The following assertion concerning the norm of a vector-valued polynomial is a consequence of the above lemma.

Lemma 3.3. *Consider a vector-valued function $u(x,t) \in C([0,T),H^2) \cap C^1([0,T),H^1)$ with $T > 0$, $x \in \mathbb{R}^n$, $n \leq 3$. Let $g(u)$ be a vector-valued polynomial with $\alpha \geq 1$. Then there are $\kappa \in (0,1)$ and $C_G > 0$ such that for all $u(x,t)$ such that*

$$||g(u)||_{H^2} \leq 2C_G||u||_{H^2}^{\alpha}. \tag{30}$$

Proof. Indeed, from the inequalities (28) and (29) it follows that

$$||g(u)||_{H^2} \leq \text{const}\,|u|_0^{\alpha-1}(1+|u|_0^{\beta})||u||_{H^2}.$$

By the embedding theorem,

$$||g(u)||_{H^2} \leq C_G||u||_{H^2}^{\alpha}(1+||u||_{H^2}^{\beta}).$$

Hence the required inequality (30) holds for sufficiently small κ.

Whence we obtain

Lemma 3.4. *Let $u(x,t)$ and $v(x,t)$ be vector-valued functions such that $u(x,t),v(x,t) \in C([0,T),H^2) \cap C^1([0,T),H^1)$ for some $T > 0$. Assume that $x \in \mathbb{R}^n$ and $n \leq 3$, $f(u)$ is a vector-valued polynomial with $\alpha \geq 2$. Then there are $\kappa \in (0,1)$ and $C_* > 0$ such that for all $u(x,t),v(x,t)$ the inequalities $||u||_{H^2} < \kappa$, $||v||_{H^2} < \kappa$ imply the inequality*

$$||f(u) - f(v)||_{H^2} \leq C_*(||u||_{H^2}^{\alpha-1} + ||v||_{H^2}^{\alpha-1})||u-v||_{H^2}. \tag{31}$$

Lemma 3.5. *Suppose that $t > 0$ and $P(\tau)$ is a continuous function such that the inequality $P(\tau) \geq 0$ for all $\tau \in [0,t]$. Let $d > 0$. Then there is a constant $C_P > 0$ such that*

$$\int_0^t (1+(t-\tau)^d)^2 P(\tau) d\tau \le C_P (1+t^d)^2 \int_0^t (1+\tau^d)^2 P(\tau) d\tau. \qquad (32)$$

Proof. We have

$$\int_0^t (1+(t-\tau)^d)^2 P(\tau) d\tau \le C_1 \int_0^t (1+\tau^{2d}+t^{2d}) P(\tau) d\tau \le$$

$$\le C_1 \left(t^{2d} \int_0^t P(\tau) d\tau + \int_0^t (1+\tau^{2d}) P(\tau) d\tau \right) \le C_1 (1+t^{2d}) \int_0^t (1+\tau^{2d}) P(\tau) d\tau \le$$

$$\le C_1 (1+t^d)^2 \int_0^t (1+\tau^d)^2 P(\tau) d\tau.$$

3.2 Method of Successive Approximations

We look for a solution to the system (26) with the initial data $u|_{t=0} = \phi(x)$ for small $\phi(x)$ by the method of successive approximations. We set $u_0 = 0$,

$$\partial_t u_k + \sum_{j=1}^n \mathscr{A}_j \partial_j u_k + B u_k = f(u_{k-1}), \quad u_k|_{t=0} = \phi(x). \qquad (33)$$

Introduce the notation $\Lambda = \sum_{j=1}^n \mathscr{A}_j i \xi_j + B$, $l_1 = \inf_\xi \min_{\lambda \in \sigma(\Lambda)} \operatorname{Re}\lambda$. We denote by $\mathscr{F}(\cdot)$ the Fourier transform with respect to the spatial variables. We estimate using the statements from above the solution u_k to the problem (33).

Lemma 3.6. *Let $l_1 > 0$. Then there exist constants $\kappa \in (0,1)$ and $C_1^* > 0$, $C_2^* > 0$ such that the solution u_k to the problem (33) with the initial data ϕ such that $\|\phi\|_{H^2} < \kappa$ satisfies the following inequality for any $t \ge 0$:*

$$\|u_k\|_{H^2} \le C_1^* (1+t^{d_\Lambda}) e^{-l_1 t} (\|\phi\|_{H^2} + C_2^* \sqrt{t} \|\phi\|_{H^2}^\alpha). \qquad (34)$$

Proof. We first prove that for sufficiently small initial data

$$\|u_k\|_{H^2} \le C_\Lambda (1+t^{d_\Lambda}) e^{-l_1 t} (\|\phi\|_{H^2} + C_k \sqrt{t} \|\phi\|_{H^2}^\alpha), \qquad (35)$$

where the constants C_k depend on k. We write an explicit expression for C_k. For this purpose, we use the method of mathematical induction. Let $k = 1$. Then the problem (33) takes the form

$$\partial_t u_1 + \sum_{j=1}^n \mathscr{A}_j \partial_j u_1 + B u_1 = 0, \quad u_1|_{t=0} = \phi.$$

The solution to this problem is written in terms of the Fourier images as follows: $\mathscr{F}(u_1) = e^{-\Lambda t} \mathscr{F}(\phi)$. By Lemma 3.6

$$||u_1||^2 = ||\mathscr{F}(u_1)||^2 = \int_{\mathbb{R}^n} |e^{-\Lambda t}\mathscr{F}(\phi)|^2 d\xi \leq C_\Lambda^2 \int_{\mathbb{R}^n} (1+t^{d_\Lambda})^2 e^{-2l_1 t}|\mathscr{F}(\phi)|^2 d\xi =$$

$$= C_\Lambda^2 (1+t^{d_\Lambda})^2 e^{-2l_1 t}||\mathscr{F}(\phi)||^2 = C_\Lambda^2 (1+t^{d_\Lambda})^2 e^{-2l_1 t}||\phi||^2.$$

A similar inequality holds for the derivatives of u_1. Thus,

$$||u_1||_{H^2} \leq C_\Lambda (1+t^{d_\Lambda})e^{-l_1 t}||\phi||_{H^2},$$

and the inequality (35) is true with $C_1 = 0$. Now, we write an explicit formula for the solution to the problem (33) in terms of the Fourier images:

$$\mathscr{F}(u_k) = e^{-\Lambda t}\mathscr{F}(\phi) + \int_0^t e^{\Lambda(\tau-t)}\mathscr{F}(f(u_{k-1}(\tau)))d\tau. \tag{36}$$

We set $I_{\sigma,k} = (i\xi)^\sigma \int_0^t e^{\Lambda(\tau-t)}\mathscr{F}(f(u_{k-1}(\tau)))d\tau$ and find

$$||I_{\sigma,k}|| \leq C_\Lambda C_k \sqrt{t}(1+t^{d_\Lambda})e^{-l_1 t}||\phi||_{H^2}^\alpha,$$

where the constant C_k is independent of σ. The proof of this assertion is similar to that of the inequality (35) for all k. Indeed, we have the auxiliary estimates

$$||I_{\sigma,k}||^2 \leq t \int_0^t ||(i\xi)^\sigma e^{\Lambda(\tau-t)}\mathscr{F}(f(u_{k-1}(\tau)))||^2 d\tau \leq$$

$$\leq t C_\Lambda^2 \int_0^t (1+(t-\tau)^{d_\Lambda})^2 e^{2l_1(\tau-t)} \int_{\mathbb{R}^n} |(i\xi)^\sigma \mathscr{F}(f(u_{k-1}(\tau)))|^2 d\xi d\tau \leq$$

$$\leq t C_\Lambda^2 \int_0^t (1+(t-\tau)^{d_\Lambda})^2 e^{2l_1(\tau-t)}||f(u_{k-1}(\tau))||_{H^2}^2 d\tau.$$

Using Lemmas 3.3 and 3.5, we find

$$||I_{\sigma,k}||^2 \leq 4C_\Lambda^2 C_F^2 C_P t(1+t^{d_\Lambda})^2 e^{-2l_1 t} \int_0^t (1+\tau^{d_\Lambda})^2 e^{2l_1 \tau}||u_{k-1}(\tau)||_{H^2}^{2\alpha}d\tau. \tag{37}$$

Let $k = 2$. Using the estimate (35) for $k = 1$, we find

$$||I_{\sigma,2}||^2 \leq 4C_\Lambda^{2+2\alpha} C_F^2 C_P t(1+t^{d_\Lambda})^2 e^{-2l_1 t} \int_0^t e^{2l_1(1-\alpha)\tau}(1+\tau^{d_\Lambda})^{2+2\alpha}||\phi||_{H^2}^{2\alpha}d\tau \leq$$

$$\leq 4C_\Lambda^{2+2\alpha} C_F^2 C_P t(1+t^{d_\Lambda})^2 e^{-2l_1 t} \int_0^{+\infty} e^{2l_1(1-\alpha)\tau}(1+\tau^{d_\Lambda})^{2+2\alpha}||\phi||_{H^2}^{2\alpha}d\tau.$$

Note that for $\alpha \geq 2$ the integral is convergent. Setting

$$C_2^2 = 4C_\Lambda^{2\alpha} C_F^2 C_P \int_0^{+\infty} e^{2l_1(1-\alpha)\tau}(1+\tau^{d_\Lambda})^{2+2\alpha}d\tau,$$

we obtain an inequality of the required form for $||I_{\sigma,2}||$.

Assume that the inequality (35) is valid for all $k \leq r$, where $r \geq 2$. Then for $k = r+1$, by the inequality (37)

$$||I_{\sigma,r+1}||^2 \leq C_\Lambda^2 t(1+t^{d_\Lambda})^2 e^{-2l_1 t} ||\phi||_{H^2}^{2\alpha} \Big(C'C_2^2 +$$

$$+4C_\Lambda^{2\alpha} C_F^2 C_P C' C_r^{2\alpha} \int_0^{+\infty} e^{2l_1(1-\alpha)\tau}(1+\tau^{d_\Lambda})^{2+2\alpha}\tau^\alpha ||\phi||_{H^2}^{2\alpha(\alpha-1)} d\tau\Big).$$

Setting

$$J = \int_0^{+\infty} e^{2l_1(1-\alpha)\tau}(1+\tau^{d_\Lambda})^{2+2\alpha}\tau^\alpha d\tau,$$

we obtain the required estimate (35) with

$$C_{r+1}^2 = C_2^2 C' + 4C_\Lambda^{2\alpha} C_F^2 C_P C' J ||\phi||_{H^2}^{2\alpha(\alpha-1)} C_r^{2\alpha}.$$

We note that $C' > 1$. We choose $\kappa > 0$ such that for $||\phi|| < \kappa$

$$4C_\Lambda^{2\alpha} C_F^2 C_P C' J ||\phi||_{H^2}^{2\alpha(\alpha-1)} < \frac{1}{(C_2^2 C' + 1)^\alpha}.$$

Let $q_r = C_r^2$. Then $q_2 < C_2^2 C' + 1$. We note that for κ, as above, $q_r < C_2^2 C' + 1$ for all $r \geq 2$. Indeed,

$$q_{r+1} = C_2^2 C' + 4C_\Lambda^{2\alpha} C_F^2 C_P C' J ||\phi||_{H^2}^{2\alpha(\alpha-1)} q_r^\alpha < C_2^2 C' + \frac{1}{(C_2^2 C' + 1)^\alpha} q_r^\alpha < C_2^2 + 1.$$

Thus, $C_r < \sqrt{C_2^2 C' + 1}$ and the inequality (35) with small κ implies (34).

From this result follows

Lemma 3.7. *Suppose that $l_1 > 0$ and $||\phi||_{H^2} < \kappa$ in (33) with sufficiently small κ. Then the solutions u_k to the system (33) converge in $C((0,+\infty);H^2)$.*

3.3 Construction of a Nonlinear Chapman Projection

3.3.1 Weak Nonlinearity

We consider the system

$$\partial_t u + \mathscr{A}_{11}\partial_x u + \mathscr{A}_{12}\partial_x v + B_{11}u + B_{12}v = 0,$$

$$\partial_t v + \mathscr{A}_{21}\partial_x u + \mathscr{A}_{22}\partial_x v + B_{21}u + B_{22}v = G(u)v, \qquad (38)$$

with the initial data $u|_{t=0} = \phi_1(x)$, $v|_{t=0} = \phi_2(x)$. We set $\phi = \begin{pmatrix} \phi_1 \\ \phi_2 \end{pmatrix}$. Suppose that $u(x,t): \mathbb{R} \times \mathbb{R}_+ \to \mathbb{R}^m$ and $v(x,t): \mathbb{R} \times \mathbb{R}_+ \to \mathbb{R}^{N-m}$. Assume that the data of the problem (38) satisfy all the assumptions of Lemma 3.7. We also assume that the following condition is satisfied.

Condition 3.1. *The linearized part of the problem (26) and the initial data satisfy all the assumptions of Theorem 2.7. Moreover,* $l_j = \inf_{\xi} \min_{\lambda \in \Gamma_j} \mathrm{Re}\, \lambda$ *and* $l_1 > 0$, $l_2 - \alpha l_1 < 0$. *We denote by* P_{21} *the symbol of the Chapman-Enskog projection for the linearized problem (38). If the initial data* ϕ *are sufficiently smooth and*

$$||\phi||^2_{H^2} + ||P_{21}(\partial_x)\phi||^2_{H^2} < \kappa \ll 1$$

then, according to the method of successive approximations, there exists a solution $\begin{pmatrix} w \\ z \end{pmatrix}$ *to the problem (38) with the initial data* $w|_{t=0} = \Upsilon(\phi_1, \phi_2)$, $z|_{t=0} = P_{21}(\partial_x)\Upsilon(\phi_1, \phi_2)$, *where* Υ *is the operator of the initial data corresponding to the sum of* U_{Ch} *and* U_{Cor} *in the linear case. The goal of this section is to construct a nonlinear operator* $\mathscr{P}_{21}(w, \partial_x)$ *such that* $z = \mathscr{P}_{21}(w, \partial_x)w$.

Let $M = S \Lambda S^{-1}$, where

$$S = \begin{pmatrix} E & 0 \\ -P_{21} & E \end{pmatrix}.$$

We write the system (38) in terms of the Fourier images and use the fact that P_{21} is the symbol of the Chapman projection for the linearized problem. Then

$$\partial_t \mathscr{F}(w) + M_{11}\mathscr{F}(w) + M_{12}\mathscr{F}(v') = 0,$$

$$\partial_t \mathscr{F}(v') + M_{22}\mathscr{F}(v') = \mathscr{F}(G(w)z),$$

where $z = P_{21}w + v'$. Based on this fact, we look for z in the following form of the state equation

$$z = P_{21}w + \sum_{j=1}^{\infty} v_j, \tag{39}$$

where v_j is a solution to the equation

$$\partial_t \mathscr{F}(v_j) + M_{22}\mathscr{F}(v_j) = \mathscr{F}(G(w)v_{j-1}) \tag{40}$$

with the initial data $v_j|_{t=0} = 0$ and $v_0 = P_{21}w$. Using the method of variation of constants, we find

$$v_j = \mathscr{F}^{-1}\left(e^{-M_{22}t}\int_0^t e^{M_{22}\tau}\mathscr{F}(G(w(\tau))v_{j-1}(\tau))d\tau\right).$$

This representation shows that $v_j = \Pi_j(w, \partial_x)w$. It remains to prove that for small ϕ the series (39) is convergent.

From Lemma 3.6 and the method of successive approximations it follows that

$$||w||_{H^2} \le C_0 e^{-l_1 t}(1 + t^d)(||\phi||_{H^2} + ||P_{21}\phi||_{H^2}).$$

Furthermore, for $P_{21}w$ we have the similar estimate

$$||P_{21}w|| \leq C_1 e^{-l_1 t}(1+t^d)||\phi||_{H^s}$$

with some s. Based on these two inequalities and the embedding theorem, we find the following estimate for v_1:

$$||v_1||^2 \leq t \int_0^t ||e^{M_{22}(\tau-t)}\mathscr{F}(G(w(\tau))P_{21}w(\tau))||^2 d\tau$$

$$\leq te^{-2l_2 t} \int_0^t e^{2l_2\tau}q_0(t,\tau)|w|_\infty^{2\alpha-2}||P_{21}w(\tau)||^2 d\tau,$$

where $q_0(t,\tau)$ is a polynomial depending only on the structure of M. Further,

$$||v_1||^2 \leq C_2 te^{-2l_2 t} \int_0^t e^{2(l_2-\alpha l_1)\tau}q_0(t,\tau)(1+\tau^d)^{2\alpha}(||\phi||_{H^2}+||P_{21}\phi||_{H^2})^{2\alpha-2}||\phi||_{H^s}^2 d\tau.$$

Hence, under the above conditions on the system (38), there are constants $d_1 \geq 0$ and $K_1 > 0$ such that

$$||v_1||^2 \leq K_1 t(1+t^{d_1})e^{-2l_2 t}||\phi||_{H^s}^{2\alpha}. \tag{41}$$

Moreover, d_1 depends only on the structure of the matrix M and K_1 is independent of ϕ.

Similarly, for v_2 we find

$$||v_2||^2 \leq t \int_0^t ||e^{M_{22}(\tau-t)}\mathscr{F}(G(w(\tau))v_1(\tau))||^2 d\tau \leq$$

$$\leq C_3 te^{-2l_2 t} \int_0^t e^{2l_2\tau}q_0(t,\tau)(1+\tau^d)^{2\alpha-2}e^{-(2\alpha-2)l_1\tau}||v_1(\tau)||^2 h_2^{2\alpha-2} d\tau,$$

$$h_2 := ||\phi||_{H^2} + ||P_{21}\phi||_{H^2}.$$

Using the above estimate for v_1, we find

$$||v_2||^2 \leq K_2 te^{-2l_2 t}(1+t^{d_1})||\phi||_{H^s}^{4\alpha-2},$$

where the constant K_2 is independent of ϕ, because $\int_0^t \tau^r e^{-\gamma\tau}d\tau \leq \int_0^\infty \tau^r e^{-\gamma\tau}d\tau =$ const.

Arguing in the same way, it is easy to obtain the inequality

$$||v_j||^2 \leq K_j t(1+t^{d_1})e^{-2l_2 t}||\phi||_{H^s}^{2j\alpha-2j+2},$$

where the constants K_j are independent of ϕ and $K_j \leq K_0^j$, $K_0 =$ const. Hence for sufficiently small ϕ the series (39) is convergent.

We note that the smallness of the norm of the initial data in some space H^s and the estimate

$$|G(w)|_\infty \le C_G |w|_\infty^{\alpha-1}(1+|w|_\infty^\beta)$$

imply

$$|G(w)|_\infty \le C'_G |w|_\infty^{\alpha-1},$$

where $C'_G > C_G$. Furthermore, d_1 is the degree of the polynomial $q_0(t,\tau)$ in the variable t. Since q_0 is a polynomial, there exist constants I_1 and I_2 such that

$$\int_0^{+\infty} e^{2(l_2-\alpha l_1)\tau} q_0(t,\tau)(1+\tau^d)^{2\alpha} d\tau \le I_1(1+t^{d_1}),$$

and

$$\int_0^{+\infty} e^{-(2\alpha-2)l_1\tau} q_0(t,\tau)(1+\tau^d)^{2\alpha-2}(1+\tau^{d_1})\tau d\tau \le I_2(1+t^{d_1}).$$

Indeed, both integrals on the left-hand sides of these inequalities are polynomials in t of degree d_1, which implies the required estimates.

To prove the assertions concerning the constants K_j, we need the following lemma.

Lemma 3.8. *Assume that all the assumptions of Lemma 3.7 and Condition 3.1 are satisfied. Then the solution v_j to the problem (40) with the initial data $v_j|_{t=0}=0$ for $j \ge 1$ satisfies the inequality*

$$||v_j||^2 \le K_j t(1+t_1^d)e^{-2l_2 t}||\phi||_{H^s}^{2j\alpha-2j+2},$$

where $K_j \le (C'_G)^j C_W^{2j\alpha-2j+2} I_2^{j-1} I_1$ and $C_W = \max\{C_0,C_1\}$.

Proof. We use the method of mathematical induction. As was already shown, the required estimate is valid for $||v_1||^2$. Furthermore, it is easy to see that

$$K_1(1+t^{d_1}) = C_2 \int_0^{+\infty} e^{2(l_2-\alpha l_1)\tau} q_0(t,\tau)(1+\tau^d)^{2\alpha} d\tau \le C_2 I_1(1+t^{d_1}),$$

where $C_2 = C'_G C_W^{2\alpha}$. Thus, the corresponding inequality for K_1 also holds.

Assume that the assertion holds for $j \le k$. Then for $j = k+1$ we have

$$||v_{k+1}||^2 \le t \int_0^t ||e^{M_{22}(\tau-t)}\mathscr{F}(G(w(\tau))v_k(\tau))||^2 d\tau \le$$

$$\le t e^{-2l_2 t} \int_0^t e^{2l_2\tau} q_0(t,\tau)|w(\tau)|_\infty^{2\alpha-2}||v_k(\tau)||^2 d\tau \le K_{k+1} t(1+t^{d_1})e^{-2l_2 t}||\phi||_{H^s}^{2(k+1)\alpha-2k},$$

where $C_3 = C'_G C_W^{2\alpha-2}$, and from the inequality

$$\int_0^t q_0(t,\tau)(1+\tau^d)^{2\alpha-2}e^{-(2\alpha-2)l_1\tau}(1+\tau^{d_1})\tau d\tau \le$$

$$\le \int_0^{+\infty} q_0(t,\tau)(1+\tau^d)^{2\alpha-2}e^{-(2\alpha-2)l_1\tau}(1+\tau^{d_1})\tau d\tau \le I_2(1+t^{d_1})$$

we obtain the required estimate for K_{k+1}.

3.3.2 General Case

We consider the system

$$\partial_t u + \mathscr{A}_{11}\partial_x u + \mathscr{A}_{12}\partial_x v + B_{11}u + B_{12}v = G_{11}(u)u,$$
$$\partial_t v + \mathscr{A}_{21}\partial_x u + \mathscr{A}_{22}\partial_x v + B_{21}u + B_{22}v = G_{21}(u)u + G_{22}(u)v, \qquad (42)$$

with the same initial data as above. Then we construct a nonlinear operator $\mathscr{P}_{21}(\partial_x, w)$ that determines the solution $\begin{pmatrix} w \\ z \end{pmatrix}$. We look for z in the form (39), where v_j are solutions to the problem

$$\partial_t \mathscr{F}(v_1) + M_{22}\mathscr{F}(v_1) = \mathscr{F}(P_{21}(G_{11}(w)w) + G_{21}(w)w + G_{22}(w)P_{21}w),$$

$$\mathscr{F}(v_1)|_{t=0} = 0, \quad \partial_t \mathscr{F}(v_j) + M_{22}\mathscr{F}(v_j) = \mathscr{F}(G_{22}(w)v_{j-1}), \quad \mathscr{F}(v_j)|_{t=0} = 0, \; j \geq 2.$$

Then we can estimate $\|v_1\|$ as follows:

$$\|v_1\|^2 \leq t \int_0^t C_M^2 (1 + (t-\tau)^{d_M})^2 e^{2l_2(\tau - t)} \|h_3\|^2(\tau) d\tau \leq$$

$$\leq \mathrm{const}\, t (1 + t^{d_M})^2 e^{-2l_2 t} \|\phi\|_{H^2}^{2\alpha} \int_0^{+\infty} e^{2(l_2 - \alpha l_1)\tau}(1 + \tau^{d_M})^2 (1 + \tau^{d_A})^{2\alpha}(1 + \sqrt{\tau})^{2\alpha} d\tau$$

$$= K_1' t (1 + t^{d_M})^2 e^{-2l_2 t} \|\phi\|_{H^2}^{2\alpha}, \qquad h_3 := P_{21}(G_{11}(w)w) + G_{21}(w)w + G_{22}(w)P_{21}w.$$

Thus, for v_1 we have an estimate of the form (41). We note that the equation for v_j, $j > 1$, is the same as in the previous subsection. Furthermore, for estimating from above v_j, $j > 1$, we used the estimate (41), but not an explicit form of v_1. Consequently, Lemma 3.8 remains valid. Therefore, the series (39) converges in the L_2-norm for small initial data, which means the existence of a nonlinear projection $\mathscr{P}_{21}$.

3.4 Properties of Nonlinear Projections

We study properties of the nonlinear operator $\mathscr{P}_{21}$ constructed in the previous section.

Lemma 3.9. *Let the data of the problem (42) satisfy all the assumptions of Lemma 3.7 and Condition 3.1. Assume that $\phi \in H^3$, $|P_{21}|_0 \leq \mathrm{const}(1 + |\xi|^s)$, $s \leq 2$. Then for every term v_j of the series (39) the following inequality holds:*

$$\|v_j\|_{H^1}^2 \leq K_0^j t (1 + t^{d_M})^2 e^{-2l_2 t} \|\phi\|_{H^3}^{2j\alpha - 2j + 2}. \qquad (43)$$

Proof. We estimate derivatives $\|\partial_k v_j\|$. For this purpose, we note that $\|\partial_k v_j\|$ satisfies the problem

$$\partial_t \mathscr{F}(\partial_k v_j) + M_{22}\mathscr{F}(\partial_k v_j) = \mathscr{F}(\partial_k(F_j(w, v_{j-1}))), \quad \mathscr{F}(\partial_k v_j)|_{t=0} = 0,$$

where

$$F_1(w, v_0) = F_1(w) = -P_{21}(G_{11}(w)w) + G_{12}(w)w + G_{22}(w)P_{21}w,$$

$$F_j(w, v_{j-1}) = G_{22}(w)v_{j-1}, \; j \geq 2.$$

Using Lemmas 3.1 and 3.5, we find

$$||\partial_k v_1||^2 \leq \text{const} \, t (1 + t^{d_M})^2 e^{-2l_2 t} \int_0^{+\infty} e^{2l_2 \tau} (1 + \tau^{d_M})^2 ||F_1(w)||^2_{H^1} d\tau,$$

Using Lemma 3.2 and the inequality $|P_{21}| \leq \text{const}(1 + |\xi|^s)$, $s \leq 2$ we finally find

$$||\partial_k v_1||^2 \leq \text{const} \, t (1 + t^{d_M})^2 e^{-2l_2 t} ||\phi||^{2\alpha}_{H^3}.$$

Further

$$||\partial_k v_j||^2 \leq C_M t (1 + t^{d_M})^2 e^{-2l_2 t} \int_0^{+\infty} e^{2l_2 \tau} (1 + \tau^{d_M})^2 ||\partial_k (G_{22}(w)v_{j-1})||^2 d\tau.$$

Taking into account that G_{22} is a matrix polynomial and arguing as in Lemma 3.8, we obtain the required estimates.

For the sake of brevity, we introduce the notation $L_1 = \sup_\xi \max_{\lambda \in \Gamma_1} \text{Re}\, \lambda$.

Theorem 3.1. *Let* $\begin{pmatrix} w \\ z \end{pmatrix}$ *be a solution to the system (42) with the initial data* $w|_{t=0} = \phi_1$, $z|_{t=0} = P_{21}\phi_1$. *Let* $\phi \in H^3$, *and let all the assumptions of Lemma 3.9 be satisfied. Denote by* $\begin{pmatrix} w_0 \\ z_0 \end{pmatrix}$ *the solution to the linearized problem (42) with the same initial data. If* $\alpha > \frac{L_1}{l_1}$, $||\phi||_{H^3} < \kappa \ll 1$, *then the following estimate holds:*

$$||e^{M_{11}t} \mathscr{F}(w - w_0)||^2 \leq \frac{1}{\gamma} \text{const}(||\phi||^{2\alpha}_{H^2} + ||\phi||^{\alpha+1}_{H^3}),$$

where

$$0 < \gamma < \min\{\frac{l_2 - L_1}{2}, 2\alpha l_1 - 2L_1\}.$$

Proof. We note that the Fourier images satisfy the equality

$$\partial_t \mathscr{F}(w) + M_{11}\mathscr{F}(w) + M_{12}\mathscr{F}(z - P_{21}w) = \mathscr{F}(G_{11}(w)w).$$

Hence

$$\mathscr{F}(w) = e^{-M_{11}t}\left(\mathscr{F}(\phi_1) + \int_0^t e^{M_{11}\tau}(\mathscr{F}(G_{11}(w)w) - M_{12}\mathscr{F}(z - P_{21}w))d\tau\right).$$

Thus,

$$||e^{M_{11}t}(w - w_0)|| \leq ||\int_0^t e^{M_{11}\tau}\mathscr{F}(G_{11}(w)w)d\tau|| + ||\int_0^t e^{M_{11}\tau}M_{12}\mathscr{F}(z - P_{21}w)d\tau||.$$

Further,

$$\left|\left| \int_0^t e^{M_{11}\tau} \mathscr{F}(G_{11}(w)w) d\tau \right|\right|^2 \le \frac{\text{const}}{\gamma} \int_0^{+\infty} e^{(2L_1+\gamma)\tau}(1+\tau^d)^2 ||w||_{H^2}^{2\alpha} d\tau.$$

Using Lemma 3.6, we find

$$\left|\left| \int_0^t e^{M_{11}\tau} \mathscr{F}(G_{11}(w)w) d\tau \right|\right|^2 \le$$

$$\le \frac{\text{const}}{\gamma} \int_0^{+\infty} e^{(2L_1+\gamma-2\alpha l_1)\tau}(1+\tau^d)^2(1+\tau^{d_\Lambda})^{2\alpha}(||\phi||_{H^2}+||\phi||_{H^2}^{\alpha}\sqrt{\tau})^{2\alpha} d\tau.$$

By the conditions on α, γ, and ϕ, it follows that $\left|\left| \int_0^t e^{M_{11}\tau} \mathscr{F}(G_{11}(w)w) d\tau \right|\right|^2 \le \frac{\text{const}}{\gamma} ||\phi||_{H^2}^{2\alpha}$. Estimating the second term, we find

$$\left|\left| \int_0^t e^{M_{11}\tau} M_{12} \mathscr{F}(z - P_{21}w) d\tau \right|\right|^2 \le$$

$$\le \int_0^{+\infty} e^{-\gamma\tau} d\tau \int_0^{+\infty} e^{(2L_1+\gamma)\tau}(1+\tau^d)^2 ||M_{12}\mathscr{F}(z-P_{21}w)||^2 d\tau.$$

We note that $M_{12} = \Lambda_{12}$. Thus, $||M_{12}\mathscr{F}(z-P_{21}w)|| = ||z-P_{21}w||_{H^1}$. Using Lemma 3.9 and taking ϕ with sufficiently small H^3-norm, we find

$$\left|\left| \int_0^t e^{M_{11}\tau} M_{12} \mathscr{F}(z - P_{21}w) d\tau \right|\right|^2 \le \frac{\text{const}}{\gamma} ||\phi||_{H^3}^{\alpha+1}.$$

which implies the required assertion.

Remark 3.1. *Applications of the obtained results to models of continuum mechanics can be found in [11, 13, 14, 16].*

Acknowledgement. This work was supported by the Russian Foundation of Basic Researches (grant no. 09-01-00288).

References

1. Chen, G.Q., Levermore, C.D., Lui, T.-P.: Hyperbolic conservation laws with stiff relaxation terms and entropy. Commun. Pure Appl. Math. 47(6), 787–830 (1994)
2. Boltzmann, L.: Rep. Brit. Assoc. 579 (1894)
3. Zhura, N.A.: Hyperbolic first order systems and quantum mechanics (in Russian). Mat. Zametki (submitted)
4. Bardos, C., Levermore, C.D.: Fluid dynamic of kinetic equation II: convergence proofs for the Boltzmann equation. Comm. Pure Appl. Math. 46, 667–753 (1993)
5. Bardos, C., Golse, F., Levermore, C.D.: Fluid dynamics limits of discrete velocity kinetic equations. In: Advances in Kinetic Theory and Continuum Mechanics, pp. 57–71. Springer, Berlin (1991)

6. Caffish, R.E., Papanicolaou, G.C.: The fluid dynamical limit of nonlinear model Boltz-
    mann equations. Comm. Pure Appl. Math. 32, 103–130 (1979)
7. Chen, G.Q., Frid, H.: Divergence-measure fields and hyperbolic conservation laws. Arch.
    Ration. Mech. Anal. 147, 89–118 (1999)
8. Chapman, S.: On Certain Integrals Occurring in the Kinetic Theory of Gases. Manchester
    Mem. 66 (1922)
9. Chapman, S.C., Cowling, T.C.: The Mathematical Theory of Non-Uniform Gases. Cam-
    bridge Univ. Press, Cambridge (1970)
10. Grad, H.: On the kinetic theory of rarefied gases. Commun. Pure Appl. Math. 2(4), 331–
    406 (1949)
11. Radkevich, E.V.: Irreducible Chapman projections and Navier-Stokes approximations.
    In: Instability in Models Connected with Fluid Flows, vol. II, pp. 85–151. Springer, New
    York (2007)
12. Radkevich, E.V.: Kinetic equations and the Chapman projection problem (in Russian).
    Tr. Mat. Inst. Steklova 250, 219–225 (2005); English transl.: Proc. Steklov Inst. Math.
    250, 204–210 (2005)
13. Radkevich, E.V.: Mathematical Aspects of Nonequilibrium Processes (in Russian).
    Tamara Rozhkovskaya Publisher, Novosibirsk (2007)
14. Palin, V.V.: On the solvability of quadratic matrix equations (in Russian). Vestn. MGU,
    Ser. 1(6), 36–42 (2008)
15. Palin, V.V.: On the solvability of the Riccati matrix equations (in Russian). Tr. Semin. I.
    G. Petrovskogo 27, 281–298 (2008)
16. Palin, V.V.: Dynamics separation in conservation laws with relaxation (in Russian).
    Vestn. SamGU 6(65), 407–427 (2008)
17. Palin, V.V., Radkevich, E.V.: Hyperbolic regularizations of conservation laws. Russian J.
    Math. Phys. 15(3), 343–363 (2008) (Submitted date: January 9, 2009)
18. Palin, V.V., Radkevich, E.V.: On the Maxwell problem. Journal of Mathematical Sci-
    ences 157(6) (2009); The date for the Table of Contents is March 28 (2009)
19. Radkevich, E.V.: Problems with insufficient information about initial-boundary data.
    In: Fursikov, A., Galdi, G.P., Pukhnachov, V. (eds.) Advances in Mathematical Fluid
    Mechanics (AMFM), Special AMFM Volume in Honour of Professor Kazhikhov.
    Birkhauser Verlag, Basel (2009)
20. Hsiao, L., Liu, T.-P.: Convergence to nonlinear diffusion waves for solutions of a system
    of hyperbolic conservation laws with damping. Commun. Math. Phys. 143, 599–605
    (1992)
21. Dreyer, W., Struchtrup, H.: Heat pulse experiments revisted. Continuum Mech. Thermo-
    dyn. 5, 3–50 (1993)
22. Palin, V.V.: Dynamics separation in conservation laws with relaxation (in Russian).
    Vestn. MGU, Ser. 1 (2009) (to appear)

Continuum-Molecular Modeling of Nanostructured Materials

Gwidon Szefer and Dorota Jasińska

Abstract. The paper presents the mixed structural mechanics - molecular approach to analysis of stresses and deformations at the nanoscale range. Plots of Young modulus, for the individual bonds, as well as for the bulk nanotubes are given, taking molecular interactions into account. Numerical calculations for carbon nanotubes of different chiralities are performed.

1 Introduction

Investigating and modeling of materials and devices in the nanoscale range constitutes an up to date topic in mechanics and material science. In recent years different approaches have been proposed to describe their mechanical phenomena and properties. Among them the idea to couple the molecular models with continuum and structural mechanics description is one of the most effective and promising methods of analysis. The continuum models with some equivalent molecular properties have been proposed (see e.g.[5], [6]). They lead mostly to mechanical models (rods, plates, shells etc) with equivalent, but constant, mechanical parameters. It contradicts the molecular situation, in which the potentials are essentially nonharmonical, and hence the stiffness parameters are non-constant. In the present paper an alternative approach will be presented, with constitutive model taking directly molecular interactions into account. Our considerations include potentials depending not only on the intermolecular distances, but also on the angles between the bonds. The paper is organized as follows: we start with the short description of the discrete molecular system, followed by the the concept of the mixed continuum-molecular model,

Gwidon Szefer · Dorota Jasińska
Cracow University of Technology, ul. Warszawska 24, 31-155 Kraków
e-mail: {szefer,jasinska}@limba.wil.pk.edu.pl

called nanocontinuum. This model is applied to carbon nanotubes (CNT), for which the modified Morse potential and Tersoff-Brenner potential are used. The role of the bond-angles is shown in details. Numerical examples for different CNTs are presented.

2 Field Quantities in the Discrete Systems

To introduce field quantities in the discrete molecular systems let us consider a set of material points A_i, with position vectors r_i, $i = 1,..N$ referred to a fixed Cartesian frame $\{0x_k\}$, $k = 1,2,3$. Let the initial position vector be denoted by r_{oi}. The system is subjected to the action of forces resulting from the given potential $U(r_1,..r_N) = U(r_{ij}, \theta_{ijk})$, where $r_{ij} = |\mathbf{r_{ij}}|$, $\mathbf{r_{ij}} = \mathbf{r_j} - \mathbf{r_i}$ and θ_{ijk} - angle between adjacent bonds. The influence of the angle θ_{ijk} and possibly further environment plays an important role in bonds of many materials (like carbon, silicon etc.). It concerns many devices in modern nanoengineering, like carbon nanotubes and silicon wafers.

The gradient

$$\mathbf{f_{ij}} = \frac{\partial U}{\partial r_{ij}} \cdot \mathbf{e_{ij}}; \quad \mathbf{e_{ij}} = \frac{1}{r_{ij}} \cdot \mathbf{r_{ij}}, \tag{1}$$

describes the interaction forces between molecules, which induce the motion of the molecules according to the equations

$$m_i \cdot \ddot{\mathbf{r}}_{\mathbf{ij}} = \sum_j \mathbf{f_{ij}} + \mathbf{f_i^{ext}}; \quad i = 1..N, \tag{2}$$

where m_i- mass of the molecule A_i, and $\mathbf{f_i^{ext}}$ stands for the external force acting on a point. The solution of the system (2) (with suitable initial conditions), for different potentials $U(r_{ij})$, is the standard task in molecular dynamics (for sets of millions of particles considered in chemistry and material science).

In analysis of mechanical phenomena like deformation, fracture, crack propagation, buckling etc., such quantities as strain, stress and stress rate are of interest. The expression

$$\sigma_i = \frac{1}{V} \sum_j \mathbf{r_{ij}} \otimes \mathbf{f_{ij}} = \frac{1}{V} \sum_j \mathbf{r_{ij}} \otimes \frac{1}{r_{ij}} \frac{\partial U}{\partial r_{ij}} \mathbf{r_{ij}} \tag{3}$$

defines the molecular virial stress tensor (see [7], [11]), which results from the generalized Clausius virial theorem (see [4], [8]), known in statistics and molecular mechanics. Above V means the volume of representative volume element assigned to the point A_i. It is convenient to present the molecular stresses in the form with a clear structural mechanics interpretation. For this purpose let's assume molecular bonds being rods with circular cross-sections having area A, and diameter d, resulting from computational chemical data verified by molecular simulations.

Assuming furthermore the bond's volume being constant $V^{bond} = A \cdot r_{ij} = A^o \cdot r_{ij}^o$, where r_{ij}^o, A^o mean the initial values of rod lenght and crossection area, one can write

$$\sigma_i = \frac{V^{bond}}{V} \sum_j \frac{1}{A \cdot r_{ij}} \mathbf{r_{ij}} \otimes \mathbf{f_{ij}} = \frac{V^{bond}}{V} \sum_j \sigma_{ij}^{bond} \mathbf{e_{ij}} \otimes \mathbf{e_{ij}} \qquad (4)$$

where

$$\sigma_{ij}^{bond} = \frac{1}{A} \frac{\partial U}{\partial r_{ij}} = \frac{1}{A^o r_{ij}^o} \cdot r_{ij} \frac{\partial U}{\partial r_{ij}}. \qquad (5)$$

It can be seen, that any component of the virial stress tensor at a point A_i is a sum of stresses arising from all bonds at a particle.

3 Hyperelastic Nanocontinuum

Molecular dynamics experiments and simulations (particularly for crystalline materials) show, that in a wide range of deformation (before bond breaking) the material behavior is elastic. Thus, using for generality the Seth's and Hencky's [3] strain measure, which in the case of uniaxial stretch test takes the form

$$\varepsilon_m = \begin{cases} \frac{1}{m} \left[\left(\frac{r}{r^o} \right)^m - 1 \right] & \text{for} \quad m \neq 0, \\ \ln \left(\frac{r}{r^o} \right) & \text{for} \quad m = 0, \end{cases} \qquad (6)$$

one can write according to (5)

$$\sigma^{bond} = \frac{f}{A} = \frac{1}{A} \frac{\partial U}{\partial r} = E^{bond} \cdot \varepsilon_m. \qquad (7)$$

Hence the definition

$$E^{bond}(r) = \frac{1}{A^o r^o} \cdot r \cdot \frac{\partial U}{\partial r} \cdot \frac{1}{\varepsilon_m} \qquad (8)$$

stands for the generalized Young modulus for the bond. It takes the form

$$E^{bond}(r) = \begin{cases} \frac{m}{A^o r^o} \cdot r \cdot \frac{\partial U}{\partial r} \left[\left(\frac{r}{r^o} \right)^m - 1 \right]^{-1} & \text{for} \quad m \neq 0, \\ \frac{1}{A^o r^o} \cdot r \cdot \frac{\partial U}{\partial r} \cdot \ln \left(\frac{r}{r^o} \right)^{-1} & \text{for} \quad m = 0. \end{cases} \qquad (9)$$

Substituting (9) into (4) leads to

$$\sigma_i = \frac{V^{bond}}{V} \sum_j \frac{E^{bond}(r) \cdot \varepsilon_m}{r_{ij}^2} \cdot \mathbf{r_{ij}} \otimes \mathbf{r_{ij}}, \qquad (10)$$

that expresses the elastic features of the molecular stresses. Between any pair of points $A_i A_j$ the bond modulus is different, depending on the distance.

As mentioned earlier, we pay particular attention to modified Morse and Tersoff-Brenner potentials, since they have been widely applied in nanoengineering. They are described by functions:

$$U(r,\theta) = D\left\{\left[1 - e^{(-\beta(r-r^o))}\right]^2 - 1\right\} + \frac{1}{2}k_\theta(\theta - \theta_o)^2 \cdot [1 + k_s(\theta - \theta_o)^4] \quad (11)$$

for modified Morse bonds (see [1]), and

$$U(r,\theta) = U_R(r) - \bar{B}(r,\theta) \cdot U_A(r) \quad (12)$$

for the Tersoff-Brenner type [2], where

$$U_R = f_c(r)\frac{D_e}{s-1}e^{\sqrt{2S}\beta(r-r_e)} \quad \text{is the repulsive part,}$$

$$U_A = f_c(r)\frac{D_e \cdot s}{s-1}e^{\sqrt{\frac{2}{s}}\beta(r-r_e)} \quad \text{is the atracting part of the potential,}$$

$$f_c(r) = \begin{cases} 1 & \text{for} \quad r < r_1, \\ \frac{1}{2} \cdot \left(1 + \cos\frac{\pi(r-r_1)}{r_2-r_1}\right) & \text{for} \quad r_1 \leq r \leq r_2, \\ 0 & \text{for} \quad r > r_1, \end{cases}$$

$$\bar{B}(r,\theta) = \frac{1}{2}(B_{ij} + B_{ji}), \qquad B_{ij} = \left[1 + \sum_{k \neq ij} G(\theta_{ijk})f_c(r_{ik})\right]^{-\delta}$$

$$G(\theta) = a_o\left[1 + (\frac{c_o}{d_o})^2 - \frac{c_o^2}{d_o^2 + (1 + \cos\theta)^2}\right].$$

Assuming the bond diameter $d = 0.147nm$, and taking data for carbon from [1], for the modified Morse potential: $D = 6.03 \cdot 10^{-9}N\text{Å}$, $r_0 = 1.42\text{Å}$, $\beta = 2.625 \text{ Å}^{-1}$, $k_\theta = 0.9 \cdot 10^{-9}N\text{Å}rad^{-2}$, $k_s = 0.754rad^{-4}$, and for the Tersoff-Brenner potential: $D_e = 9.648 \cdot 10^{-9}N\text{Å}$, $r_e = 1.4507 \text{ Å}$, $r_1 = 1.7\text{Å}$, $r_2 = 2.0 \text{ Å}$, $\beta = 2.1 \text{ Å}^{-1}$, $s = 1.22$, $\delta = 0.5$, $a_o = 0.00020813$, $c_o = 330$, $d_o = 3.5$, one obtains plots for the modulus functions $E^{bond}(r)$ shown in figures 1, 2 and 3.

Figures 4 and 5 show plots of the relation (7), which is the constitutive equation of bonds. The presented figures show the sensitivity of the Young modulus to the distance r_{ij}. The evidently assymetric strain-stress response appears due to difference between repulsive and attractive force, resulting directly from the structure of the potential. The characteristic camel-back effect for the Tersoff-Brenner potential is also easily noticeable.

The mixed continuum-atomistic models of CNT based on a harmonic approximation of molecular potentials (see [9]) incline to supposition, that the constitutive formula (7) may be extended to the three-dimensional, isotropic, hyperelastic nanocontinuum, that is the continuum model with molecular interactions, in the form

$$\sigma = 2 \cdot G(\varepsilon)\varepsilon_\mathbf{m} + \Lambda(\varepsilon) \cdot tr\varepsilon_\mathbf{m} \cdot \mathbf{1}, \quad (13)$$

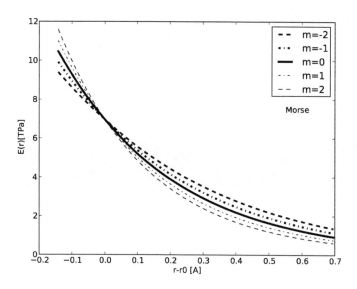

Fig. 1 Young modulus of the bond for Morse potential, for different strain measures

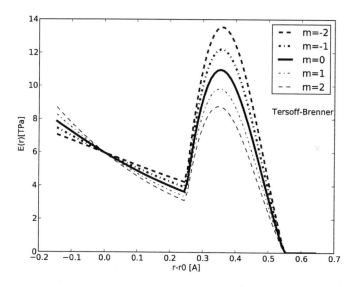

Fig. 2 Young modulus of the bond for Tersoff-Brenner potential, for $\theta = 120^o$, for different strain measure

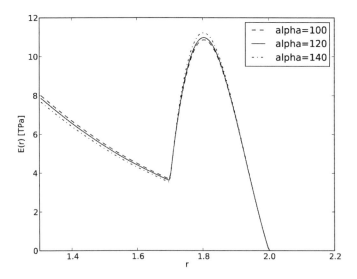

Fig. 3 Young modulus of the bond for TB potential, for m=0, for different θ angles

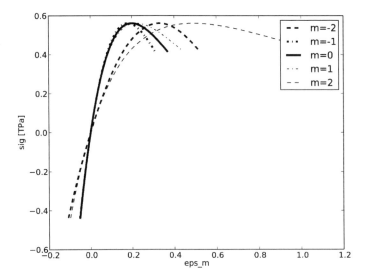

Fig. 4 Stresses of the bond for Morse potential, for different strain measures

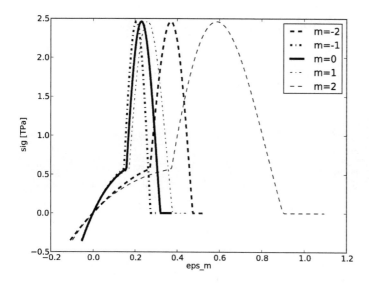

Fig. 5 Stresses of the bond for Tersoff-Brenner potential, for $\theta = 120^o$, for different strain measures

with $G(\varepsilon) = \frac{E(\varepsilon)}{2(1+v)}$, $\Lambda(\varepsilon) = \frac{v}{1+v} \cdot \frac{E(\varepsilon)}{1-2v}$, v - Poisson's ratio.

$$\varepsilon_m = \begin{cases} \frac{1}{m} \cdot (\mathbf{U}^m - \mathbf{1}) & \text{for} \quad m \neq 0, \\ \sum_{k=1}^{3} \ln \lambda_k \mathbf{w}_k \otimes \mathbf{w}_k & \text{for} \quad m = 0, \end{cases}$$

$\mathbf{U}$ means the Cauchy Green right stretch tensor, $\lambda_k, \mathbf{w}_k$- eigenvalues and eigenvectors of $\mathbf{U}$ respectively. In determination of $E(\varepsilon)$ ε takes the maximum value of λ_k.

Successful simulations of CNT's and graphens with mixed continuous-molecular plate, shell, and tube models allow us to presume, that the proposed formula, (13) with variable modulus $E(\varepsilon)$, will be more accurate in continuum-molecular modeling, than the models with equivalent, but constant E. Thus using the Voronoi tesselation of the nanocontinuum, and hence the Delannay tetrahedrons, to build the mesh of finite elements - the standard FE technique for the nanostructure can be applied. Expressing the displacements field $\mathbf{u}(\mathbf{x},t)$ of the nanocontinuum in the usual form

$$u_k(\mathbf{x},t) = \sum_{\alpha} N_{k\alpha}(\mathbf{x}) q_\alpha(t),$$

one obtains the governing systems of equations in the matrix form:

$$\left[M_{\alpha\beta} \right] \left[\ddot{q}_\beta \right] + \left[F_\alpha^{int} \right] = \left[F_\alpha^{ext} \right], \tag{14}$$

where

$$M_{\alpha\beta} = \int_V \rho N_{k\alpha}(\mathbf{x}) N_{k\beta}(\mathbf{x}) dV,$$

$$F_\alpha^{int} = \int_V \sigma_{kl}(\mathbf{x}) \frac{\partial N_{k\alpha}}{\partial x_l} dV,$$

$$F_\alpha^{ext} = \int_V \rho b_k N_{k\alpha}(\mathbf{x}) dV + \int_S p_k(\mathbf{x}) N_{k\alpha}(\mathbf{x}) dS,$$

and $N_{k\alpha}(\mathbf{x})$-shape functions, $q_\alpha(t)$ - nodal displacements, $\sigma_{kl}(\mathbf{x}) = \sigma_{kl}(r_{ij})$ - components of the stress tensor (13), b_k, p_k - external body forces and surface traction respectively, $r_k^{ij} = r_{ok}^{ij} + u_k^{ij}(\mathbf{x})$, $u_k^{ij}(\mathbf{x}) = u_k^j - u_k^i$. In the range of inelastic deformations (e.g. dislocation cores in crystals) where the constitutive equation (13) is not valid, more general formula

$$\sigma_{kl} = \sum_i \sigma_{kl}^i = \sum_i \sum_j \frac{1}{V_i} \chi_{ij} r_k^{ij} r_l^{ij}, \qquad \chi_{ij} = \frac{1}{r_{ij}} \frac{\partial U}{\partial r_{ij}} \qquad (15)$$

resulting directly from (3), should be substituted into the term F_α^{int}.

4   Computational Method

To validate the effectiveness and accuracy of the presented description, we consider carbon nanotubes with different chiralities and boundary conditions. The earlier considered Morse (11), and Tersoff-Brenner (12) potentials are the most suitable for this analysis. It's worth mentioning, that without bond -angle potential energy in (11), the stable configuration cannot be found for the nanotube. This is also evident from the mechanical point of view, since the honeycomb truss structure of the CNT constitutes the unstable system. The angle between adjacent bonds contributes to the equilibrium configuration, as follows from the calculations presented below (see Fig. 6).

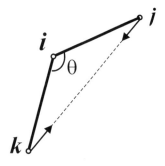

Fig. 6 Diagram of interactions in modified Morse potential

$$\theta = \arccos \frac{r_{ij}^2 + r_{ik}^2 - r_{jk}^2}{2 r_{ij} r_{ik}},$$

$$dU^m = \frac{\partial U}{\partial r_{ij}} dr_{ij} + \frac{\partial U}{\partial \theta} \frac{\partial \theta}{\partial r_{ij}} dr_{ij} + \frac{\partial U}{\partial r_{ik}} dr_{ik} + \frac{\partial U}{\partial \theta} \frac{\partial \theta}{\partial r_{ik}} dr_{ik} + \frac{\partial U}{\partial \theta} \frac{\partial \theta}{\partial r_{jk}} dr_{jk} =$$

$$= \left(\frac{\partial U}{\partial r_{ij}} + \frac{\partial U}{\partial \theta} \frac{\partial \theta}{\partial r_{ij}} \right) dr_{ij} + \left(\frac{\partial U}{\partial r_{ik}} + \frac{\partial U}{\partial \theta} \frac{\partial \theta}{\partial r_{ik}} \right) dr_{ik} + \frac{\partial U}{\partial \theta} \frac{\partial \theta}{\partial r_{jk}} dr_{jk} =$$

$$= f_{ij} dr_{ij} + f_{ik} dr_{ik} + f_{jk}^{con} dr_{jk}. \quad (16)$$

The force

$$f_{ij} = \frac{\partial U}{\partial r_{ij}} + \frac{\partial U}{\partial \theta} \frac{\partial \theta}{\partial r_{ij}} \quad (17)$$

contributes to the bond stress σ_{ij}^{bond}, whereas the term

$$f_{jk}^{con} = \frac{\partial U}{\partial \theta} \frac{\partial \theta}{\partial r_{jk}}, \quad (18)$$

defined as a connector force, expresses additional contribution to the equilibrium configuration of the system. The additional angle terms should be calculated for all bonds of the point A_i, and for all points in the bulk system. Analogous considerations, but with more complex calculations, (see Fig. 7) must be performed for the potential (12)

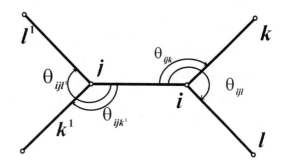

Fig. 7 Diagram of interactions in Tersoff-Brenner potential

$$dU^{T-B} = \frac{\partial U}{\partial r_{ij}} dr_{ij} + \sum_{\alpha=1}^{4} \frac{\partial U}{\partial \bar{B}} \frac{\partial \bar{B}}{\partial \theta_\alpha} d\theta_\alpha. \quad (19)$$

In the considered static case of the rod structure the FEM equation set takes the form

$$\sum_i \mathbf{F}_i^{int} = \sum_i \mathbf{F}_i^{ext} - \sum_i \mathbf{F}_i^{con}, \quad (20)$$

with

$$\mathbf{F}_i^{int} = \sum_j \left(\int_v \sigma_{ij}^{bond} \frac{\partial \varepsilon_m}{\partial q} dV \right) \mathbf{e}_{ij}, \quad \mathbf{F}_i^{con} = \sum_k \mathbf{f}_{ik}^{con},$$

where ε_m is described by (6), and q stands for the nodal displacement. In numerical calculations the logarithmic strain measure has been used.

Remark. The connector forces $\mathbf{f}_{jk}^{con}$ in (18), and hence in (20), appear as the result of the applied truss model. They can be eliminated by considering other models, like spatial frames, or micro-polar theories. Since the connector forces don't represent atomic bonds, they cannot be expressed by the bond stresses. Therefore the spring formula

$$\mathbf{f}_{jk}^{con} = k(r_{jk})\Delta_{jk} \cdot \mathbf{e}_{jk}, \quad (21)$$

where

$$k(r_{jk}) = \frac{1}{r_{jk} - r_{jk}^o} \frac{\partial U}{\partial \theta} \frac{\partial \theta}{\partial r_{jk}}, \quad \Delta_{jk} = r_{jk} - r_{jk}^o,$$

has been applied in both the Morse and Tersoff-Brenner potentials respectively.

5 Numerical Results

A series of numerical calculations were performed. Single wall carbon nanotubes of chirality $(5,5), (5,0), (20,20)$ and $(20,0)$ were subjected to stretching. The edge nodes of CNT were displaced axially with constant increment of elongation, and fixed transverse position (see fig. 8,9). Such conditions correspond to experiments of YU et al [10], where individual nanotubes were mounted between two opposing AFM tips. The length of the tubes has been set at 5 and 15 nm, and the diameter varied between 0.7 and 2.7 nm.

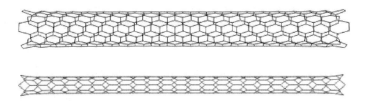

Fig. 8 Deformations of armchair (5,5), and zig-zag (5,0) CNT (lenght $\approx$ 5nm), for 20 % stretch

The aim of this test was to determine the values of the Young modulus for the bulk nanotube, and it dependence on strain. For stress calculation nanotube was treated as a cylindrical rod with thickness 0.34nm. The results are presented on figures 8-11.

They are generally in good agreement with results of molecular simulations and experiments quoted in literature. One can see that arm-chair tubes are stiffer than the zig-zag ones.

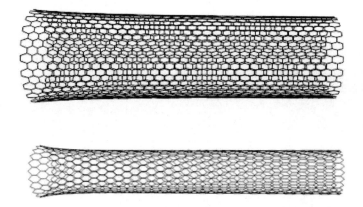

Fig. 9 Deformations of half of armchair (20,20), and zig-zag (20,0) CNT (lenght ≈ 15nm), for 20 % stretch

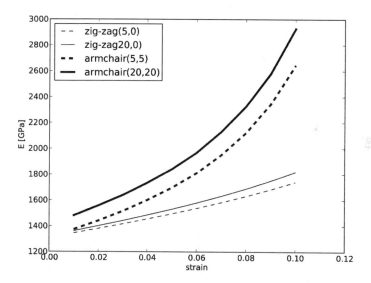

Fig. 10 Young modulus of CNT versus strain for different chiralities

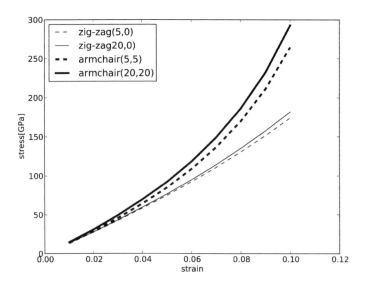

Fig. 11 Tensile stress of CNT versus strain for different chiralities

6 Conclusions

The plots of Young modulus and stresses depending on molecular interactions show effectiveness of constitutive modeling at the nanoscale range. The values of E_{CNT} for the bulk carbon nanotubes, calculated with the elastic rod model, were compared with the values reported by Belytschko et al [1] . They are in good agreement in terms of small deformations. Our description shows also the dependance of the modulus of E_{CNT} on strain (see fig.10), that leads to more accurate analysis of the nanostructures.

References

1. Belytschko, T., Xiao, S., Schatz, G., Ruoff, R.: Atomistic simulations of nanotubes fracture. Phys. Rev. B. 65, 235430 (2002)
2. Brenner, D.: Empirical potential for hydrocarbons for use in simulating the chemical vapour deposition of diamond films. Phys. Rev. B. 42, 9458–9471 (1990)
3. Dłużewski, P., Traczykowski, P.: Numerical simulation of atomic position in quantum dot by means of molecular statics. Arch. Mech. 55(5-6), 393–406 (2003)
4. Hoover, W.: Molecular Dynamics. Lect. Not. Phys., vol. 258. Springer, Berlin (1986)
5. Li, C., Chou, T.: A structural mechanics approach for the analysis of carbon nanotubes. Int. J. Solid. Struct. 40, 2487–2499 (2003)
6. Odegard, G., Gates, T., Nicholson, L., Wise, K.: Equivalent-continuum modeling of nano-structured materials. Compos. Sci. Technol. 62, 1869–1870 (2002)
7. Ribersy, M., Landman, U.: Dynamical simulation of stress, strain and finite deformation. Phys. Rev. B. 38(14), 9522–9537 (1988)

8. Szefer, G., Jasińska, D.: Modeling of strains and stresses of material nanostructures. Bull. Pol. Acad. Tech. 57(1), 41–46 (2009)
9. Tserpes, K., Papanikos, P.: Finite element modeling of single walled carbon nanotubes. Compos. B36, 468–479 (2005)
10. Yu, M.-F., Louire, O., Dyer, M.J., Moloni, K., Kelly, T.F., Ruoff, R.S.: Strength and breaking mechanism of multiwalled carbon nanotubes under tensile load. Science 287, 637–640 (2000)
11. Zhou, M.: A new look at the atomic level virial stress: on continuum-molecular system equivalence. Proc. Roy. Soc. Lond. A459, 2347–2392 (2003)

Linear Wave Propagation in Unsaturated Rocks and Soils

Bettina Albers

Abstract. In this contribution an overview of the continuum mechanical modeling of linear elastic partially saturated porous media and the application of such a model to linear wave propagation is given. First the involved microstructural variables are discussed and the construction of the model is presented. The macroscopic parameters used in the model are obtained by micro-macro-transition procedure from the measurable microscopic quantities. The linear elastic wave propagation analysis is demonstrated exemplarily for sandstone, sand and clayey loam. The properties of the four appearing waves – three compressional and one shear wave – are compared. Phase speeds and attenuations of these waves depend both on the frequency and on the degree of saturation.

1 Introduction

The classification of soils is nonuniform. Nearly every country has its own standards in which the characterization is regulated. In the German standards alone at least two different classifications of soil types do exist. Here we use the classification of DIN 4220 [9] named "Pedologic site assessment – Designation, classification and deduction of soil parameters (normative and nominal scaling)" which does not only contain pedologic quantities but also some parameters which are needed in the wave analysis. Thirty-one soil types are given in this standard. They consist of different fractions of the three main soil types sand, silt and clay. In the present paper, beneath

Bettina Albers
Technische Universität Berlin, Institute for Geotechnical Engineering and Soil Mechanics, Sekr. TIB1-B7, Gustav-Meyer-Allee 25, 13355 Berlin, Germany
e-mail: albers@grundbau.tu-berlin.de

sandstone, exemplarily, two of them (a sand mS and a clayey loam $Lt3$) have been chosen for a detailed investigation.

2 Microstructural Variables, Microscopic Material Parameters

2.1 Porosity and Mass Densities

The porosity, n, is a microstructural variable of a macroscopic model in which components are immiscible. It is interpreted as the ratio of the volume of the voids (whether filled by a single pore fluid or by a mixture of pore fluids) over the entire representative elementary volume, REV, of a porous medium which is substantially smaller than the whole flow region but decisively larger than the size of a single pore. The size of REV depends on the considered problem.

Soils and rocks are composed of a mixture of grains of different sizes and of voids which may be filled by a gas, a fluid or a mixture of fluids. The particles are either loose (e.g. sands, then we speak about *granular media*) or held together by compression and cementing material (e.g. sandstone, then we call them a *porous material*).

In Table 31 of DIN 4220 air capacity, usable field capacity and field capacity for the 31 soil types for different values of the oven-dry density are given. The oven-dry density is classified in Table 12 of DIN 4220. Further a medium value of this quantity is assumed, namely the class $\rho t3$. In the standard the calculation formula for the total void fraction and thus for the porosity (as volume fraction in %) is proposed:

$$\text{total void fraction} = \text{field capacity} + \text{air capacity}. \tag{1}$$

In this manner, the porosity for different soil types can be determined. It lies in the range $0.42 \leq n_0 \leq 0.46$ in which the values for the sands are smallest and those of clays are biggest.

The notion of porosity is closely connected to those of compactness and grain density commonly used in geotechnics, namely

$$\text{void fraction} = 1 - \frac{\text{compactness}}{\text{grain density}}. \tag{2}$$

The Relation (2) is used – with somewhat different denotations – to determine the reference mass density of the skeleton, ρ_0^S, which is equivalent to the notion of compactness of soil mechanics. Namely, if we denote the grain density by its counterpart of continuum mechanics – the initial true mass density of the skeleton, ρ_0^{SR}, then (2) yields with n_0 being the initial porosity

$$\rho_0^S = (1 - n_0)\,\rho_0^{SR}. \tag{3}$$

The mass densities of the considered soil types which are obtained under the assumption of $\rho_0^{SR} = 2650 \text{ kg/m}^3$ are given in one column of Table 1.

2.2 Compressibility of the Skeleton, Poisson's Number, Shear Modulus

For the compressibility modulus of the skeleton for sandstone the value $K_s = 48$ GPa is used and for sand and loam the value $K_s = 35$ GPa has been chosen. The latter may be too high if the soil is clayey. However VANORIO, PRASAD & NUR in [17] report in their article of theoretical values for the bulk modulus of clay between 20 and 50 GPa. These values have been widely adopted in the literature because measurements of clay minerals had proven to be difficult. In contrast, the at that time only published experimental measurement of Young's modulus in a clay mineral gave a much lower value of 6.2 GPa. The measurements described in [17] provide a value in between, namely $K_s = 12$ GPa.

The value of Poisson's number has been estimated according to the fractions of the three main soil types. The fractions of clay, silt and sand are given in Table 6 of DIN 4220. The following Poisson numbers for the main soil types are proposed:

$$v_{sand} = 0.31, \qquad v_{silt} = 0.37, \qquad v_{clay} = 0.43. \qquad (4)$$

Using these values and approximately the median of the fractions of the main soil types (but taking into account that the fractions do have to sum up to 100%) the two granular soil types under consideration possess the values of Poisson's number given in Table 1.

In ALBERS & WILMANSKI [7] it had been discussed for a two-component porous medium that the shear modulus, μ^S, does not depend on any coupling parameters but only on Poisson's number and on the compressibility modulus of the dry matrix, K_d. Thereby, it depends on the initial porosity n_0 since K_d is used in this work in the form proposed by GEERTSMA [11] $K_d = K_s/(1+50n_0)$. As done in almost all approaches on partially saturated porous media it is assumed that the shear modulus does not depend on the degree of saturation. Thus, the shear modulus can be expressed by

$$\mu^S = \frac{3}{2}\frac{1-2v}{1+v}\frac{K_s}{1+50n_0}. \qquad (5)$$

Table 1 Porosity, initial mass density, compressibility of the skeleton, Poisson's number and medians of the grain size fractions of three chosen soil types.

short cut	porosity	ρ_0^S $\left[\frac{kg}{m^3}\right]$	compressibility modulus [GPa]	Poisson's number	median of the grain size fraction clay	silt	sand
sandstone	0.25	1988	48	0.3	–	–	–
mS	0.42	1537	35	0.32	3	5	92
Lt3	0.44	1484	35	0.38	40	40	20

This form of the shear modulus stems from elasticity. Of course, there are other approaches in predicting it. WHITE [19], for example, suggests for granular soils another form of the shear modulus which goes back on a sphere pack model and which yields the shear wave speeds of approximately one half of those calculated in this work using the elasticity approach. A comparison of both approaches in application to several soil types can be found in ALBERS [1].

2.3 Saturation and Capillary Pressure

If multicomponent fluid mixtures fill the pores one needs, apart from porosity, at least one additional microstructural variable – a fraction of contributions of these components. If we consider a three-component medium consisting of a solid and two pore fluids (e.g. water and air) the saturation or degree of saturation (S) is defined as the fraction of the volume of the first fluid over the volume of all voids within a representative elementary volume REV. The sum of the fractions of the two pore fluids, of course, is equal to one, i.e. that the two pore fluids completely fill out the void volume.

Between two immiscible fluids, of which one may be gaseous, a discontinuity in pressure exists across the interface separating them. The difference is the capillary pressure p_c. It is a measure of the tendency of a porous medium to suck in the wetting fluid or to repel the nonwetting phase. In soil science, the negative of the capillary pressure (expressed as the pressure head) is called suction. The capillary pressure depends on the geometry of the void space, on the nature of the solids and fluids (e.g. on the contact angle) and on the degree of saturation. Laboratory experiments are probably the only method to derive the relationship $p_c = p_c(S)$.

Measured capillary pressure curves for several soil types differ considerably. The entire curve, according to SCHICK [13], can only be obtained by combining several measuring techniques because the range of capillary pressures can cover up to seven orders of magnitude while each technique is applicable only up to three orders of magnitude. Particularly, the range of the saturation which for a special soil can appear is rather distinct. In Figure 1 three curves for sandstone, sand (mS) and loam ($Lt3$) are shown. The latter two result from values given in DIN 4220.

For the presentation of capillary pressure curves by formulas, in the present work the approach of van Genuchten is chosen because the necessary parameters are presented in the German standard DIN 4220. They are given there for the original formula with non-normalized quantities and in units [hPa]. The normalized form of the van Genuchten formula is

$$S = (1/[1+(\alpha p_c)^n])^m, \qquad (6)$$

where the parameters α, n and m are the van Genuchten parameters. The corresponding values of the van Genuchten parameters for the three soil types are presented in Table 2. Parameter m is not given in the tables of DIN 4220 but calculated from n by $m = 1 - 1/n$.

Fig. 1 Capillary pressure
curves for sandstone, sand
(*mS*) and loam (*Lt*3).

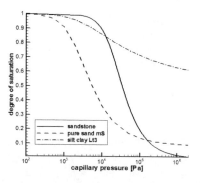

Table 2 Van Genuchten parameters in dependence on soil type (after Table 37 of DIN 4220 [9]; supplemented by the third van-Genuchten-parameter $m = 1 - \frac{1}{n}$); medium permeability in water-saturated soil for a medium value of the oven-dry density (see Table 35 of DIN 4220).

soil type	VAN GENUCHTEN parameters			saturated water permeability K
short cut	α [Pa]	n	m	in m/s
sandstone	0.00005	2	0.5	$2.45 \cdot 10^{-4}$
Lt3	0.0003005	1.28574	0.22224	$8.10 \cdot 10^{-7}$
mS	0.0004732	1.75418	0.42993	$5.67 \cdot 10^{-5}$

2.4 Compressibilities of Fluid and Gas, Viscosity, Permeability

The microscopic compressibilities of water and air do not change by consideration of different soil types. Thus, they are independent of the soil type they have the values

$$K_f = 2.25 \cdot 10^9 \mathrm{Pa}, \qquad K_g = 1.01 \cdot 10^5 \mathrm{Pa}. \tag{7}$$

For the wave analysis the incorporation of relative resistances is necessary. They are proposed by

$$\pi^{FS} = \pi^F / k_f, \qquad \pi^{GS} = \pi^G / k_g, \tag{8}$$

where

$$k_f = S^{\frac{1}{2}} \left[1 - \left(1 - S^{\frac{1}{m}} \right)^m \right]^2, \qquad k_g = (1-S)^{\frac{1}{3}} \left(1 - S^{\frac{1}{m}} \right)^{2m}. \tag{9}$$

The parameter m, again, is the van Genuchten parameter which is arranged in Table 2. The viscosities of water and air are implicitly incorporated in the parameters π^F and π^G. While the permeability or resistance with respect to the gas is not influenced by the soil type, the permeability with respect to the fluid differs. Thus, the value $\pi^G = 1.82 \cdot 10^5$ Pa is the same for all soil types while π^F is different. This value can be deduced from the hydraulic conductivity in water-saturated soil which is given in Table 2. The conversion of the permeability, K, into the resistance, π^F, happens according to

$$\mathrm{K}/(\rho g) \sim 1/\pi^F, \qquad \text{i.e.} \qquad \pi^F = (n_0 \cdot 1000 \cdot 9.81)/\mathrm{K} \quad [\mathrm{kg}/(\mathrm{m}^3 \mathrm{s})]. \tag{10}$$

3   Governing Equations

We construct a macroscopic model which is intended to describe the wave propagation in partially saturated – three-component – elastic porous and granular media. The three-component model contains features of two models for the description of saturated porous media namely the Simple Mixture Model by WILMANSKI (see: WILMANSKI [21], [22]) and the BIOT Model (see: TOLSTOY [14]). One important property of Biot's model is the interaction between components by partial volume changes which yields an additional contribution to partial stresses characterized by the material parameter Q. In contrast to the Biot Model the present model incorporates not only one coupling (between skeleton and fluid) but three of them (solid-fluid Q^F, solid-gas Q^G and fluid-gas Q^{FG}). These parameters account for changes of the stress state (the effect of surface tension between the fluid and the gas). In the Simple Mixture Model the porosity belongs to the set of fields and satisfies an own balance equation.

A systematic method of derivation of relations between macroscopic (average) material parameters (compressibilities and coupling parameters) and their counterparts for true (microscopic) materials is in ALBERS [3] and will be summarized here. The basis for the construction of this method is the work concerning two-component porous media by WILMANSKI [20]. The transition from the true properties to the averaged macroscopic properties is based on volume averaging over the representative elementary volume REV. The results of the averaging depend on both porosity and saturation and also on the microscopic compressibilities.

3.1   *Linear Three-Component Model for Elastic Porous and Granular Media*

The linear elastic three-component model for isothermal processes is based on the following set of unknown fields

$$\{\rho^S, \rho^F, \rho^G, \mathbf{v}^S, \mathbf{v}^F, \mathbf{v}^G, \mathbf{e}^S, n\}, \tag{11}$$

where ρ^S, ρ^F, ρ^G are the macroscopic current partial mass densities of the skeleton, of the fluid and of the gas, respectively, $\mathbf{v}^S, \mathbf{v}^F, \mathbf{v}^G$ are macroscopic velocity fields of these components. $\mathbf{e}^S$ is the macroscopic deformation tensor and n denotes the current porosity. The current saturation of the fluid, S, is not included in the series of fields. Different from the incorporation of the porosity into the model we do not state an own balance equation for it but we introduce a constitutive law in form of the van Genuchten approach for the current saturation.

The fields (11) are functions of the spatial variable $\mathbf{x} \in \mathscr{B}$, and time $t \in \mathscr{T}$. They must satisfy field equations which follow from partial balance equations by a linear closure. The following partial balance equations are used:

- mass conservation laws

$$\frac{\partial \rho^S}{\partial t} + \rho_0^S \operatorname{div} \mathbf{v}^S = 0, \quad \frac{\partial \rho^F}{\partial t} + \rho_0^F \operatorname{div} \mathbf{v}^F = 0, \frac{\partial \rho^G}{\partial t} + \rho_0^G \operatorname{div} \mathbf{v}^G = 0, \tag{12}$$

• momentum balance equations

$$\rho_0^S \frac{\partial \mathbf{v}^S}{\partial t} = \operatorname{div} \mathbf{T}^S + \hat{\mathbf{p}}^S, \quad \rho_0^F \frac{\partial \mathbf{v}^F}{\partial t} = \operatorname{div} \mathbf{T}^F + \hat{\mathbf{p}}^F,$$

$$\rho_0^G \frac{\partial \mathbf{v}^G}{\partial t} = \operatorname{div} \mathbf{T}^G + \hat{\mathbf{p}}^G, \tag{13}$$

• balance equation of porosity
$$\frac{\partial (n - n_E)}{\partial t} + \operatorname{div} \mathbf{J} = \hat{n}, \tag{14}$$

• integrability condition for the deformation tensor

$$\frac{\partial \mathbf{e}^S}{\partial t} = \operatorname{sym} \operatorname{grad} \mathbf{v}^S. \tag{15}$$

Quantities with subindex zero are initial values of the corresponding current quantity. The partial stress tensors satisfy the following constitutive relations

$$\begin{aligned}
\mathbf{T}^S &= \mathbf{T}_0^S + \lambda^S e \mathbf{1} + 2\mu^S \mathbf{e}^S + Q^F \varepsilon^F \mathbf{1} + Q^G \varepsilon^G \mathbf{1}, \quad e := \operatorname{tr} \mathbf{e}^S, \\
\mathbf{T}^F &= -p^F \mathbf{1}, \quad p^F = p_0^F - \rho_0^F \kappa^F \varepsilon^F - Q^F e - Q^{FG} \varepsilon^G, \\
\mathbf{T}^G &= -p^G \mathbf{1}, \quad p^G = p_0^G - \rho_0^G \kappa^G \varepsilon^G - Q^G e - Q^{FG} \varepsilon^F,
\end{aligned} \tag{16}$$

where the constant tensor $\mathbf{T}_0^S$ is the initial partial stress in the skeleton, p_0^F is the initial partial pressure in the fluid and p_0^G the initial partial pressure in the gas. These quantities as well as the material parameters $\lambda^S, \mu^S, \kappa^F, \kappa^G, Q^F, Q^G, Q^{FG}$ are functions of an initial porosity n_0 and an initial saturation S_0. The parameters λ^S and μ^S correspond to the classical Lamé constants and κ^F, κ^G correspond to compressibility coefficients of an ideal fluid and of the gas, respectively. The volume changes of fluid and of gas are denoted by ε^F and ε^G, respectively.

The linear constitutive relations for the flux of porosity and for the sources have the following form

$$\hat{\mathbf{p}}^S = \pi^{FS} \left(\mathbf{v}^F - \mathbf{v}^S \right) + \pi^{GS} \left(\mathbf{v}^G - \mathbf{v}^S \right), \quad \hat{\mathbf{p}}^F = -\pi^{FS} \left(\mathbf{v}^F - \mathbf{v}^S \right),$$

$$\hat{\mathbf{p}}^G = -\pi^{GS} \left(\mathbf{v}^G - \mathbf{v}^S \right), \quad \hat{n} = -\frac{n - n_E}{\tau_n}, \tag{17}$$

$$\mathbf{J} = \Phi^F \left(\mathbf{v}^F - \mathbf{v}^S \right) + \Phi^G \left(\mathbf{v}^G - \mathbf{v}^S \right), \quad n_E = n_0 \left(1 + \delta e \right),$$

where n_E denotes values of the porosity in the thermodynamical equilibrium which corresponds to vanishing sources $\hat{\mathbf{p}}^S, \hat{\mathbf{p}}^F, \hat{\mathbf{p}}^G$ and $\hat{n}$. In the linear model the material parameters $\Phi^F, \Phi^G, \pi^{FS}, \pi^{GS}, \tau_n, \delta$ are constants depending on the initial porosity n_0 and the initial saturation S_0.

It is convenient to use a description of the equations which instead on the partial mass densities depends on the volume changes of the components. To this aim we eliminate the mass density ρ^S from the set of fields. We integrate the mass balance equation $(12)_1$ and get under consideration of (15) the relation for volume changes of the skeleton

$$\frac{\partial e}{\partial t} = \operatorname{div} \mathbf{v}^S \quad \Rightarrow \quad \frac{\partial \rho^S}{\partial t} = -\rho_0^S \frac{\partial e}{\partial t} \quad \Rightarrow \quad e = \frac{\rho_0^S - \rho^S}{\rho_0^S}. \tag{18}$$

Similar relations hold for the volume changes of fluid and gas, ε^F and ε^G

$$\varepsilon^F := \frac{\rho_0^F - \rho^F}{\rho_0^F}, \quad \varepsilon^G := \frac{\rho_0^G - \rho^G}{\rho_0^G}. \tag{19}$$

The porosity balance equation (14) can be solved. We obtain in different notation

$$\frac{\partial (n - n_E)}{\partial t} + \frac{n - n_E}{\tau_n} = \Phi^F \frac{\partial (e - \varepsilon^F)}{\partial t} + \Phi^G \frac{\partial (e - \varepsilon^G)}{\partial t}, \tag{20}$$

and the solution

$$n = n_0 \left[1 + \delta e + \frac{\Phi^F}{n_0}(e - \varepsilon^F) + \frac{\Phi^G}{n_0}(e - \varepsilon^G) - \right.$$
$$\left. -\frac{\Phi^F}{n_0 \tau_n} \int_0^t (e - \varepsilon^F)\big|_{t-s} e^{-s/\tau_n} ds - \frac{\Phi^G}{n_0 \tau_n} \int_0^t (e - \varepsilon^G)\big|_{t-s} e^{-s/\tau_n} ds \right]. \tag{21}$$

The last two contributions describe memory effects. Like in the two-component case they are neglected which corresponds to the assumption $\tau_n \to \infty$.

The linear thermodynamical model without memory effects can now be described in the following way:

$$\{\mathbf{v}^S, \mathbf{v}^F, \mathbf{v}^G, \mathbf{e}^S, \varepsilon^F, \varepsilon^G\}, \tag{22}$$

are the essential fields which have to satisfy the field equations

$$\rho_0^S \frac{\partial \mathbf{v}^S}{\partial t} = \operatorname{div} \{\lambda^S e \mathbf{1} + 2\mu^S \mathbf{e}^S + Q^F \varepsilon^F \mathbf{1} + Q^G \varepsilon^G \mathbf{1}\} +$$
$$+ \pi^{FS}(\mathbf{v}^F - \mathbf{v}^S) + \pi^{GS}(\mathbf{v}^G - \mathbf{v}^S),$$
$$\rho_0^F \frac{\partial \mathbf{v}^F}{\partial t} = \operatorname{grad} \{\rho_0^F \kappa^F \varepsilon^F + Q^F e + Q^{FG} \varepsilon^G\} - \pi^{FS}(\mathbf{v}^F - \mathbf{v}^S), \tag{23}$$
$$\rho_0^G \frac{\partial \mathbf{v}^G}{\partial t} = \operatorname{grad} \{\rho_0^G \kappa^G \varepsilon^G + Q^G e + Q^{FG} \varepsilon^F\} - \pi^{GS}(\mathbf{v}^G - \mathbf{v}^S),$$

and

$$\frac{\partial \mathbf{e}^S}{\partial t} = \operatorname{sym} \operatorname{grad} \mathbf{v}^S, \quad \frac{\partial \varepsilon^F}{\partial t} = \operatorname{div} \mathbf{v}^F, \quad \frac{\partial \varepsilon^G}{\partial t} = \operatorname{div} \mathbf{v}^G, \quad e \equiv \operatorname{tr} \mathbf{e}^S,$$
$$n = n_0 \left[1 + \delta e + \frac{\Phi^F}{n_0}(e - \varepsilon^F) + \frac{\Phi^G}{n_0}(e - \varepsilon^G) \right]. \tag{24}$$

If the third component, i.e. the gas, is neglected then the set of equations (23) coincides with the set of Biot's equations, however, as we have seen above, the classical Biot Model does not contain any counterpart to the relation (14) for porosity.

3.2 Micro-macro Transition

In this Section the micro-macro transition method is summarized. For a detailed discussion see: ALBERS [3].

Homogeneous and spherically symmetric deformations are assumed so that mechanical reactions of the system reduce to pressures. For such a system first the macroscopic constitutive relations for partial pressures are considered. In the static case the full pressure change must be in equilibrium with a given excess pressure. The latter four macroscopic equations contain only unknown quantities, namely volume changes $e, \varepsilon^F, \varepsilon^G$, partial pressures p^S, p^F, p^G, porosity n and saturation S as well as material parameters $\lambda^S, \mu^S, \kappa^F, \kappa^G, Q^F, Q^G, Q^{FG}$.

On the other hand, microscopic constitutive relations are assumed which are set up using microscopic variables like volume changes on the microscopic level, e^R, ε^{FR} and ε^{GR}. and corresponding pressures p^{SR}, p^{FR} and p^{GR}. In these relations the measurable microscopic material parameters K_s, K_f and K_g, i.e. real (true) compressibility moduli of the solid component (granulae), of the fluid and of the gas, appear.

A relationship between macroscopic and microscopic properties has to be found. Therefore two sets of compatibility conditions are at hand. These are dynamical and geometrical compatibility relations.

Consequently, for the 14 unknown quantities of spherical homogeneous deformations $\left\{ e, \varepsilon^F, \varepsilon^G, p^S, p^F, p^G, e^R, \varepsilon^{FR}, \varepsilon^{GR}, p^{SR}, p^{FR}, p^{GR}, n, S \right\}$, 13 equations are at the disposal: 1 equilibrium condition, 3 macroscopic constitutive relations, 3 microscopic constitutive relations, 3 dynamical compatibility relations, and 3 geometrical compatibility conditions. Thus, one further equation is needed to solve the problem. This is the microscopic relation for the capillary pressure $p_c(S) = p^{GR} - p^{FR}$ which reflects the dependence of the capillary pressure on the saturation.

The combination of the above given 14 equations yields the explicit solution of the problem, i.e. the volume changes e, ε^F and ε^G as well as all other quantities are given in terms of the given excess pressure. However, the excess pressure still contains the macroscopic material parameters $\left\{ \lambda^S + \frac{2}{3}\mu^S, \kappa^F, \kappa^G, Q^F, Q^G, Q^{FG} \right\}$. Due to the presence of six unknown parameters we need six further conditions to solve the problem completely. For static, homogeneous, spherically symmetric problems such conditions can be easily formulated. For the two-component medium, first by BIOT & WILLIS [8] and later by WILMANSKI [20] three simple tests were considered which concern the investigation of the volume changes and the pressures under certain conditions. They are called "jacketed drained", "jacketed undrained" and "unjacketed" tests. However, it became clear that the incorporation of three tests yields too many conditions which are partly equivalent. It turned out that the two-component problem can be solved using only two of the aforementioned tests: either both jacketed tests or, equivalently, the jacketed undrained and the unjacketed test. It was shown by SANTOS, DOUGLAS & CORBERÓ [12] that also in the case of unsaturated porous media only two tests are necessary to set up an adequate number of boundary conditions. They proposed a "generalized jacketed compressibility test" and a "generalized partially jacketed compressibility test".

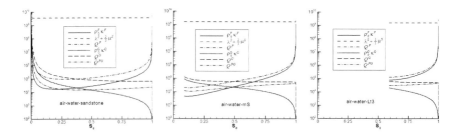

Fig. 2 Material parameters of the three chosen soil types.

The six conditions follow from these compressional tests and using them the problem for the six unknown material parameters $\left\{\lambda^S + \frac{2}{3}\mu^S, \kappa^F, \kappa^G, Q^F, Q^G, Q^{FG}\right\}$ can be solved. One of the solutions to the six equations which are highly nonlinear in the six unknowns is the same as this provided by SANTOS, CORBERÓ & DOUGLAS [12]. The relation between microscopic and macroscopic parameters is not as simple as the GASSMANN relations [10] of the two-component body and, therefore, not quotable here in a compact form. Moreover, for their determination another approach using the total stress tensor of the bulk material and microscopic pressure changes of the two fluid components and another notation are used. However, some conversions yield the dependence of the material parameters on the saturation (for details see: ALBERS [3]).

The macroscopic material parameters of the three soil types under consideration are given in Figure 2. Note that for sand mS and clayey loam $Lt3$ they are not defined for each degree of saturation since these soil types do not exist for low degrees of saturation (also the capillary pressure curves do not cover these regions of saturation; see Figure 1).

4 Wave Propagation in Partially Saturated Rocks

In this Section the propagation of bulk waves in partially saturated elastic media is investigated by means of the above introduced model.

In a one-component medium there exists a single longitudinal wave P and a transversal wave S. The additional component in saturated porous media yields – compared to the one-component body – the existence of an additional bulk wave, the $P2$-wave. Due to the existence of the second pore-fluid (the gas) in the unsaturated medium besides the two compressional waves ($P1$ and $P2$) and the shear wave (S) which appear in the saturated porous medium, an additional compressional wave ($P3$) emerges. In the present work, speeds and attenuations of these waves are shown for three different types of skeleton filled by an water-air-mixture in dependence on the frequency ω and on the initial saturation S_0.

Instead of presenting the wave analysis here (for details see: ALBERS [4]) in the next Section only some representative results are shown.

5   Numerical Analysis

Three types of the skeleton are considered: sandstone, sand mS and silt clay $Lt3$. In the pores a water-air-mixture is present.

The true initial pressures of skeleton, p_0^{SR}, and fluid, p_0^{FR}, are negligible in comparison to K_s and K_f . Therefore they are chosen to be zero. For the corresponding pressure in the gas this is not the case. It is of the same order as K_g. It is determined by the capillary pressure $p_c(S_0)$. Since $p_c(S_0) = p_0^{GR} - p_0^{FR}$ (see page 213) it follows for the gas

$$p_0^{GR} = p_c(S_0); \qquad p_0^{SR} = 0, \quad p_0^{FR} = 0. \tag{25}$$

The following material parameters for water (w) and air (a) have been used:

- true mass densities $\rho_0^{wR} = 1000 \frac{kg}{m^3}$ and $\rho_0^{aR} = 1.2 \frac{kg}{m^3}$,
- microscopic compressibilities $K_w = 2.25 \cdot 10^9$ Pa and $K_a = 1.01 \cdot 10^5$ Pa,
- viscosities (20°C) $\mu_w = 1$ mPas and $\mu_a = 0.0182$ mPas and
- resistances $\pi^w = 10^7 \frac{kg}{m^3 s}$ and $\pi^w = 1.82 \cdot 10^5 \frac{kg}{m^3 s}$.

5.1   Discussion of Numerical Results

Below the numerical results of the wave analysis are discussed. The dispersion relations for the shear wave and for longitudinal waves for three soil types have been solved for the complex wave number k. The results for k reveal both the phase speeds of the body waves $c = \omega/(\mathrm{Re}\,k)$ and the corresponding attenuations $\mathrm{Im}\,k$.

Speeds and attenuations are given for relative low frequencies $\omega = 10, 100, 500,$ 1000 and 5000 Hz. These are frequencies which have a practical bearing in geophysical applications.

The saturation axis covers the whole region $0 \le S_0 \le 1$, however, the analysis shows that the model is not applicable for very low degrees of saturation. That was to be expected because in this region presumptions of the model are not fulfilled anymore. Instead of a continuous fluid with air inclusions in this region of saturation a frothy structure of the the pore fluids is encountered. A second reason for acceptance of the limitation is that smaller values of the saturation do not appear in reality (a residual amount of fluid is trapped in the channels).

Figures 3 and 4 provide a comparison of the wave speeds and attenuations in sandstone filled by an air-water-mixture (black curves) with those on sand (mS, grey curves) and silty clay (Lt3, light grey curves).

Phase speeds of S- and P1-wave are independent of the frequency and – except for the rapid increase near water saturation – nearly constant with change of saturation, however, for the three soil types they differ in size. The reason for this behavior is that the shear modulus, μ^S, for a given Poisson number is constant and the wave speed depends besides it only on the mass densities of the components. Thus, the speed of the S-waves decreases linearly with increasing saturation. The low frequency limit of the S-wave is determined by the inverse of the sum of the mass densities of all three components. In nearly the whole range of saturations also the other elastic constant of the skeleton, λ^S, changes only marginally (see Figure 2). However, for a degree of saturation which is lies close to the state of water saturation it increases abruptly and reaches approximately the double of the preceding value. Since the $P1$-wave is mainly influenced by the compressibility of the skeleton, $\lambda^S + \frac{2}{3}\mu^S$, its speed proceeds nearly constant for almost all values of the saturation and increases only for very large values of S_0. Experimental observations found in the literature support the occurrence of this effect. It may be an important feature for applications in geotechnics. Namely, it provides the hope for the development of a non-destructive testing method to warn against land slides. For both waves, S and $P1$, the speed in sandstone is biggest while this of clay is the smallest. The shear wave speed of sand is approximately one half of the sandstone speed, for silty clay it is around a fourth. For the $P1$ -wave the curves of sand and clay are closer. These waves propagate with approximately 65% (Lt3) and 70% (mS) of the $P1$-speed in sandstone.

Figure 3 shows, that both $P1$- and S-wave have a very low attenuation. S- and $P1$-wave attenuation expose at least one minimum for a certain degree of saturation.

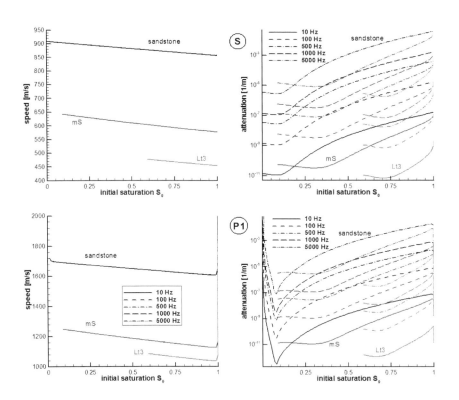

Fig. 3 Comparison of the phase speeds (left column) and the attenuations (right column) of S- and $P1$waves appearing in sandstone (black), sand (grey) and silt clay (light grey) filled by an air-water-mixture in dependence on the initial saturation.

The occurrence of those minima is hard to explain. WEI & MURALEETHARAN [18] attribute the peak near saturation to an influence of residual saturation where the nonwetting phase becomes trapped in the water phase. Since, however, in the modeling the residual saturation for simplification has been neglected, this interpretation is questionable. Various continuum models of multicomponent systems indicate such instabilities near the points of some characteristic frequencies. This is the case for surface waves, as indicated in ALBERS [5] and also apparent for adsorption processes as shown in the work ALBERS [6]. Anyway, the attenuations of both $P1$- and S-waves are almost negligible. In contrast to the speeds they do not differ in same extent between the three soil types.

Figure 4 illustrates the dependence on the saturation of the speeds and attenuations of the slower longitudinal waves, $P2$ and $P3$. For these waves the influence of the degree of saturation is higher. They are effected by the existence of the fluid and

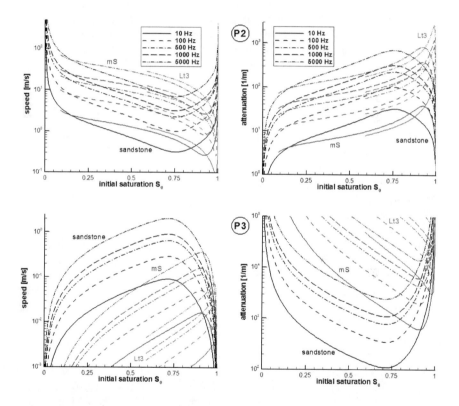

Fig. 4 Comparison of the phase speeds (left column) and the attenuations (right column) of $P2$- and $P3$-waves appearing in sandstone (black), sand (grey) and silt clay (light grey) filled by an air-water-mixture in dependence on the initial saturation.

the gas. In order to make the dependence clearer these diagrams are plotted logarithmically. The speed of the P2-wave decreases as the degree of saturation increases and then increases until the full saturation is reached. The speeds of the P2-wave are higher for higher frequencies.

It is opposite for the P3-wave: it first increases with increasing S_0 and then decreases. Both for gas saturation and for water saturation this wave disappears. This shows that it is driven by the capillary pressure and thus only exists if a second pore fluid is existent. Its speed is much smaller than this of the other waves. However, its attenuation is much higher. The fact that a third longitudinal wave appears in partially saturated porous media is known for almost 15 years. Most likely the first in predicting this form of P3-waves were TUNCAY & CORAPCIOGLU [15] in 1996.

Speeds and attenuations of the P2-wave are of same order for all three soil types. However, both for speed and for attenuation it is obvious that the curves for sandstone are much smoother than these for sand or even clay. The minimum in the speed and the maximum in the attenuation move in comparison to sandstone towards higher values of the saturation for sand and even more for clay. The more the extremum scrolls towards water saturation the more pronounced it is. The reason for this seems to be the bigger difference in the resistances of water and gas for clay and sand ($\pi^G = 1.82 \cdot 10^5 \frac{kg}{m^3 s}$; sandstone: $\pi^F = 10^7 \frac{kg}{m^3 s}$, mS: $\pi^F = 7.27 \cdot 10^7 \frac{kg}{m^3 s}$, Lt3: $\pi^F = 5.33 \cdot 10^9 \frac{kg}{m^3 s}$).

The remark concerning the smoothness of the curves and the intensity and location of the extrema is also valid for both speed and attenuation of the P3-wave. These are shown in the last row of Figure 4. However, for this wave the order of the values for the three soil types are different. If we check, for example, the extremum P3-speeds for $\omega = 10$ Hz, it is obvious that it is approximately 0.09 m/s for sandstone, 0.016 m/s for mS and 0.003 m/s for Lt3. I.e. they nearly differ in one order of magnitude. For the attenuation similar relations are obtained. Since the P3-wave is attributed to the capillary pressure this seems to be an effect of different retention curves for different soil types.

In Figure 4 also the attenuations of P2- and P3-waves are shown. From the analysis of wave propagation in saturated poroelastic media it is already known that the second longitudinal wave is highly damped and therefore hard to observe especially in non-artificial materials. This problem is even more pronounced for the P3-wave because its attenuation is even higher than this of the P2-wave. Thus, the observation of this wave will be nearly impossible. The attenuations of these two waves behave in the opposite way to their speeds. For degrees of saturation for which the speed has a minimum, in the attenuation a maximum appears and vice versa.

The above comments indicate two limitations of the three-component model. Firstly, it cannot be applied in the region of small saturations because the structure of the pore fluids becomes frothy and the presumptions of the model are not fulfilled anymore. On the other hand, as explained above, the model is not applicable to very high frequencies if the permeability parameters are not incorporated in a frequency dependent way.

6   Final Remarks

In the last section the phase speeds and attenuations of two unconsolidated soils have been investigated and compared to the acoustic behavior of sandstone filled by a water-air-mixture. The wave analysis has been conducted with a new model for partially saturated porous materials whose construction has also been indicated in this work. It has been shown in which way macroscopic material parameters can be determined by a micro-macro transition procedure.

It turned out that the S- and $P1$-waves in sandstone are the fastest waves followed by the waves in sand. These two waves in $Lt3$ are the slowest. The reason is that both waves are dependent on the shear modulus which in turn varies with porosity, Poisson's number and mass density and the $P1$-wave additionally to the elasticity of the skeleton. Calculated values are qualitatively concordant with measurements (see: ALBERS [3, 2]).

The $P2$- and $P3$-wave are attributed to the two pore fluids and behave in the opposite way. Where in the $P2$-wave a minimum appears, in the $P3$-wave a maximum occurs. The values of speeds and of the attenuations depend mainly on the resistances and on the capillary pressure.

Due to the complexity of the analysis for the three-component medium only bulk waves have been studied so far. Of course, also for these media a surface wave analysis is aspired. Surface waves have commonly a lower damping than body waves and therefore their investigation will sooner lead to an experimental verification. The model for three-component media does not cover the range of low saturations since in this region the structure of the material is frothy and the presumptions of the model are not anymore valid.

The application of acoustics on soils described in this work demonstrates the practical potential of the micro-macro-transition method for partially saturated soils: it can be used for the development of nondestructive testing methods: for instance, in monitoring and predicting landslides or dam failures. Further applications are imaginable as for example in piping, oil industry, landfilled waste management or for the remediation of compacted residual soils.

References

1. Albers, B.: Micro-macro transition and linear waves in compact granular materials. In: AIP Conference Proceedings of the IUTAM-ISIMM Symposium on Mathematical Modeling and Physical Instances of Granular Flows (to be published, 2009)
2. Albers, B.: Linear elastic wave propagation in unsaturated sands, silts, loams and clays. Submitted to Transport in Porous Media (2009)
3. Albers, B.: Modeling and Numerical Analysis of Wave Propagation in Saturated and Partially Saturated Porous Media, habilitation thesis, Veröffentlichungen des Grundbauinstitutes der Technischen Universität Berlin. Shaker (2010)
4. Albers, B.: Analysis of the propagation of sound waves in partially saturated soils by means of a macroscopic linear poroelastic model. Transport in Porous Media 80(1), 173–192 (2009)

5. Albers, B.: Modelling of surface waves in poroelastic saturated materials by means of a two component continuum. In: Lai, C., Wilmanski, K. (eds.) Surface Waves in Geomechanics: Direct and Inverse Modelling for Soils and Rocks, CISM Courses and Lectures, pp. 277–323. Springer, Wien (2005)

6. Albers, B.: Relaxation analysis and linear stability vs. adsorption in porous materials. Continuum Mech. Thermodyn. 15(1), 73–95 (2003)

7. Albers, B., Wilmanski, K.: On modeling acoustic waves in saturated poroelastic media. J. Engrg. Mech. 131(9), 974–985 (2005)

8. Biot, M.A., Willis, D.G.: The Elastic Coefficients of the Theory of Consolidation. J. Appl. Mech. 24, 594–601 (1957)

9. DIN 4220: Pedologic site assessment – Designation, classification and deduction of soil parameters (normative and nominal scaling). DIN Deutsches Institut für Normung e.V., Beuth Verlag GmbH (draft, in German), German title: Bodenkundliche Standortbeurteilung – Kennzeichnung, Klassifizierung und Ableitung von Bodenkennwerten (normative und nominale Skalierungen) (2005)

10. Gassmann, F.: Über die Elastizität poröser Medien. Vierteljahresschrift der Naturforschenden Gesellschaft in Zürich 96(1), 1–23 (1951)

11. Geertsma, J.: The effect of fluid pressure decline on volumetric changes of porous rocks. Trans. AIME 210, 331–340 (1957)

12. Santos, J., Douglas, J., Corbero, J.: Static and dynamic behaviour of a porous solid saturated by a two-phase fluid. JASA 87(4), 1428–1438 (1990)

13. Schick, P.: Ein quantitatives Zwei-Komponenten-Modell der Porenwasser-Bindekräfte in teilgesättigten Böden. habilitation-thesis, Universität der Bundeswehr, München (2003); In German, Heft 17, Mitteilungen des Instituts für Bodenmechanik und Grundbau

14. Tolstoy, I.: Acoustics, Elasticity and Thermodynamics of Porous Media: Twenty-One Papers by Biot, M.A. Acoustical Society of America (1991)

15. Tuncay, K., Corapcioglu, M.Y.: Body waves in poroelastic media saturated by two immiscible fluids. J. Geophys. Res. 101(B11), 25149–25159 (1996)

16. Van Genuchten, M.T.: A Closed-form Equation for Predicting the Hydraulic Conductivity of Unsaturated Soils. Soil Sci. Soc. Am. J. 44, 892–898 (1980)

17. Vanorio, T., Prasad, M., Nur, A.: Elastic properties of dry clay mineral aggregates, suspensions and sandstones. Geophysical Journal International 155(1), 319–326 (2003)

18. Wei, C., Muraleetharan, K.K.: Acoustical Waves in Unsaturated Porous Media. In: Proceedings 16th Engineering Mechanics Conference, ASCE, Seattle (2003)

19. White, J.E.: Underground sound. In: Application of seismic waves, Methods in Geochemistry and Geophysics, vol. 18, Elsevier, Amsterdam (1983)

20. Wilmanski, K.: On microstructural tests for poroelastic materials and corresponding Gassmann-type relations. Geotechnique 54(9), 593–603 (2004)

21. Wilmanski, K.: Waves in Porous and Granular Materials. In: Hutter, K., Wilmanski, K. (eds.) Kinetic and Continuum Theories of Granular and Porous Media, CISM 400, pp. 131–186. Springer, Wien (1999)

22. Wilmanski, K.: Thermomechanics of Continua. Springer, Berlin (1998)

Explicit Solution Formulas for the Acoustic Diffraction Problem with a Slit in a Hard and a Soft Screen

Matthias Kunik

Dedicated to Professor Wilmanski on the occasion of his 70th birthday.

Abstract. We consider the boundary operators for acoustic slit diffraction problems with a hard and soft screen, respectively. This model can also be applied to Sommerfeld's boundary operators with a Hankel kernel for the diffraction of light through a slit. For a logarithmic approximation of the Hankel kernel we use the Fourier method to derive explicit solutions together with certain regularity results.

1 Introduction

For the spatially two-dimensional acoustic diffraction problem we consider the slit region $|x| < a$ in the $x-y$-plane for $z = 0$. Then solutions $u = u(x,z)$ of the homogeneous Helmholtz-equation $(k^2 + \Delta)u = 0$ in the upper half-space $z > 0$ which are constant along the edges of the slit (i.e. u is independent on y) depend only on the product ak with a wave-number $k > 0$. Thus we may choose $a := 1$ without loss of generality, and in the sequel $I := (-1, 1)$ denotes the slit interval.

In the following section we present the boundary equations for the acoustic slit diffraction problem with a soft and a hard screen, respectively. The mathematical models are the same as in Sommerfeld's electromagnetic diffraction theory, reduced to the slit problem, see the textbooks [27] and [4] and the articles [26], [15].

The Sommerfeld diffraction problem for the slit poses much more difficult mathematical problems than the diffraction problems for some other special geometries, like the half-plane or wedges. Regarding mathematical methods for treating concrete diffraction problems we want to cite the works [5], [6], [7], [17], [18], [19], [21], [22] and [28].

Matthias Kunik

Otto-von-Guericke-Universität, Institut für Analysis und Numerik, Universitätsplatz 2, 39106 Magdeburg, Germany

e-mail: matthias.kunik@ovgu.de

Regarding attempts for obtaining explicit solutions of the boundary integral equation for the hard screen we want to refer to the papers [2], [3], [8], [9], [10], [11],
[13], [16], [20], [24], [25] and [29].

There are well known existence theorems which imply that the corresponding
diffraction problems are uniquely solvable, see for example the preprint [12], but the
general solutions are usually distributions in certain Sobolev spaces. In our study we
replace the Hankel kernel by a simplier logarithmic approximation, which is useful
for a small wave-number $k > 0$, and under appropriate regularity conditions we are
able to solve the slit diffraction problem in terms of classical functions instead of
distributions. The two model equations for the acoustic diffraction problem with a
soft and a hard screen are proposed in the Section 2, whereas in Section 3 we present
there explicit solutions.

2   The Acoustic Diffraction Problem

2.1   The Soft Screen with a Slit

Here the diffraction problem is governed by the boundary integro-differential equation

$$f(x) = \left(k^2 + \frac{d^2}{dx^2} \right) \int_{-1}^{1} u(y) H_0(k|x-y|)\, dy. \tag{1}$$

The first kind Hankel function of order zero is
$H_0 = H_0^{(1)} : \mathbb{C} \setminus (-\infty, 0] \to \mathbb{C}$ with $H_0 = J_0 + iY_0$, given by the Bessel functions
$J_0 : \mathbb{C} \to \mathbb{C}$ and $Y_0 : \mathbb{C} \setminus (-\infty, 0] \to \mathbb{C}$, see the mathematical handbook [1] for more
details.

In the case that the wave number k is very small, we can use a well known logarithmic approximation of the Hankel kernel. For this purpose we define the function
H_0^{ln} by

$$H_0^{ln}(z) = 1 + \frac{2i}{\pi}(\gamma - \ln 2) + \frac{2i}{\pi} \ln z \quad \text{for } z \in \mathbb{C} \setminus \{0\} \text{ with } |\arg(z)| < \pi, \tag{2}$$

where $\gamma = \lim\limits_{n\to\infty} \left(\sum\limits_{j=1}^{n} \frac{1}{j} - \ln n \right)$ is Euler's constant. We have

$$H_0(z) = H_0^{ln}(z) + O(|z|^2 |\ln z|) \quad \text{for } z \to 0 \text{ with } |\arg(z)| < \pi \tag{3}$$

([14], 8.405, 1.; 8.441, 1. and 8.403, 2.).

Now we replace in (1) u by the new unknown $g = \frac{2i}{\pi} u$ and the Hankel function
H_0 by H_0^{ln}, and obtain the simplier logarithmic boundary equation

$$f(x) = \frac{d^2}{dx^2} \int_{-1}^{1} g(u) \ln|x-u|\, du. \tag{4}$$

This equation will be solved in Section 3 for prescribed $f \in C^2[-1,1]$.

2.2 The Hard Screen with a Slit

Here the diffraction problem is governed by the boundary integral equation

$$f(x) = \int_{-1}^{1} u(y) H_0(k|x-y|) \, dy, \tag{5}$$

where $f \in C^1[-1,1]$ is prescribed in the slit interval I, and u is an unknown function in I.

If we replace here again the Hankel kernel by its logarithmic approximation in (3), then we obtain with a constant $\alpha \in \mathbb{C}$ the simplier boundary integral equation

$$f(x) = \int_{-1}^{1} g(y)(\alpha + \ln|x-y|) \, dy, \tag{6}$$

This equation will also be solved in Section 3 by reducing it to the problem (4).

3 Explicit Solution Formulas for the Operators with the Logarithmic Kernel

3.1 Diffraction through a Slit in a Soft Screen

Lemma 3.1. *We define the function* $F : [-1,1]^2 \setminus \{(s,s) \in [-1,1]^2\} \to \mathbb{R}$ *by*

$$F(s,t) := \ln \frac{1 - st + \sqrt{1-s^2}\sqrt{1-t^2}}{|s-t|}. \tag{7}$$

Then we obtain for $s,t \in (-1,1)$ *with* $s \neq t$ *that*

$$F(s,t) = F(t,s), \quad \frac{\partial F}{\partial s}(s,t) = \frac{\sqrt{1-t^2}}{\sqrt{1-s^2}(t-s)}, \quad \lim_{s \to \pm 1} F(s,t) = 0. \tag{8}$$

Moreover, for $s,t \in [-1,1]$ *with* $s \neq t$ *there holds the inequalities*

$$0 \leq F(s,t)|s-t| \leq \sqrt{1-s^2}\sqrt{1-t^2}. \tag{9}$$

Proof. Equations (8) can be checked easily. The first inequality in (9) results from

$$F(s,t) = \frac{1}{2} \ln \frac{1 + \frac{\sqrt{1-s^2}\sqrt{1-t^2}}{1-st}}{1 - \frac{\sqrt{1-s^2}\sqrt{1-t^2}}{1-st}} \geq 0. \tag{10}$$

On the other hand there holds the following inequality for any two positive numbers $b \geq a > 0$,

$$(\ln b - \ln a)\sqrt{ab} \leq b - a. \tag{11}$$

Namely, if we set $x := \frac{b}{a} \geq 1$, we can rewrite this inequality in the form

$$\ln x \leq \frac{x-1}{\sqrt{x}}, \tag{12}$$

which is obvious, because the left- and right-hand side vanish at $x = 1$, whereas for $x > 1$

$$\frac{d}{dx} \ln x = \frac{1}{x} \leq \frac{1}{x} \frac{\sqrt{x} + \sqrt{\frac{1}{x}}}{2} = \frac{d}{dx} \frac{x-1}{\sqrt{x}}. \tag{13}$$

If we put $b := 1 - st + \sqrt{1-s^2}\sqrt{1-t^2}$, $a := 1 - st - \sqrt{1-s^2}\sqrt{1-t^2}$, then $\sqrt{ab} = |s-t|$, and from (11) we obtain the second inequality in (9). ∎

Lemma 3.2. *For all $s \in [-1,1]$ the function $F(s,\cdot) : (-1,1) \setminus \{s\} \to \mathbb{R}$ is in $L^2(-1,1)$ and hence in $L^1(-1,1)$.*

Proof. We have already stated in Lemma 3.1 that F is non-negative and vanishes at $s = \pm 1$. For $s \in (-1,1)$ the non-negative function $F(s,\cdot)$ has a logarithmic singularity at $t = s \in (-1,1)$, so that it is locally square integrable in a certain neighbourhood of s. Due to (9) $F(s,\cdot)$ is also square integrable on $(-1,1)$. ∎

Theorem 3.1. *For $s,t \in [-1,1]$ we define*

$$K(s,t) = -\frac{1}{\pi} \int_{-1}^{1} \ln(|s-u|) F(u,t) \, du. \tag{14}$$

Then we obtain for all $s,t \in [-1,1]$

$$K(s,t) = \begin{cases} (1+\ln 2)\sqrt{1-t^2} - (t-s)\arccos(t), & s \leq t, \\ (1+\ln 2)\sqrt{1-t^2} - (s-t)(\pi - \arccos(t)), & t \leq s. \end{cases} \tag{15}$$

Proof. Suppose that $s,t \in [-1,1]$ with $s \neq t$ and define

$$F_\pm(s,t) := \ln(1 - st \pm \sqrt{1-s^2}\sqrt{1-t^2}). \tag{16}$$

Putting $s = \sin \sigma$, $t = \sin \vartheta$, $u = \sin v$ with $|\sigma|, |\vartheta| \leq \frac{\pi}{2}$, $|v| < \frac{\pi}{2}$, we have $\sqrt{1-s^2} = \cos \sigma$, $\sqrt{1-t^2} = \cos \vartheta$, $\sqrt{1-u^2} = \cos v$ and obtain that

$$F_\pm(\sin v, \sin \vartheta) = -\ln 2 - \sum_{n=1}^{\infty} \frac{2(\mp 1)^n}{n} \cos(n(v \pm \vartheta)). \tag{17}$$

Equation (17) can be derived as follows: For all $z \in \mathbb{C}$ with $|z| \leq 1$ and $z \neq 1$ we use the identity

$$\sum_{n=1}^{\mu} \frac{z^n}{n} = \frac{z}{1-z} \left(1 - \frac{z^\mu}{\mu}\right) - \frac{1}{1-z} \sum_{n=2}^{\mu} \frac{z^n}{n(n-1)} \quad \text{for all } \mu \in \mathbb{N} \tag{18}$$

in order to conclude the well-known result

$$\sum_{n=1}^{\infty} \frac{z^n}{n} = \ln \frac{1}{1-z}.$$
(19)

Putting $z = \mp e^{i(v \pm \vartheta)}$ and forming the real part of (19), we find

$$\sum_{n=1}^{\infty} \frac{2(\mp 1)^n}{n} \cos(n(v \pm \vartheta)) = 2\ln \frac{1}{|1 \pm e^{i(v \pm \vartheta)}|}$$
$$= -\ln 2 - \ln(1 \pm \cos(v \pm \vartheta)).$$
(20)

Regarding that $F_{\pm}(\sin v, \sin \vartheta) = \ln(1 \pm \cos(v \pm \vartheta))$ we obtain (17) from (20). Since $F(u,t) = \frac{1}{2}(F_+(u,t) - F_-(u,t))$, it holds that

$$F(\sin v, \sin \vartheta) = 2 \sum_{n=1}^{\infty} \frac{\sin(2nv)\sin(2n\vartheta)}{2n}$$
$$+ 2 \sum_{n=1}^{\infty} \frac{\cos((2n-1)v)\cos((2n-1)\vartheta)}{2n-1}.$$
(21)

On the other hand we have $\ln|s - u| = \frac{1}{2}(F_+(s,u) + F_-(s,u))$, and conclude

$$\ln|\sin \sigma - \sin v| = -\ln 2 - 2 \sum_{n=1}^{\infty} \frac{\cos(2n\sigma)\cos(2nv)}{2n}$$
$$- 2 \sum_{n=1}^{\infty} \frac{\sin((2n-1)\sigma)\sin((2n-1)v)}{2n-1}.$$
(22)

From (21) we derive the four identities

$$\int_{-\pi/2}^{\pi/2} F(\sin v, \sin \vartheta) \cos v\, dv = \pi \cos \vartheta,$$
(23)

$$\int_{-\pi/2}^{\pi/2} F(\sin v, \sin \vartheta) \cos(2nv) \cos v\, dv$$
$$= \frac{\pi}{2} \left[\frac{\cos((2n-1)\vartheta)}{2n-1} + \frac{\cos((2n+1)\vartheta)}{2n+1} \right] \quad \text{for all } n \in \mathbb{N},$$
(24)

$$\int_{-\pi/2}^{\pi/2} F(\sin v, \sin \vartheta) \sin(v) \cos v\, dv = \frac{\pi}{4} \sin(2\vartheta),$$
(25)

$$\int_{-\pi/2}^{\pi/2} F(\sin v, \sin \vartheta) \sin((2n-1)v) \cos v\, dv$$
$$= \frac{\pi}{2} \left[\frac{\sin(2(n-1)\vartheta)}{2(n-1)} + \frac{\sin(2n\vartheta)}{2n} \right] \quad \text{for all } n \in \mathbb{N}, n \geq 2.$$
(26)

These identities and (22) imply that

$$K(\sin\sigma, \ \sin\vartheta) = -\frac{1}{\pi}\int_{-\pi/2}^{\pi/2} F(\sin v, \sin\vartheta)\ln(|\sin\sigma - \sin v|)\cos v\, dv$$

$$= \ln 2\cos\vartheta + \frac{1}{2}\sin(\sigma)\sin(2\vartheta)$$

$$+ \sum_{n=1}^{\infty} \frac{\cos(2n\sigma)}{2n}\left[\frac{\cos((2n-1)\vartheta)}{2n-1} + \frac{\cos((2n+1)\vartheta)}{2n+1}\right] \tag{27}$$

$$+ \sum_{n=2}^{\infty} \frac{\sin((2n-1)\sigma)}{2n-1}\left[\frac{\sin(2(n-1)\vartheta)}{2(n-1)} + \frac{\sin(2n\vartheta)}{2n}\right].$$

For all $\sigma, \vartheta \in [-\pi/2, \pi/2]$ the expansion (27) converges uniformly as well as absolutely. Especially $K : [-1, 1]^2 \to \mathbb{R}$ is a continuous function.

For $f_\pm : [-\pi, \pi] \times [-\pi/2, \pi/2] \to \mathbb{R}$ with

$$f_\pm(\sigma, \vartheta) := \cos\vartheta + (\vartheta \pm \frac{\pi}{2})(\sin\vartheta - \sin\sigma) \tag{28}$$

we define the *continuous* kernel $K_* : [-\pi, \pi] \times [-\pi/2, \pi/2] \to \mathbb{R}$ by

$$K_*(\sigma, \vartheta) := \begin{cases} f_-(\sigma, \vartheta), & \vartheta \geq 0 \ \text{and} \ \sigma \leq \vartheta, \\ f_+(\sigma, \vartheta), & \vartheta \geq 0 \ \text{and} \ \vartheta \leq \sigma \leq \pi - \vartheta, \\ f_-(\sigma, \vartheta), & \vartheta \geq 0 \ \text{and} \ \pi - \vartheta \leq \sigma, \\ f_+(\sigma, \vartheta), & \vartheta \leq 0 \ \text{and} \ \sigma \leq -\pi - \vartheta, \\ f_-(\sigma, \vartheta), & \vartheta \leq 0 \ \text{and} \ -\pi - \vartheta \leq \sigma \leq \vartheta, \\ f_+(\sigma, \vartheta), & \vartheta \leq 0 \ \text{and} \ \vartheta \leq \sigma. \end{cases} \tag{29}$$

If we calculate for fixed $\vartheta \in [-\pi/2, \pi/2]$ the 2π-periodic Fourier expansion of $K_*(\sigma, \vartheta)$ with respect to the variable $\sigma \in [-\pi, \pi]$ and compare with (27), then we obtain

$$K(\sin\sigma, \sin\vartheta) = \ln 2\cos\vartheta + K_*(\sigma, \vartheta) \tag{30}$$

on the restricted range $\sigma, \vartheta \in [-\pi/2, \pi/2]$. This equation is an equivalent form of (15). ∎

From (23) and $F(s,t) = F(t,s)$ we obtain immediately the following theorem.

Theorem 3.2. *For all $s \in [-1, 1]$ we have $\int_{-1}^{1} F(s,t)\, dt = \pi\sqrt{1 - s^2}$.*

Now we give the solution formula for the logarithmic approximation with a soft screen.

Theorem 3.3. Solution formula for the soft screen. *For any C^1-function $f : [-1, 1] \to \mathbb{C}$ define $g : [-1, 1] \to \mathbb{C}$ by*

$$g(u) = \frac{1}{\pi^2}\int_{-1}^{1} F(u,t) f(t)\, dt. \tag{31}$$

Then g is continuous with $g(-1) = g(1) = 0$, and for all $s \in [-1, 1]$ we have

$$\int_{-1}^{1} \ln(|s - u|) g(u)\, du = -\frac{1}{\pi} \int_{-1}^{1} K(s,t) f(t)\, dt \tag{32}$$

and

$$f(s) = \frac{d^2}{ds^2} \int_{-1}^{1} \ln(|s - u|) g(u)\, du. \tag{33}$$

There exists the derivative $g' : (-1, 1) \to \mathbb{C}$, such that $u \mapsto \sqrt{1 - u^2} g'(u)$ has a continuous extension to the compact interval $[-1, 1]$. This derivative satisfies for $|u| < 1$ the growth condition

$$\left| g'(u) \sqrt{1 - u^2} + \frac{1}{\pi} u f(u) \right| \leq \frac{1}{2\pi} \max_{-1 \leq x \leq 1} |f'(x)|. \tag{34}$$

Proof. From Theorem 3.2 we obtain

$$g(u) = \frac{f(u) \sqrt{1 - u^2}}{\pi} + \frac{1}{\pi^2} \int_{-1}^{1} F(u,t)(t - u) \frac{f(t) - f(u)}{t - u}\, dt. \tag{35}$$

Since $(u,t) \mapsto F(u,t)(t - u)$ has a continuous extension to the compact square $[-1, 1]^2$, we conclude that g is continuous, and by (9) we obtain

$$|g(u)| \leq \frac{\sqrt{1 - u^2}}{\pi} \left(|f(u)| + \frac{1}{2} \max_{-1 \leq x \leq 1} |f'(x)| \right) \tag{36}$$

for all $u \in [-1, 1]$. Especially g has zero boundary values at $u = \pm 1$.

Equation (32) results from the definition of the kernel K with Fubini's theorem, and (33) from (32) and Theorem 3.1 according to

$$\frac{d}{ds} \qquad \int_{-1}^{1} \ln(|s - u|) g(u)\, du$$

$$= -\frac{1}{\pi} \frac{d}{ds} \int_{-1}^{s} K(s,t) f(t)\, dt - \frac{1}{\pi} \frac{d}{ds} \int_{s}^{1} K(s,t) f(t)\, dt$$

$$= -\frac{1}{\pi} K(s,s) f(s) + \frac{1}{\pi} K(s,s) f(s) - \frac{1}{\pi} \int_{-1}^{1} \frac{\partial K}{\partial s}(s,t) f(t)\, dt \tag{37}$$

$$= -\frac{1}{\pi} \int_{-1}^{s} (\arccos(t) - \pi) f(t)\, dt - \frac{1}{\pi} \int_{s}^{1} \arccos(t) f(t)\, dt$$

$$= \int_{-1}^{s} f(t)\, dt - \frac{1}{\pi} \int_{-1}^{1} \arccos(t) f(t)\, dt.$$

It remains to justify the existence of $g' : (-1, 1) \to \mathbb{C}$ satisfying the estimate (34) and the stated extensibility of $u \mapsto \sqrt{1 - u^2} g'(u)$. The estimate (34) follows from the representation (35) regarding the second equation in (8) and Theorem 3.2,

$$g'(u) = \frac{f'(u)\sqrt{1-u^2}}{\pi} - \frac{uf(u)}{\pi\sqrt{1-u^2}} + \frac{1}{\pi^2}\frac{d}{du}\int_{-1}^{1} F(u,t)\left(f(t)-f(u)\right)dt$$

$$= -\frac{uf(u)}{\pi\sqrt{1-u^2}} + \frac{1}{\pi^2}\int_{-1}^{1}\frac{\sqrt{1-t^2}}{\sqrt{1-u^2}}\frac{f(t)-f(u)}{t-u}dt. \qquad (38)$$

From the last representation it also follows that g' is continuous and that $u \mapsto \sqrt{1-u^2}\,g'(u)$ has a continuous extension to the compact interval $[-1,1]$. ∎

3.2 Diffraction through a Slit in a Hard Screen

The following result shows that the solution of the diffraction problem for a small wavenumber k with a soft screen given in Theorem 3.3 can be used to construct the corresponding solution for the hard screen.

Lemma 3.3. *For* $-1 < x < 1$ *we have*

$$\int_{-1}^{1}\frac{\ln|x-y|}{\sqrt{1-y^2}}dy = -\pi\ln 2. \qquad (39)$$

Proof. We put $x = \cos\alpha$ and apply the substitution $y = \cos\beta$ for $\alpha,\beta \in (0,\pi)$ and $\beta \neq \alpha$. Using that

$$\cos\alpha - \cos\beta = 2\sin\left(\frac{\beta+\alpha}{2}\right)\sin\left(\frac{\beta-\alpha}{2}\right), \qquad (40)$$

we conclude that

$$\int_{-1}^{1}\frac{\ln|x-y|}{\sqrt{1-y^2}}dy = \int_{0}^{\pi}\ln|\cos\alpha - \cos\beta|\,d\beta = \frac{1}{2}\int_{0}^{2\pi}\ln|\cos\alpha - \cos\beta|\,d\beta$$

$$= \pi\ln 2 + \frac{1}{2}\int_{0}^{2\pi}\ln\left|\sin\frac{\beta+\alpha}{2}\right|\,d\beta + \frac{1}{2}\int_{0}^{2\pi}\ln\left|\sin\frac{\beta-\alpha}{2}\right|\,d\beta. \qquad (41)$$

The last two and hence all integrals in (41) are independent of $\alpha \in \mathbb{R}$, because $\beta \mapsto \left|\sin\frac{\beta}{2}\right|$ is a 2π-periodic function of $\beta \in \mathbb{R}$, such that (put $\alpha = \frac{\pi}{2}$)

$$J := \int_{-1}^{1}\frac{\ln|x-y|}{\sqrt{1-y^2}}dy = \int_{0}^{\pi}\ln|\cos\beta|\,d\beta \qquad (42)$$

as well as (put $\alpha = 0$)

$$\frac{1}{2}\int_{0}^{2\pi}\ln\left|\sin\frac{\beta\pm\alpha}{2}\right|\,d\beta = \frac{1}{2}\int_{0}^{2\pi}\ln\left(\sin\frac{\beta}{2}\right)\,d\beta$$

$$= \int_{0}^{\pi}\ln\left(\sin\frac{\beta}{2}\right)\,d\beta = 2\int_{0}^{\pi/2}\ln\left(\sin u\right)\,du = J. \qquad (43)$$

With (41), (42) and (43) we conclude the proof of Lemma 3.3.

Theorem 3.4. Solution formula for the hard screen. *We assume that* $\tilde{f} : [-1, 1] \to$ $\mathbb{C}$ *is twice continuously differentiable, and that* $\alpha \neq \ln 2$ *is a given complex number. Let be*

$$\kappa := \frac{1}{\pi(\alpha - \ln 2)} \int_{-1}^{1} \frac{\tilde{f}(s)}{\sqrt{1 - s^2}} \, ds \,, \tag{44}$$

and define $G : [-1, 1] \to \mathbb{C}$ *by*

$$G(s) := \frac{\kappa}{\pi} \arcsin s + \frac{1}{\pi^2} \int_{-1}^{1} F(s, t) \tilde{f}'(t) \, dt \,. \tag{45}$$

Then there exists the derivative $g : (-1, 1) \to \mathbb{C}$ *of* G, *i.e.* $G' = g$, *such that* $s \mapsto$ $\sqrt{1 - s^2} g(s)$ *has a continuous extension to the compact interval* $[-1, 1]$. *For* $|s| < 1$ *we have that* g *is solution of the boundary integral equation*

$$\int_{-1}^{1} (\alpha + \ln|s - u|) \, g(u) \, du = \tilde{f}(s) \,. \tag{46}$$

Proof. If in Theorem 3.3 we replace f by $\tilde{f}'$, then for $g(u)$ in Theorem 3.3 we obtain the expression $G(u) - \frac{\kappa}{\pi} \arcsin(u)$, which has zero boundary values at $u = \pm 1$. Therewith we immediately obtain from Theorem 3.3 that $s \mapsto \sqrt{1 - s^2} G'(s)$ has a continuous extension to $[-1, 1]$. From (37) and Lemma 3.3 now we obtain with $g = G'$ for $|s| < 1$ that

$$\begin{aligned}
& \tilde{f}(s) - \tilde{f}(-1) - \frac{1}{\pi} \int_{-1}^{1} \arccos(t) \, \tilde{f}'(t) \, dt \\
&= \frac{d}{ds} \int_{-1}^{1} \ln(|s - u|) \left[G(u) - \frac{\kappa}{\pi} \arcsin(u) \right] du \\
&= -\frac{d}{ds} \int_{-1}^{1} \int_{s}^{u} \ln|s - v| dv \left[g(u) - \frac{\kappa}{\pi} \frac{1}{\sqrt{1 - u^2}} \right] du \\
&= \frac{d}{ds} \int_{-1}^{1} \left[(s - u) \ln(|s - u|) - (s - u) \right] \left[g(u) - \frac{\kappa}{\pi} \frac{1}{\sqrt{1 - u^2}} \right] du \quad (47) \\
&= \int_{-1}^{1} \ln(|s - u|) g(u) \, du - \frac{\kappa}{\pi} \int_{-1}^{1} \frac{\ln|s - u|}{\sqrt{1 - u^2}} \, du \\
&= \int_{-1}^{1} \ln(|s - u|) g(u) \, du + \kappa \ln 2 \,.
\end{aligned}$$

Next we conclude with partial integration and the definition of κ that

$$\begin{aligned}
& \kappa \ln 2 + \tilde{f}(-1) + \frac{1}{\pi} \int_{-1}^{1} \arccos(t) \, \tilde{f}'(t) \, dt \\
&= \kappa \ln 2 + \frac{1}{\pi} \int_{-1}^{1} \frac{\tilde{f}(u) \, du}{\sqrt{1 - u^2}} = \alpha \kappa \,.
\end{aligned} \tag{48}$$

Inserting this in (47) gives

$$\tilde{f}(s) = \alpha\kappa + \int_{-1}^{1} \ln(|s-u|)g(u)\,du\,, \qquad |s| < 1. \tag{49}$$

Now multiplication of (49) by $\frac{1}{\sqrt{1-s^2}}$ and integration with respect to s over $(-1,1)$ gives in view of (44) and Lemma 3.3

$$\pi\kappa(\alpha - \ln 2) = \int_{-1}^{1} \frac{\tilde{f}(s)}{\sqrt{1-s^2}}\,ds = \pi\alpha\kappa - \pi\ln 2 \int_{-1}^{1} g(t)\,dt\,,$$
$$\kappa = \int_{-1}^{1} g(t)\,dt\,. \tag{50}$$

From (49) and the second formula in (50) we conclude (46). ∎

Remark. *We differentiate (45) for $|s| < 1$ with respect to s, and obtain from (8) in terms of a Cauchy principal value integral for $|s| < 1$*

$$g(s) = \frac{\kappa}{\pi\sqrt{1-s^2}} - \frac{1}{\pi^2\sqrt{1-s^2}}\,\text{p.v.}\int_{-1}^{1} \frac{\tilde{f}'(t)\sqrt{1-t^2}}{s-t}\,dt\,. \tag{51}$$

The boundary integral equation (46) with $\alpha = 0$ is called Carleman's or Symm's integral equation. It is treated e.g., for the integration interval $(0,1)$ instead of $(-1,1)$, in the textbook [23] in 6, Example 6, where for $\alpha = 0$ the solution formula analogous to (51) is obtained in a different way than in our considerations above.

References

1. Abramowitz, M., Stegun, I.: Handbook of mathematical functions. Dover Publications, Inc., Mineola (1968)
2. Bastos, M.A., dos Santos, A.F.: Convolution equations of the first kind on a finite interval in Sobolev spaces. Integral Equations and Operator Theory 13, 638–659 (1990)
3. Belward, J.A.: The solution of an integral equation of the first kind on a finite interval. Quart. Appl. Math. 27, 313–321 (1969)
4. Born, M., Wolf, E.: Principles of Optics: Electromagnetic Theory of Propagation, 7th edn. Interference and Diffraction of Light. Cambridge University Press, Cambridge (1999)
5. Castro, L., Kapanadze, D.: Pseudo-differential Operators in a Wave Diffraction Problem with Impedance Conditions. Fractional Calculus & Applied Analysis 11(1), 15–26 (2008)
6. Castro, L., Kapanadze, D.: The Impedance Boundary-Value Problem of Diffraction by a Strip. J. Math. Anal. Appl. 337, 1031–1040 (2008)
7. Castro, L., Kapanadze, D.: Dirichlet-Neumann-impedance boundary-value problems arising in rectangular wedge diffraction problems. Proceedings of the American Mathematical Society 136, 2113–2123 (2008)
8. Dörr, J.: Zwei Integralgleichungen erster Art, die sich mit Hilfe Mathieuscher Funktionen lösen lassen. ZAMP III, 427–439 (1952)

9. Gorenflo, N.: A new explicit solution method for the diffraction through a slit - Part 2. ZAMP 58, 16–36 (2007)
10. Gorenflo, N.: A new explicit solution method for the diffraction through a slit. ZAMP 53, 877–886 (2002)
11. Gorenflo, N.: A characterization of the range of a finite convolution operator with a Hankel kernel. Integral Transforms and Special Functions 12, 27–36 (2001)
12. Gorenflo, N., Kunik, M.: A new and self-contained presentation of the theory of boundary operators for slit diffraction and their logarithmic approximations, Fakultät für Mathematik, Otto-von-Guericke Universität Magdeburg (preprint, 4/2009)
13. Gorenflo, N., Werner, M.: Solution of a finite convolution equation with a Hankel kernel by matrix factorization. SIAM J. Math. Anal. 28, 434–451 (1997)
14. Gradshteyn, I.S., Ryzhik, I.M.: Table of Integrals, Series, and Products, Corrected and Enlarged Edition. Academic Press, New York (1980)
15. Kunik, M., Skrzypacz, P.: Diffraction of light revisited. Math. Methods in the Appl. Sci. 31(7), 793–820 (2008)
16. Latta, G.E.: The solution of a class of integral equations. J. Rational Mech. Anal. 5, 821–834 (1956)
17. Meister, E.: Einige gelöste und ungelöste kanonische Probleme der mathematischen Beugungstheorie. Expositiones Mathematicae 5, 193–237 (1987)
18. Meister, E., Santos, P.A., Teixeira, F.S.: A Sommerfeld-type diffraction problem with second-order boundary conditions. Z. Angew. Math. Mech. 72(12), 621–630 (1992)
19. Noble, B.: Methods based on the Wiener-Hopf technique for the solution of partial differential equations. Pergamon Press, New York (1958)
20. Pal'cev, B.V.: A generalization of the Wiener-Hopf method for convolution equations on a finite interval with symbols having power-like asymptotics at infinity. Math. USSR Sbornik 41(3), 289–328 (1982)
21. dos Santos, A.F., Teixeira, F.S.: The Sommerfeld problem revisited: Solution spaces and the edge conditions
22. Moura Santos, A., Speck, F.-O.: Sommerfeld diffraction problems with oblique derivatives. Math. Methods in the Appl. Sci. 20(7), 635–652 (1997)
23. Schmeidler, W.: Integralgleichungen mit Anwendungen in Physik und Technik, Band I, Lineare Integralgleichungen, Akademische Verlagsgesellschaft, Geest & Portig K.-G., Leipzig (1950)
24. Shanin, A.V.: Three theorems concerning diffraction by a strip or a slit. Q. J. Mech. Appl. Math. 54, 107–137 (2001)
25. Shanin, A.V.: To the problem of diffraction on a slit: Some properties of Schwarzschild's series. In: Proceedings of the International Seminar Day on Diffraction Millennium Workshop, St. Petersburg, Russia, pp. 143–155 (2000)
26. Sommerfeld, A.: Mathematische Theorie der Diffraction. Math. Ann. 47(2,3), 317–374 (1896)
27. Sommerfeld, A.: Vorlesungen über Theoretische Physik, Band IV, Optik. Verlag Harri Deutsch (1989)
28. Spahn, R.: The diffraction of a plane wave by an infinite slit. I. Quart. Appl. Math. 40(1), 105–110 (1982/83)
29. Williams, M.H.: Diffraction by a finite strip. Q. J. Mech. Appl. Math. 35(1), 103–124 (1982)

On the Stability of the Inversion of Measured Seismic Wave Velocities to Estimate Porosity in Fluid-Saturated Porous Media

Carlo G. Lai and Jorge G.F. Crempien de la Carrera

Abstract. The theory of linear poroelasticity developed by Biot [4] in the low-frequency limit can be profitably used to estimate porosity in a fluid-saturated continuum from measured transversal (shear) and longitudinal (compression) wave velocities. Porosity is an important state parameter which controls the hydro-mechanical response of porous materials such as soils. While in fine-grained geomaterials such as silts and clays porosity can be easily measured in laboratory on undisturbed samples, in coarse-grained soils such as sands and gravels undisturbed sampling is difficult and the possibility of estimating porosity from direct inversion of seismic velocities, which are routinely measured using geophysical seismic prospecting methods, is appealing. The methodology has been proposed and successfully applied by Foti et al., 2002 [13] and it is based on the apparent insensitivity of the inversion procedure on the Poisson ratio of the (evacuated) soil skeleton which is *a-priori* an unknown quantity. This paper attempts to thoroughly investigate the stability of the inversion of measured seismic wave velocities for porosity estimation and to assess the degree of well-posedness of the procedure.

1 Introduction

In soil mechanics, porosity is a state parameter and it is essential for the description of the natural state of a soil deposit. It can be considered the simplest volumetric variable describing particle arrangement in soils and it is formally represented by the first (scalar) invariant of the fabric tensor [33]. Experimental evidence shows that porosity exerts a strong influence on the mechanical response of porous media

Carlo G. Lai
EUCENTRE, Via Ferrata 1, 27100 Pavia, Italy
e-mail: carlo.lai@eucentre.it

Jorge G.F. Crempien de la Carrera
NORSAR, Gunnar Randers vei 15, P.O. Box 53, NO-2007 Kjeller, Norway
e-mail: jorge.crempien@norsar.no

both under static and dynamic loading (e.g. the phenomenon of dilatancy during a seismic excitation).

In experimental soil mechanics, porosity is routinely determined in fine-grained materials (such as silts and clays) through standard laboratory tests carried out on undisturbed soil specimen. In uncemented, coarse-grained soils (e.g. sands and gravels) however, the experimental measurement of porosity is not straightforward as undisturbed sampling is difficult and rarely performed for it requires the use of sophisticated and very expensive ground freezing techniques [1, 18, 27, 16, 30, 14].

In principle, porosity could also be determined through the use of empirical correlations relating this parameter (or its surrogates void ratio and relative density) to either the low-strain shear modulus of soils measured in laboratory or the penetration resistance determined from CPT and SPT field tests [17]. However these correlations are site-specific and strongly dependent upon the particular characteristics (e.g. mineralogy, grain size distribution, confining pressure, stress/strain history, etc.) of the material under investigation. In addition they suffer from both model-based and aleatory uncertainty.

In the geophysical literature several empirical correlations have been proposed to relate porosity in fluid-saturated porous materials with the speeds of propagation of transversal and longitudinal waves V_s and V_p measured from seismic prospecting [34, 10, 8, 11, 24, 2, 20, 21, 22, 25, 15]. Depending on their founding assumptions, there are different categories of velocity-porosity relationships. They have been mostly developed for rock-like materials. A complete review of these correlations is beyond the scope of this paper.

A relatively recent, apparently promising approach for determining porosity from the field measurement of the speeds of propagation of transversal and longitudinal waves is that proposed by Foti et al., 2002 [13]. The methodology aims to estimate porosity using the results of the theory of linear poroelasticity in the low-frequency limit which was developed by Biot [4]. As such, the method assumes the soil deposit as a fluid-saturated porous medium where no relative movement occurs between the fluid and the solid phase (closed system). This is a reasonable assumption considering the frequency content of typical mechanical disturbances generated in seismic prospecting and the hydraulic conductivity of most soils [23].

The theory of linear poroelastodynamics, formulated by Biot [4, 5], predicts the existence, in fluid-saturated media, of two dilatational waves (dilatational wave of the first and second kind) and one rotational or transversal wave. All bulk waves predicted by Biot's theory are *dispersive* i.e. their speed of propagation is frequency-dependent. However, considering the frequency content of stress waves radiated by geophysical sources typically used in seismic prospecting and the values of hydraulic conductivity in coarse and fine-grained soils, it can be reasonably assumed [6, 9, 23] that wave propagation in fluid-saturated soils occur below the frequency at which solid skeleton and fluid move in phase (characteristic frequency). Under these conditions the speed of propagation of both transversal and dilatational wave of the first kind can be determined using the low-frequency approximation of Biot theory [4].

A peculiar aspect of the method proposed by Foti et *al.*, 2002 [13] is that its application only requires measurement of V_s and V_p of the medium, which are quantities that can be easily measured using standard seismic prospecting methods such as cross-hole and down-hole tests. The other parameters required for the implementation of the method are the Poisson ratio of the (evacuated) soil skeleton v^{sk} and four physical constants which assume standard numerical values. Usually the parameter v^{sk} is *a-priori* unknown however Foti et *al.*, 2002 [13] and Foti and Lancellotta, 2004 [12] claim that numerical simulations and experimental data show that the computed values of porosity are little sensitive to this parameter. As far as numerical simulations are concerned the apparent insensitivity of porosity to variations of v^{sk}, all other parameters being constant, was demonstrated for three different pairs $\{V_s, V_p\}$ and for v^{sk} ranging in the interval $[0.10, 0.40]$. Although the method proposed by Foti et *al.*, 2002 [13] to determine porosity is very appealing because this parameter can be estimated directly *in-situ* (and this is particularly useful in fluid-saturated coarse-grained soil deposits) from measured speeds of propagation of transversal and longitudinal waves, it is believed that the apparent negligible influence of the Poisson ratio of the (evacuated) soil skeleton on the estimated values of porosity should be assessed more thoroughly. This is the objective of this article that attempts to systematically investigate the stability of the inversion of measured seismic wave velocities V_s and V_p to estimate porosity and to assess the degree of well-posedness of the corresponding algorithm given the uncertainty of the parameter v^{sk}.

2 Propagation of Seismic Waves in Fluid-Saturated Porous Media

Maurice A. Biot is widely regarded as the first scientist making a thorough theoretical investigation upon the properties of wave propagation in linear, fluid-saturated porous continua. The constitutive model adopted to derive the equations of dynamic poroelasticity [4, 5] was the same introduced by Biot in 1941 in a paper devoted to illustrate the general theory of three dimensional consolidation [3]. In this model the saturated-porous medium is idealized as a binary continuum (nowadays called a *mixture*) obtained from the superposition of a fluid and a solid phase occupying simultaneously the same regions of space. The solid matrix representing the soil skeleton is modeled as an isotropic, linear elastic material. The pores are assumed filled by a non-dissipative compressible fluid. The Biot's constitutive equations can be written as follows:

$$\mathbf{T}^s = 2G \cdot sym(\nabla \mathbf{u}^s) + [(H - 2G) \cdot \mathbf{I} \cdot \nabla \cdot \mathbf{u}^s - C \cdot n \cdot \mathbf{I} \cdot \nabla \cdot (\mathbf{u}^s - \mathbf{u}^f)]$$
$$\mathbf{T}^f = p^f \mathbf{I} = M \cdot n\mathbf{I} \cdot \nabla \cdot (\mathbf{u}^s - \mathbf{u}^f) - C \cdot \mathbf{I} \cdot \nabla \cdot \mathbf{u}^s \tag{1}$$

where $\mathbf{T}^s$ and $\mathbf{T}^f$ are the partial stress tensors of the soil skeleton and fluid phase respectively, $\mathbf{u}^s$ and $\mathbf{u}^f$ are the corresponding displacement vectors, p^f is the pressure of the pore fluid, n is the porosity of the porous medium, $\mathbf{I}$ is the identity matrix, the notation $sym[\cdot] = 1/2[(\cdot) + (\cdot)^T]$ denotes the symmetry operator. Finally G is the (small-strain) elastic shear modulus of the soil skeleton and H, C and M are the Biot elastic constants defined by the following micro-macro transition relationships [28]:

$$H = \frac{(K^s - K^{sk})^2}{(D - K^{sk})} + K^{sk} + \frac{4}{3}G$$

$$C = K^s \frac{(K^s - K^{sk})}{(D - K^{sk})}$$

$$M = \frac{(K^s)^2}{(D - K^{sk})}$$

$$(2)$$

$$D = K^s [1 + n(\frac{K^s}{K^f} - 1)]$$

where K^s is the bulk modulus of the soil grains, K^f is the bulk modulus of the pore fluid, and K^{sk} is the bulk modulus of the soil skeleton.

Biot's equations of motion may be obtained from the application of Equation (1) to the balance law of linear momentum for each constituent of the two-phase medium neglecting body forces and the non-linear contribution of diffusion velocity [7]. The result is:

$$G(\Delta \mathbf{u}^s + \nabla(\nabla \cdot \mathbf{u}^s)) + (H - 2G)\nabla(\nabla \cdot \mathbf{u}^s) -$$
$$C \cdot n \cdot \nabla(\nabla \cdot (\mathbf{u}^s - \mathbf{u}^f)) + \hat{\mathbf{p}}^s = \rho^s(1 - n)\ddot{\mathbf{u}}^s$$

$$(3)$$

$$M \cdot n \cdot \nabla(\nabla \cdot (\mathbf{u}^s - \mathbf{u}^f)) - C \cdot \nabla(\nabla \cdot \mathbf{u}^s) + \hat{\mathbf{p}}^f = \rho^f \cdot n \ddot{\mathbf{u}}^f$$

where $\Delta(\cdot)$ is the Laplacian operator, ρ^s and ρ^f are the mass densities of soil particles and pore fluid respectively, $\hat{\mathbf{p}}^\alpha$ ($\alpha = S, F$) is the momentum supply vector of the α-th component of the mixture [7] and it accounts for the local interaction of the two constituents of the mixture. It is defined as the rate of increase, per unit mass of the mixture, of momentum of the α-th constituent (for instance the solid skeleton) due to the interaction with the other (for instance the pore fluid).

The compatibility of Equation (3) with the balance of linear momentum for the mixture as a whole requires that $\hat{\mathbf{p}}^s = -\hat{\mathbf{p}}^f$. Finally, the dot notation for the displacement vectors $\mathbf{u}^s$ and $\mathbf{u}^f$ is used to denote time-differentiation.

It should be mentioned here that the original Biot's equations of motion included a term ρ^a named *apparent* mass density which was introduced to account for the inertial coupling caused by the relative motion between the solid and fluid phases [4]. The value of ρ^a is a function of the topology of porous network through a parameter called *tortuosity* [9]. It is remarked however that more recent formulations of porous media theory have shown that the concept of apparent mass density is not only unnecessary [7, 31], but it even violates the fundamental principle of material frame indifference [32]. By searching monochromatic solutions of the Biot's equations of motion in the form:

$$\mathbf{u}^s = \mathbf{U}^s \cdot e^{i(k\mathbf{n}\cdot\mathbf{x}-\omega t)}$$
$$\mathbf{u}^f = \mathbf{U}^f \cdot e^{i(k\mathbf{n}\cdot\mathbf{x}-\omega t)}$$

(4)

Equation (3) becomes:

$$\left(\omega^2\mathbf{I} - \frac{H-G-2nC}{\rho^s(1-n)}k^2\mathbf{n}\otimes\mathbf{n} - \frac{G}{\rho^s(1-n)}k^2\mathbf{I} + i\frac{\pi\omega}{n\rho^f}\mathbf{I}\right)\mathbf{U}^s$$

$$-\left(\frac{2nC}{\rho^s(1-n)}k^2\mathbf{n}\otimes\mathbf{n} + i\frac{\pi\omega}{\rho^s(1-n)}\mathbf{I}\right)\mathbf{U}^f = 0$$

(5)

$$-\left(\frac{2nC}{n\rho^f}k^2\mathbf{n}\otimes\mathbf{n} + i\frac{\pi\omega}{n\rho^f}\mathbf{I}\right)\mathbf{U}^s$$

$$+\left(\omega^2\mathbf{I} - \frac{C(1-2n)}{n\rho^f}k^2\cdot\mathbf{n}\otimes\mathbf{n} + i\frac{\pi\omega}{n\rho^f}\mathbf{I}\right)\mathbf{U}^f = 0$$

where $\mathbf{U}^s$ and $\mathbf{U}^f$ are constant amplitudes of particle motion, k is the complex-valued wavenumber, ω is the (real-valued) frequency of excitation, t is time, $i = \sqrt{-1}$, the symbol $\otimes$ denotes the tensor product operator, $\mathbf{x}$ is the position vector and $\mathbf{n}$ is a unit vector defining the direction of wave propagation. Equations (5) define a linear eigenvalue problem with the wavenumber being the eigenvalue and $\mathbf{U}^s$ and $\mathbf{U}^f$ being the eigenvectors.

Since the porous medium is assumed isotropic, the rotational and dilatational motions are uncoupled and the above eigenvalue problem may be split by decomposing the particle motion of both solid and fluid phases, into a component parallel and perpendicular to the direction of wave propagation $\mathbf{n}$ as follows:

$$\mathbf{U}^S = \mathbf{U}^S_\perp + (\mathbf{U}^S\cdot\mathbf{n})\mathbf{n}$$
$$\mathbf{U}^F = \mathbf{U}^F_\perp + (\mathbf{U}^F\cdot\mathbf{n})\mathbf{n}$$

(6)

This decomposition allows to obtain two independent dispersion relations [31]:

$$\omega^3 + i\pi\left(\frac{1}{\rho^s(1-n)} + \frac{1}{n\rho^f}\right)\omega^2 - \frac{k^2 G}{\rho^s(1-n)}\left(\omega + \frac{i\pi}{n\rho^f}\right) = 0 \qquad (7)$$

$$\left(\omega^2 - k^2\frac{H-2nC}{\rho^s(1-n)} + i\frac{\pi\omega}{\rho^s(1-n)}\right)$$

(8)

$$\cdot\left(\omega^2 - k^2\frac{C(1-2n)}{n\rho^f} + i\frac{\pi\omega}{n\rho^f}\right) - \frac{(2nCk^2 + i\pi\omega)^2}{\rho^s\rho^f n(1-n)} = 0$$

Seeking the low frequency limit of the above two equations corresponds to assume that no relative motion occurs between the solid and the fluid phases (i.e. $\mathbf{U}^s = \mathbf{U}^f$). In this case Equations (7) and (8) yield the following result:

$$\lim_{\omega \to 0} \left(\frac{\omega}{\mathrm{Re}(k)} \right)^2 = V_s^2 = \frac{G}{\rho^s(1-n)+n\rho^f} \tag{9}$$

$$\lim_{\omega \to 0} \left(\frac{\omega}{\mathrm{Re}(k)} \right)^2 = V_p^2 = \begin{cases} \dfrac{H}{\rho^s(1-n)+n\rho^f} \\[2ex] 0 \end{cases} \tag{10}$$

where $\mathrm{Re}(k)$ denotes the real part of the wavenumber. Equation (9) represents the speed of propagation V_s in the low frequency approximation of transversal (shear) waves in a fluid-saturated porous medium whereas Equation (10) indicate that at low frequency the dilatational wave of the second kind ceases to exist since its speed of propagation goes to zero. Therefore V_p denotes the velocity of propagation of the dilatational wave of the first kind.

As mentioned in the introduction, V_s and V_p can be easily determined using techniques from seismic prospecting which are standard in geophysics and experimental geotechnical engineering (e.g. cross-hole and down-hole tests, seismic tomography and many others).

3   Estimate of Porosity from Measured Seismic Wave Velocities

Equations (9) and (10) were used by Foti et al., 2002 [13] to propose a method for the estimate of porosity in fluid-saturated soil deposits knowing V_s and V_p that is the speed of propagation of transversal and dilatational waves at low frequency. By substituting Equation (2) into Equation (10) Foti et al., 2002 [13] obtained:

$$\rho V_p^2 - \left[K^{sk} + \frac{4}{3}G \right] - \frac{\left(1 - \dfrac{K^{sk}}{K^s} \right)^2}{\left[\dfrac{n}{K^f} + \dfrac{(1-n)}{K^s} - \dfrac{K^{sk}}{(K^s)^2} \right]} = 0 \tag{11}$$

From the assumption that the shear modulus G^{sk} of the (evacuated) soil skeleton is equal to the shear modulus G of the mixture (since an inviscid pore fluid cannot sustain shear stresses), one can derive the following identity [13]:

$$K^{sk} = \left[\frac{2(1-v^{sk})}{(1-2v^{sk})} - \frac{4}{3} \right] \rho V_s^2 \tag{12}$$

where V_s^{sk} and V_p^{sk} are respectively the speeds of propagation of transversal and dilatational waves of the (evacuated) soil skeleton. K^{sk} and v^{sk} are respectively the soil skeleton bulk modulus and Poisson ratio. The symbol $\rho^{sk} = \rho^s(1-n)$ denotes the *dry* mass density of the soil skeleton. If Equation (9) and Equation (12) are substituted into Equation (11) one obtains:

$$[n(\rho^f - \rho^s) + \rho^s]\left[V_p^2 - 2\frac{1 - v^{sk}}{1 - 2v^{sk}}V_s^2\right] -$$

$$\frac{\left[1 - \frac{V_s^2}{K^s}[n(\rho^f - \rho^s) + \rho^s]\left[2\frac{1 - v^{sk}}{1 - 2v^{sk}} - \frac{4}{3}\right]\right]^2}{\left[\frac{n}{K^f} + \frac{1 - n}{K^s} - \frac{V_s^2}{(K^s)^2}[n(\rho^f - \rho^s) + \rho^s] \cdot \left[\frac{2(1 - v^{sk})}{(1 - 2v^{sk})} - \frac{4}{3}\right]\right]} = 0 \qquad (13)$$

In the next Section it will be shown that Equation (13) is a polynomial of second order which obviously admits, in the most general case, two solutions. However what makes Equation (13) appealing for applications is the fact that if porosity is assumed to be the unknown variable and the pair $\{V_p, V_s\}$ the input data, all the other parameters but v^{sk}, are physical constants assuming rather standard numerical values [13]. Based on these considerations and on the claim that porosity computed from Equation (13) is little sensitive to the specific value assumed for v^{sk} (within its range of variability in the interval $[0.10, 0.40]$, Foti et al., 2002 [13] proposed a methodology to estimate porosity in fluid-saturated soil deposits directly from the measured values of V_p and V_s. The apparent insensitivity of porosity to variations of v^{sk} was assessed by Foti et al., 2002 [13] for three different pairs $\{V_s, V_p\}$ and for v^{sk} ranging in the interval $[0.10, 0.40]$.

The next Section will illustrate the results of a thorough numerical investigation aimed to systematically assess the influence of the Poisson ratio of soil skeleton on porosity calculated from Equation (13).

As a corollary, it is noted that if in Equation (13) soil particles are assumed incompressible that is if $K^s \to \infty$, then this equation degenerates into a simpler quadratic equation:

$$[n(\rho^f - \rho^s) + \rho^s] \cdot \left[V_p^2 - 2\frac{1 - v^{sk}}{1 - 2v^{sk}}V_s^2\right] - \frac{K^f}{n} = 0 \qquad (14)$$

which admits the following explicit, *nonnegative* solution for n [13]:

$$n = \frac{\rho^s - \sqrt{(\rho^s)^2 - \frac{4(\rho^s - \rho^f)K^f}{V_p^2 - 2\frac{1 - v^{sk}}{1 - 2v^{sk}}V_s^2}}}{2(\rho^s - \rho^f)} \qquad (15)$$

This solution can be complex-valued for specific values of ρ^s, ρ^f, K^f, V_p, V_s and v^{sk}. More will be said about this in the next sections.

4 Stability of the Inversion Algorithm

To study the stability of the inversion algorithm used to determine porosity from Equation (13), it is convenient to rewrite this equation in terms of first order polynomials as follows [19]:

$$p_1(n)\left[V_p^2 - C_1 V_s^2\right]\left[p_2(n) - \frac{V_s^2}{(K^s)^2}p_1(n)\right] - \left[1 - \frac{V_s^2}{K^s}p_1(n)\left[C_1 - \frac{4}{3}\right]\right]^2 = 0 \quad (16)$$

where the polynomials of first order and the coefficient C_1 are defined by the following relations:

$$p_1(n) = [n(\rho^f - \rho^s) + \rho^s]$$

$$p_2(n) = \frac{n}{K^f} + \frac{1-n}{K^s}$$

$$C_1 = 2\frac{1 - v^{sk}}{1 - 2v^{sk}}$$

Then, using standard recurrence relations among polynomials, it is possible to write:

$$p_2(n) - \frac{V_s^2}{(K^s)^2}p_1(n) = p_3(n)$$

which is a polynomial of first order. Finally:

$$\left[1 - \frac{V_s^2}{K^s}p_1(n)\left[C_1 - \frac{4}{3}\right]\right]^2 = p_4(n)$$

where

$$p_1(n)p_3(n) = p_5(n)$$

are polynomials of second order. The sum or subtraction of two polynomials of second order yield a polynomial of second order, thus Equation (16) and (12) above are also polynomials of second order. The roots of Equation (16) are then well defined, and even if the formation of the coefficients of the quadratic equation may turn to be cumbersome, they can be determined using standard algorithms. Depending on the specific value assumed by the pair $\{V_s, V_p\}$ and by the physical constants ρ^f, ρ^s, K^f, K^s and v^{sk}, the two roots of Equation (16) for porosity maybe either real or complex conjugates (since all input parameters are real-valued).

Figure (1) shows the values of porosity calculated from Equation (16) for the parameter v^{sk} ranging in the interval $[0.10, 0.40]$. The pairs $\{V_s, V_p\}$ chosen for the numerical simulation are considered to span the whole range of variation of these two parameters in ordinary saturated, non-stony materials. The black dots represent the specific pairs $\{V_s, V_p\}$ adopted by Foti et al., 2002 [13] to assess the sensitivity of porosity to changes in v^{sk} in the solution of Equation (13). The results shown in Figure (1) were obtained by assuming for the parameters ρ^f, ρ^s, K^f, K^s the values of $1\ t/m^3$, $2.7\ t/m^3$, $2.25 \cdot 10^6\ kPa$, and $6.89 \cdot 10^7\ kPa$ respectively which were the same adopted by Foti et al., 2002 [13]. From Figure (1) the dependence of porosity on v^{sk} appears to be moderate at least for values of $v^{sk} < 0.30$. There are particular values of the pairs $\{V_s, V_p\}$ for which the roots provided by Equation (13) are complex-conjugates. These are indicated in the figure as white areas. Figure (2) provides a more effective representation of the sensitivity of porosity calculated from Equation (13) to changes in v^{sk}. The plots in fact show the *relative percent difference*

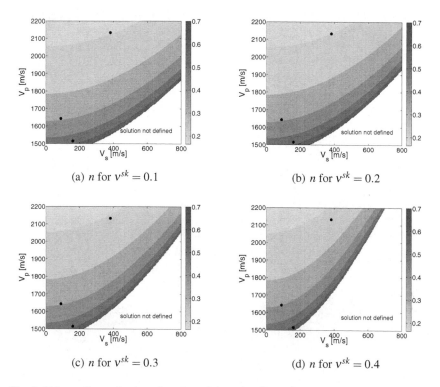

(a) n for $v^{sk} = 0.1$

(b) n for $v^{sk} = 0.2$

(c) n for $v^{sk} = 0.3$

(d) n for $v^{sk} = 0.4$

Fig. 1 Values of porosity as a function of the pairs $\{V_s, V_p\}$ obtained from the solution of Equation (13) for v^{sk} ranging in the interval $[0.10, 0.40]$ with a step increment of 0.10. The black dots correspond to specific the specific pairs $\{V_s, V_p\}$ shown by Foti et al., 2002 [13] in Figure (1), to assess the sensitivity of porosity to changes in v^{sk} (from Lai and Crempien, 2009 [19]).

of porosity (RPD), as a function of the pairs $\{V_s, V_p\}$, obtained from Equation (13) for v^{sk} ranging in the interval $[0.10, 0.45]$. The RPD is the ratio (expressed in percent) between the absolute difference of maximum and minimum values of a quantity (in this case the porosity) and their average value. The definition is similar to that of the *percent error* however for the RPD it is inappropriate to use the term *error* since the actual value of porosity is unknown due to the uncertainty of v^{sk}.

The plots shown in Figure (2) were obtained using for the parameters ρ^f, ρ^s, K^f, K^s the same values adopted to generate the plots of Figure (1). In the chromatic scale of Figure (2) the darker is the gray the greater is the uncertainty in determining porosity given the uncertainty of v^{sk}.

The white regions represent pairs $\{V_p, V_s\}$ where the roots provided by Equation (13) are complex-conjugates. The extension of this region is large, particularly for $v^{sk} = 0.45$. In Figure (2) the black dots fall in regions of the domain $\{V_s, V_p\}$ where the estimation of porosity is well-posed with respect to v^{sk} since the RPD and thus the uncertainty in determining this parameter is small. This was expected as it was already demonstrated in the article by Foti et al., 2002 [13]. However, as Figure (2) clearly shows, there are areas of the $\{V_s, V_p\}$ domain for which the inversion of Equation (13) to determine porosity is *ill-posed* in the sense that the influence of the parameter v^{sk} in the computed porosity is strong. This influence appear to be particularly pronounced as the stiffness of the soils increases and it is large for geomaterials characterized by values of V_s greater than about 500 m/s. Yet, it is important to remark that if values of $V_s > 500$ m/s are associated with values of $V_p > 1700$ m/s, still the determination of porosity remains a relatively *well-posed* problem since the influence of v^{sk} in the interval $[0.10, 0.40]$ is not too strong. However such combination of values for $\{V_s, V_p\}$ may not correspond to typical categories of soils.

Figure (3) provides a confirmation of these findings through a plot of the partial derivative of porosity with respect to the Poisson ratio of (evacuated) soil skeleton. If in Equation (13) porosity were truly independent from v^{sk}, its partial derivative with respect to this parameter, $n_{,v^{sk}}$, would be equal to zero.

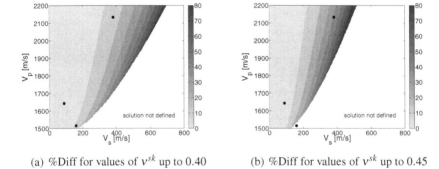

(a) %Diff for values of v^{sk} up to 0.40   (b) %Diff for values of v^{sk} up to 0.45

Fig. 2 *Relative percent difference*, RPD (i.e. ratio between the absolute difference of maximum and minimum values of a quantity and their average value) of porosity as a function of the pairs $\{V_s, V_p\}$ obtained from Equation (13) for v^{sk} ranging in the interval $[0.10, 0.45]$. Again, the black dots correspond to specific the specific pairs $\{V_s, V_p\}$ shown by Foti et al., 2002 [13] in Figure (1) to assess the sensitivity of porosity to changes in v^{sk} (from Lai and Crempien, 2009 [19]).

Figure (3) shows that the area of the $\{V_s, V_p\}$ domain where the quantity $n_{,v^{sk}}$ is small (which is chromatically represented by a light gray color) gets larger as the value of v^{sk} decreases. Conversely, for the same value of v^{sk}, the partial derivative $n_{,v^{sk}}$ increases as the value of V_s increases that is to say when the soil gets stiffer. This is particularly pronounced as the value of v^{sk} increases.

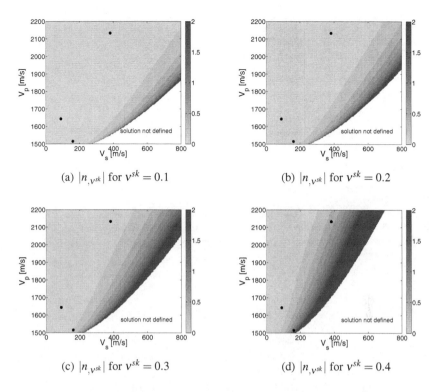

Fig. 3 Absolute values of the partial derivative of porosity with respect to v^{sk} as a function of the pairs $\{V_s, V_p\}$ obtained from the solution of Equation (13) for v^{sk} ranging in the interval $[0.10, 0.40]$ with a step increment of 0.10. The black dots correspond to specific the specific pairs $\{V_s, V_p\}$ shown by Foti et al., 2002 [13] in Figure (1) to assess the sensitivity of porosity to changes in v^{sk} (from Lai and Crempien, 2009 [19]).

5   Applications

This section shows an example of application of the method for determining porosity from direct measurements of V_s and V_p. An assessment of the uncertainty associated with this calculation and due to the (unknown) Poisson ratio of (evacuated) soil skeleton is also illustrated in light of the findings of previous section.

The case study refers to a site located at the Lagoon of Venice, Italy. At this testing site experimental measurements of V_s and V_p were available from a cross-hole seismic test performed up to a depth of 56 m below the mean sea level (msl). Several values of porosity were also determined in laboratory from undisturbed soil specimen. They were retrieved at depths ranging from 11.2 to 56.9 m below the msl.

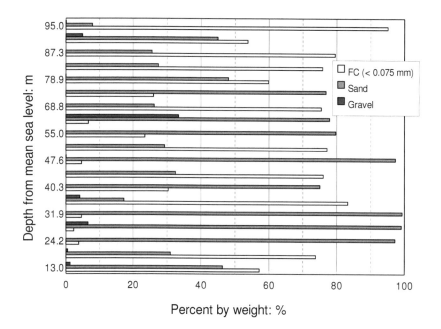

Fig. 4 Grain size distribution profile at the Lagoon of Venice as determined from borehole SC7 executed at *Bocca di Chioggia* testing site (experimental data from [26]). The symbol FC in the legend denotes the fine content and it is defined as the percent by weight passing No. 200 sieve size (0.075 *mm*) (from Lai and Crempien, 2009[19]).

Figure (4) shows the grain size distribution profile obtained from borehole SC7 executed at the *Bocca di Chioggia* testing site [26]. As shown by the figure, the soil deposit consists of alternate layers of sandy silt and weakly-clayey silty sand. Figure (5) (*left*) shows the V_p and V_s profiles measured at *Bocca di Chioggia* testing site from boreholes CH1N and SC7. The measured shear wave velocities are typical of a soft site. The minimum and maximum values of V_s are equal to 115 and 274 m/s respectively.

Figure (5) (*right*) illustrates the porosity profile determined in laboratory from undisturbed soil specimen retrieved at borehole SC7 [26]. Porosity data are not available in the depth range from 34.4 *m* through 54.07 *m* below the msl. The minimum and maximum porosity are equal to 0.315 and 0.723 respectively.

Figure (6) shows the values of porosity predicted by Equation (13) compared with those measured in the laboratory and shown in Figure (5) (*right*). In Figure (6) (*left*) the values of porosity predicted from Biot's theory have been obtained for v^{sk} ranging in the interval $[0.10, 0.40]$ whereas on Figure (6) (*right*) the values of porosity have been obtained for v^{sk} ranging in the interval $[0.10, 0.45]$. The physical parameters ρ^f, ρ^s, K^f, K^s required to compute porosity from Equation (13) were assumed equal to $1 \, t/m^3$, $2.70 \, t/m^3$, $2.25 \cdot 10^6 \, kPa$, and $2.0 \cdot 10^8 \, kPa$ respectively.

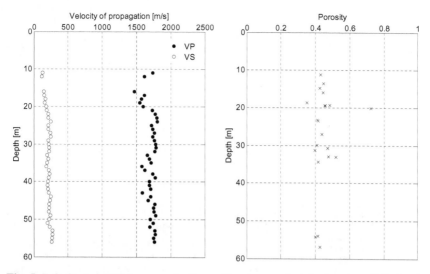

Fig. 5 *Left*: P-wave and S-wave velocity profiles measured at the Lagoon of Venice from boreholes CH1N and SC7 executed at *Bocca di Chioggia* testing site. *Right*: Porosity profile determined in laboratory from undisturbed soil specimen retrieved at borehole SC7 (experimental data from [26]) (from Lai and Crempien, 2009 [19]).

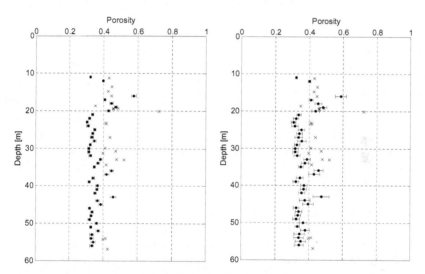

Fig. 6 Comparison between porosity predicted at the Lagoon of Venice (*Bocca di Chioggia* testing site) using Biot's theory in the low frequency limit (Equation (13) above) and data measured in the laboratory from undisturbed soil specimen. *Left*: The predicted values of porosity have been obtained from the solution of Equation (13) for v^{sk} ranging in the interval $[0.10, 0.40]$. *Right*: The predicted values of porosity have been obtained from the solution of Equation (13) for v^{sk} ranging in the interval $[0.10, 0.45]$. The error bars denote the uncertainty associated with the predictions whereas the x symbols represent the laboratory measurement of porosity (from Lai and Crempien, 2009 [19]).

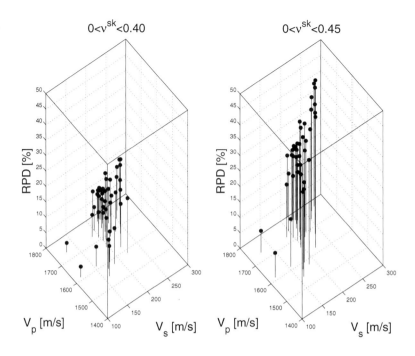

Fig. 7 *Relative percent difference* (RPD) of porosity as a function of the pairs $\{V_s, V_p\}$ obtained from Equation (13) for v^{sk} ranging in the interval $[0.10, 0.40]$ *(left)* and $[0.10, 0.45]$ *(right)*. The black dots represent the uncertainty associated to the estimate of porosity using the values of V_s and V_p measured at the Lagoon of Venice (*Bocca di Chioggia* seismic cross-hole testing site) (from Lai and Crempien, 2009 [19]).

From Figure (6) the comparison between the measured and predicted values of porosity appears to be satisfactory especially at depths between 11 and 20 *m* The dependence of porosity from the Poisson ratio of (evacuated) soil skeleton is weak if v^{sk} ranges in the interval $[0.10, 0.40]$. This is a more realistic range of variation for this parameter. However Figure (6) *(right)* shows that the sensitivity of porosity to v^{sk} increases if this parameter is allowed to vary in the broader range $[0.10, 0.45]$. Figure (7) is an alternative representation to illustrate the influence of v^{sk} in computing porosity at the Lagoon of Venice. The figure shows the uncertainty associated to the estimate of porosity using the values of V_s and V_p measured at the *Bocca di Chioggia* testing site. The relative percent difference (RPD) has been chosen as a measure of the uncertainty in estimating porosity as a function of the $\{V_s, V_p\}$ pairs. The calculation has been conducted for v^{sk} ranging in the interval $[0.10, 0.40]$ (left figure) and $[0.10, 0.45]$ (right figure). Overall Figure (7) confirms that at the Lagoon of Venice testing site the problem of determining porosity from cross-hole data is mathematically well-posed in the sense that it is weakly influenced by the specific value adopted for v^{sk} at least in the range $[0.10, 0.40]$.

6 Conclusions

This paper attempted to investigate the stability of the inversion of measured seismic wave velocities to estimate porosity in fluid-saturated porous media. Porosity is a fundamental parameter for the description of the natural state of a soil deposit. In experimental soil mechanics, the possibility of determining porosity directly in-situ using geophysical seismic testing is appealing because the measurement of this parameter in coarse-grained soil deposits is troublesome due to difficulties associated to undisturbed sampling. The method proposed by Foti et al., 2002 [13] is therefore promising since it aims to estimate porosity of fluid-saturated soil deposits from measured speeds of propagation of transversal and longitudinal waves.

The technique fully exploits the results of the Biot's theory of linear poroelasticity in the low-frequency limit. A critical aspect of the method proposed by Foti et al., 2002 [13] to determine porosity is the influence of Poisson ratio of the (evacuated) soil skeleton. This is an input parameter together with the measured V_s and V_p and it is a-priori unknown and difficult to measure. Yet, Foti et al., 2002 [13] and Foti and Lancellotta, 2004 [12] claim that the computed values of porosity are little sensitive to variations of v^{sk}. This might be true however the apparent insensitivity of porosity to variations of v^{sk} should be thoroughly assessed. Indeed the main objective of this paper was to systematically evaluate the stability and degree of well-posedness of the algorithm for the computation of porosity from the inversion of measured seismic wave velocities V_s and V_p, given the uncertainty of v^{sk}.

The results of the study show that in fluid-saturated soils (that is in non-stony materials) the values of porosity determined from measured V_s and V_p are in general weakly influenced by the Poisson ratio of the (evacuated) soil skeleton as long as v^{sk} is constrained to vary in the interval $[0.10, 0.40]$. From a physical point of view this range of variation is sufficiently large to cover all practical situations. The analysis however, allowed to identify regions of the $\{V_s, V_p\}$ domain of interest, where the influence of v^{sk} in determining porosity is more pronounced. These regions correspond to combinations of V_s and V_p that are characteristic of stiff soils, specifically geomaterials whose values of V_s are greater than about 500 m/s and simultaneously $V_p > 1700\ m/s$. The study has also allowed to identify regions of $\{V_s, V_p\}$ domain where the problem of determining porosity from measured seismic wave velocities is physically not possible since it would lead to roots that are complex-valued. Thus for these pairs of $\{V_s, V_p\}$, porosity is undefined.

Acknowledgements

The authors are grateful to Dr. Fioravante and Dr. Andrea Saccenti from the University of Ferrara, Italy for kindly providing the results of cross-hole seismic measurements and laboratory data on porosity.

References

1. Adachi, K., Tokimatsu, K.: Development in sampling of cohesionless soils in Japan. In: Proc. 13th Int. Conf. on Soil Mechanics and Foundation Engineering, New Delhi, India, January 5-10, vol. 6, pp. 239–250 (1994)

2. Berge, P.A., Bonner, B.P., Berryman, J.G.: Ultrasonic velocity-porosity relationships for sandstone analogs made from fused glass beads. Geophysics 60(1), 108–119 (1995)
3. Biot, M.A.: General theory of three dimensional consolidation. Journal of Applied Physics 12, 155–164 (1941)
4. Biot, M.A.: Theory of Propagation of Elastic Waves in a Fluid-Saturated Porous Solid, I. Lower Frequency Range. J. Acoust. Soc. Am. 28, 168–178 (1956a)
5. Biot, M.A.: Theory of Propagation of Elastic Waves in a Fluid-Saturated Porous Solid. II. Higher Frequency Range. J. Acoust. Soc. Am. 28, 179–191 (1956b)
6. Bourbie, T., Coussy, O., Zinszner, B.: Acoustics of Porous Media. Editions Technip, Paris (1987)
7. Bowen, R.M.: Compressible Porous Media Models by Use of the Theory of Mixtures. International Journal of Engineering Science 20(6), 697–735 (1982)
8. Castagna, J.P., Batzle, M.L., Eastwood, R.L.: Relationship between compressional-wave and shear-wave velocities in elastic silicate rocks. Geophysics 50(4), 571–581 (1985)
9. Coussy, O.: Mechanics of Porous Continua, p. 455. John Wiley and Sons, Chichester (1995)
10. Domenico, S.N.: Rock lithology and porosity determination from shear and compressional wave velocity. Geophysics 49(8), 1188–1195 (1984)
11. Eberhart-Phillips, D., Han, D.-H., Zoback, M.D.: Empirical relationships among seismic velocities, effective pressure, porosity and clay content in sandstones. Geophysics 54(1), 82–89 (1989)
12. Foti, S., Lancellotta, R.: Soil Porosity from Seismic Velocities. Géotechnique 54(8), 551–554 (2004)
13. Foti, S., Lai, C.G., Lancellotta, R.: Porosity of Fluid-Saturated Porous Media from Measured Seismic Wave Velocities. Géotechnique 52(5), 359–373 (2002)
14. Ghionna, V.N., Porcino, D.: Undrained monotonic and cyclic behaviour of a coarse sand from undisturbed and reconstituted samples. In: Di Benedetto, et al. (eds.) Deformation Characteristics of Geomaterials, Swets and Zeitlinger, Lisse, pp. 527–531 (2003) ISBN 9058096041
15. Han, D.-H., Batzle, M.: Gassmann's equation and fluid-saturation effects on seismic velocities. Geophysics 69(2), 398–405 (2004)
16. Hofman, B.A., Sego, D.C., Robertson, P.K.: In situ ground freezing to obtain undisturbed samples of loose sand. Journal of Geotechnical and Geoenvironmental Engineering 126(11), 979–989 (2000)
17. Ishihara, K.: Soil Behaviour in Earthquake Geotechnics, p. 350. Oxford Science Publications, Oxford (1996)
18. Kokusho, T., Tanaka, Y.: Dynamic properties of gravel layers investigated by in-situ freezing sampling. In: Proc. of Sessions on ground failures under seismic conditions. ASCE 0-7844-0055-5, Georgia, October 9-13 (1994); Special Publication (44), 121–140
19. Lai, C.G., Crempien de la Carrera, J.G.F.: Stable inversion of measured V_p and V_s to estimate porosity. Submitted to Geotechnique (2009)
20. Mavko, G., Chan, C., Mukerji, T.: Fluid substitution: estimating changes in VP without knowing V_s. Geophysics 60(6), 1750–1755 (1995)
21. Mavko, G., Mukerji, T.: Seismic pore space compressibility and Gassmanns relation. Geophysics 60(6), 1743–1749 (1995)
22. Mavko, G., Mukerji, T.: Comparison of the Krief and critical porosity models for prediction of porosity and VP/VS. Geophysics 63(3), 925–928 (1998)
23. Miura, K., Yoshida, N., Kim, Y.S.: Frequency Dependent Property of Waves in Saturated Soil. Soils and Foundations 41(2), 1–19 (2001)

24. Murphy, W., Reischer, A., Hsu, K.: Modulus decomposition of compressional and shear velocities in sand bodies. Geophysics 58(2), 227–239 (1993)
25. Nolen-Hoeksema, R.C.: Modulus-porosity relations, Gassmanns equations, and the low-frequency elastic-wave response to fluids. Geophysics 65(5), 1355–1363 (2000)
26. Saccenti, A.: On the mechanical behaviour of the soils of Venices lagoon. PhD thesis in Civil Engineering. University of Ferrara, Italy (2005) (in Italian)
27. Sego, D.C., Hofman, B.A., Robertson, P.K., Wride, C.E.: Undisturbed sampling of loose sand using in-situ ground freezing. In: Lade, P.V., Yamamuro, J.A. (eds.) Proc. Int. workshop on the physics and mechanics of soil liquefaction, Baltimore, Maryland, USA, September 10-11. Rotterdam, Balkema (1999)
28. Stoll, R.D.: Sediment Acoustics. Lecture Notes in Earth Sciences. Springer, Berlin (1989)
29. Vernik, L.: Predicting porosity from acoustic velocities in siliciclastics: a new look. Geophysics 62(1), 118–128 (1997)
30. Yoshimi, Y.: A frozen sample of sand that did not melt. In: Geotech-year 2000 on Developments in Geotechnical Engineering, Bangkok, Thailand, November 2000, pp. 27–30 (2000)
31. Wilmanski, K.: Thermomechanics of Continua, p. 273. Springer, Berlin (1998)
32. Wilmanski, K.: Some Questions on Material Objectivity Arising in Models of Porous Materials. In: Brocato, M. (ed.) Rational Continua, Classical and New. A Collection of Papers Dedicated to G. Capriz. Springer, Italy (2001)
33. Wood, D.M.: Soil Behaviour and Critical State Soil Mechanics, p. 462. Cambridge University Press, Cambridge (1990)
34. Wyllie, M.R.J., Gregory, A.R., Gardner, L.W.: Elastic wave velocity in heterogeneous and porous media. Geophysics 21(1), 41–70 (1956)

On Two Insufficiently Exploited Conservation Laws in Continuum Mechanics: Canonical Momentum and Action

G.A. Maugin and M. Rousseau

Abstract. This contribution has for main purpose to introduce the conservation laws of canonical momentum and action in the continuum-mechanics framework as these are too often ignored or neglected entities while they play an essential role in the theory of the driving forces on defects and inhomogeneities and in some wave problems. Both variational framework (with application of Noether's theorem) and direct statement of balance laws for general continua are presented. Simple classes of materials are considered for the sake of illustration.

1 Introduction

In field theory –of which continuum mechanics is only a small, but paradigmatic, part–, one distinguishes between *field equations* (*one* field equation for each field, thus one scalar field equation for each displacement component in continuum mechanics) and *conservation laws* that are deduced in a second step, via Noether's identity in a variational formulation. In continuum mechanics the field equations correspond classically to local balance laws while field conservation laws are consequences of various symmetries exhibited by the examined physical system as a *whole*. Such a typical *conservation law* is that of energy. We shall strictly adhere to these definitions.

In contrast, in contemporary continuum mechanics illustrated by well known textbooks (*e.g.*, [9, 10]) and treatises (*e.g.*, [44, 43, 11]), the standard local *balance laws* (mass, linear (physical) momentum, moment of momentum, Maxwell's equations), are deduced by localization from global balance laws written for a whole body. Some of these local balance laws may have the form of strict conservation

G.A. Maugin · M. Rousseau
UPMC Univ Paris 6, UMR CNRS 7190, Institut Jean Le Rond d'Alembert, Case 162,
Tour 55, 4 place Jussieu, 75252 Paris Cedex 05, France
e-mail: {gerard.maugin,martine.rousseau}@upmc.fr

laws in a mathematical sense (no source terms). But the only *conservation laws* concerning the whole physical (thermomechanical, electromagnetic, multi-field) system under study in a true field-theoretic sense are the already mentioned *conservation of energy* (expressed by the first principle of thermodynamics; the second law being an inequality indicating a privileged direction of evolution in time) and also much less used laws to which we devote this paper. These are the conservation of *canonical momentum* and that of *action*. The first of these has been shown to relate to the study of material inhomogeneities and various singularities (cracks, dislocations, phase-transformation fronts, shock waves) in the form of a useful expression for representing so-called *driving forces*; *cf.* [24, 25, 29], *etc.* Contrary to what is thought by some philistines, this conservation law cannot be the result of the localization of any meaningful global balance law. The second mentioned additional conservation law, of obvious usefulness in quantum mechanics (*cf.* [16]), has emerged in the continuous framework through the problem of invariance in expansion of the material framework as also in studying some types of waves (*cf.* [31, 32]). The first should not come as a surprise because canonical momentum is intimately related to the conservation of *energy-momentum* in four-dimensional (relativistic or non-relativistic) physics and thus logically complements the conservation of energy. We say in a somewhat pedantic wording that it has the same ontological status as the energy equation. Action (physical dimension: work multiplied by time) is known through the notion of Hamiltonian action in Lagrangian-like variational principles. But we also know that the elementary reduced action (represented by a barred h, *i.e.*, $\hbar$) is the cornerstone of the formulation of quantum mechanics since its introduction by Ernst Planck in the celebrated formula providing the quantum of energy as $E = \hbar v$. The quantum world disappears altogether with $\hbar$ going to zero. Here we emphasize the role of *action* in continuum physics.

It is a pity that the two essential notions that we consider in this contribution are practically ignored by all "mechanicians" of the continuum, and obviously not taught to engineering students, with very few exceptions, in spite of the obviously important role played by conservation laws in modern physics. The present contribution has for main object to discuss that matter and to illustrate with a few selected examples the usefulness of the two notions of *canonical momentum* and *action* in continuum mechanics even in the absence of variational formulation, *i.e.*, when dissipative effects are present.

Section 2 considers the special case of inhomogeneous hyperelasticity deduced from a Hamiltonian-Lagrangian variational principle. This, together with application of Noether's [36] theorem provides the motivation for this contribution as the conservation laws of interest follow naturally this application. More important is the general formulation recalled in Section 3 which proves that results along the same line can be obtained, but in a larger framework, for a general continuum. Special cases are illustrated with an emphasis on the simultaneous co-existence of the conservation, or "non-conservation" of both energy and canonical momentum. The brief Section 4 mentions more generalizations. Section 5 exhibits the dual formulation in "wave vector-frequency space" that is useful in studying some nonlinear wave-propagation problems. Then we have a conservation of wave action and canonical wave momentum.

2   Motivation: Variational Formulation and Noether's Theorem

When a variational formulation in the manner of Hamilton–Lagrange is used to introduce hyperelasticity, the field equations are none other than the components of the local balance law of linear (physical) momentum that reads as follows in the so-called *Piola–Kirchhoff format* (direct notation, no body force for the sake of simplicity)

$$\frac{\partial}{\partial t}\mathbf{p} - \text{div}_R \mathbf{T} = 0. \tag{1}$$

Here we use the following notation. If $\mathbf{x} = \bar{\mathbf{x}}(\mathbf{X},t)$ is the smooth enough placement function (so-called direct deformation mapping between a reference configuration K_R and the actual configuration K_t at Newtonian time t), ∇_R represents the (nabla) material gradient and div_R will stand for the divergence operator (acting on the first index of tensors or geometric objects considered as such for this operation). We have (T = transpose)

$$\mathbf{v} := \left.\frac{\partial \bar{\mathbf{x}}}{\partial t}\right|_X, \qquad \mathbf{F} := \left.\frac{\partial \bar{\mathbf{x}}}{\partial \mathbf{X}}\right|_t = (\nabla_R \bar{\mathbf{x}})^T, \qquad \mathbf{F}^{-1} \equiv (\mathbf{F})^{-1}, \qquad J_F := \det \mathbf{F} \tag{2}$$

respectively the physical velocity, the direct deformation gradient, the inverse deformation gradient, and the Jacobian determinant of $\mathbf{F}$. The object $\mathbf{T}$ is a two-point tensor field (not a tensor in the usual sense) called the first Piola–Kirchhoff stress. In (1), $\mathbf{p}$ is the physical linear momentum, a vector in the actual configuration, such that

$$\mathbf{p} = \rho_0 \mathbf{v}, \qquad \mathbf{T} = J_F \mathbf{F}^{-1} \boldsymbol{\sigma}, \tag{3}$$

where $\boldsymbol{\sigma}$ is the (here) supposedly symmetric Cauchy stress and ρ_0 is the matter density at material point X in K_R. The latter has to satisfy the continuity equation which simply reads in the present format:

$$\left.\frac{\partial \rho_0}{\partial t}\right|_X \neq 0, \tag{4}$$

not excluding any material inhomogeneity from the point of view of inertial properties and any creation or annihilation of matter.

Equation (1) (three components in the physical framework at K_t) represents the set of three Euler–Lagrange (field) equations that follow from the independent variation of the involved fields (here the components of the placement $\mathbf{x}$). In this variational formulation the Lagrangian density per unit reference volume to be considered reads

$$L(\mathbf{v},\mathbf{F};\mathbf{X},t) = K(\mathbf{v};\mathbf{X},t) - \widehat{W}(\mathbf{F};\mathbf{X},t), \tag{5}$$

with kinetic energy

$$K(\mathbf{v};\mathbf{X},t) = \frac{1}{2}\widehat{\rho}_0(\mathbf{X},t)\,\mathbf{v}^2. \tag{6}$$

Systems described by (5) are called *rheonomic* (here inhomogeneous) systems because of the explicit dependency on time (*cf.* [19], following Boltzmann).

The variation of the space-time parameters (X, t) yields, according to Noether's identity and Noether's theorem (see, *e.g.*, [40, 24]), the (in fact non-strict) *conservation laws of energy* and *canonical momentum* in the following form:

$$\left.\frac{\partial H}{\partial t}\right|_{\mathbf{X}} - \nabla_R \cdot \mathbf{Q} = h, \tag{7}$$

and

$$\left.\frac{\partial \mathbf{P}}{\partial t}\right|_{\mathbf{X}} - \mathrm{div}_R\,\mathbf{b} = \mathbf{f}^{\,inh}, \tag{8}$$

where the energy (Hamiltonian) density at the reference configuration, H, the material energy flux $\mathbf{Q}$, the canonical (material) momentum $\mathbf{P}$, the Eshelby (material) stress $\mathbf{b}$, and the heat source h and the so-called material inhomogeneity $\mathbf{f}^{\,inh}$ force are given by

$$H = K + W, \qquad \mathbf{Q} = \mathbf{T}\cdot\mathbf{v}, \tag{9}$$

$$\mathbf{P} := -\rho_0\,\mathbf{v}\cdot\mathbf{F}, \qquad \mathbf{b} = -(L\mathbf{1}_R + \mathbf{T}\cdot\mathbf{F}), \tag{10}$$

$$h := -\left.\frac{\partial L}{\partial t}\right|_{expl} = -\frac{K}{\rho_0}\left.\frac{\partial\widehat{\rho}_0}{\partial t}\right|_{X} + \left.\frac{\partial\widehat{W}}{\partial t}\right|_{expl}, \tag{11}$$

and

$$\mathbf{f}^{\,inh} := \frac{\mathbf{v}^2}{2}\nabla_R\overline{\rho}_0 - \left.\frac{\partial\overline{W}}{\partial\mathbf{X}}\right|_{expl}, \tag{12}$$

where the lower indication "*expl*" means that the derivatives are taken at fixed fields (here $\mathbf{F}$ and $\mathbf{v}$). Once $\widehat{\rho}_0$'s dependency is specified, equations (1), (4), (7) and (8) form the complete set of "primary" local balance laws and conservation laws in the theory of hyperelasticity in which the stress is deduced from the energy by

$$\mathbf{T} = \frac{\partial\widehat{W}}{\partial\mathbf{F}}. \tag{13}$$

Several remarks are in order:

Remark 1. According to (7) and (8), energy and canonical momentum are not strictly conserved in rheonomic materially inhomogeneous systems while physical momentum is "conserved" (in the mathematical sense) in the absence of body force. The loss of energy conservation is due to the explicit time dependence while the loss of conservation of material momentum is due to material inhomogeneities (here of both inertial and elastic origins). Obvious special cases are as follows.

(i)   In *spatially and temporally homogeneous materials*, we simply have

$$\rho_0 = \text{const.}, \qquad W = \overline{W}(\mathbf{F}). \tag{14}$$

(ii)   In materials that are only spatially *inhomogeneous from both inertial and elastic points of view*, we have

$$\rho_0 = \overline{\rho}_0(\mathbf{X}), \qquad W = \overline{W}(\mathbf{F};\mathbf{X}), \tag{15}$$

In both cases (14) and (15) the mass balance in the Piola–Kirchhoff format (4) reads as a true conservation law:

$$\left.\frac{\partial \rho_0}{\partial t}\right|_X = 0. \tag{16}$$

On account of (15) we have the material co-vector $\mathbf{f}^{inh}$

$$\mathbf{f}^{inh} := \frac{\mathbf{v}^2}{2}\nabla_R \overline{\rho}_0 - \left.\frac{\partial \overline{W}}{\partial \mathbf{X}}\right|_{expl}. \tag{17}$$

(iii)   In materials that are simultaneously space and time inhomogeneous, we would *a priori* write

$$\rho_0 = \widetilde{\rho}_0(\mathbf{X},t), \qquad W = \widetilde{W}(\mathbf{F};\mathbf{X},t), \tag{18}$$

although this is seldom done. The explicit dependence on time in the first expression opens up new horizons related to, *e.g.*, the theory of *material growth* (more material of the same type is pushed in at material point $\mathbf{X}$ in the configuration K_R; *cf.* [7]). The form (16) of the continuity equation is no longer valid. There exists a non-vanishing right-hand side (hence the form (4)). The system becomes thermodynamically open. The explicit dependence on time of the second expression in (18) leads to an evolution in time of, say, elasticity coefficients. This may represents the phenomenon of *ageing* (see [33]) of which *creep* is an example. This, of course, does not conserve energy (see below). In view of the very structure of the strict conservation law that the balance law of physical linear momentum (1) achieves in the present paragraph, an interesting sub-case of (18) is

$$\rho_0 = \widehat{\rho}_0(\mathbf{X}), \qquad W = \widehat{W}(\mathbf{F};t), \tag{19}$$

i.e., when there exists a purely inertial material inhomogeneity (*i.e.*, inhomogeneous distribution of mass in K_R) and only a time evolution of the elasticity coefficients, a situation that may be easier to realize experimentally than the general case (18). The time dependence in $(19)_2$ can only be through a relative time (Galilean invariance; so that there exists a birth time t_0 but this is not so relevant here; *cf.* [8]). Also, some of the terms will automatically come out in factor in the two terms in (1).

Remark 2. We have qualified of "primary" the above-given balance and conservation laws. Indeed, another balance law that was not recalled is that of (physical) *moment of momentum*, here reduced to the symmetry of the Cauchy stress or, in terms of **T** and **F** (T = transpose),

$$\mathbf{T}^T \mathbf{F}^T = \mathbf{F} \mathbf{T}. \tag{20}$$

This is "secondary" because no new concept arises in the formulation of this balance law if there are no applied couples (*cf.* [24]). In the same line of thought, (8) is not the only "material" conservation law. It corresponds to the invariance under material translation of the considered physical system (from an equivalent viewpoint, it is generated by an infinitesimal variation of **X**). Corresponding to (20) there is also an invariance of the physical system under rotations in material space – apart from traditional material symmetry. This is exploited in certain problems involving defects (see, *e.g.*, [27, 28]). Furthermore, and more exotic *a priori*, is the possible invariance under **expansion** in material space, an invariance which has no equivalent in physical space. At each regular material point, these two additional conservation laws can be generated by a proper application of Noether's theorem (*cf.* [24], pp. 86–95) for rotation of the material frame and change of scale. But another manner to reach such identities is to envisage multiplying vectorially and scalarly (8) by **X** – *cf.* [25]. This was done by Kienzler and Herrmann [17] in their "*mechanics in material space*". We may call *moment of material momentum* and *material scalar moment* (or "expansion moment" or, else "virial") the material vector $\mathbf{X} \times \mathbf{P}$ and the scalar $\mathbf{X} \cdot \mathbf{P}$, respectively. The resulting identities will play a role in constructing the expression of driving forces for defects (disclinations, spherical voids) related to these symmetries. The quantity $\mathbf{X} \cdot \mathbf{P}$ will necessarily appear in the transformation $\mathbf{X} \cdot \partial \mathbf{P}/\partial t = \partial(\mathbf{P} \cdot \mathbf{X})/\partial t - \mathbf{P} \cdot \mathbf{V}$. Of course the most relevant quantity will often be given by the transformation $\mathbf{X} \cdot \operatorname{div}_R \mathbf{b} = \nabla_R \cdot (\mathbf{b} \cdot \mathbf{X}) - \operatorname{tr} \mathbf{b}$. This will be the case in quasi-statics yielding the so-called M-integral of defect theory (see [24, 25]). This also relates to the derivation of the driving force acting on a *point defect* such as a center of dilatation (*cf.* [18], pp. 213–215; [38]). But in dynamics there is even better because if we consider that time is dilated by the same amount as the material coordinates, we will necessarily generate a scalar material (non) conservation law of the form

$$\frac{\partial A}{\partial t} - \nabla_R \cdot (\mathbf{b} \cdot \mathbf{X} - \mathbf{Q}t) \neq 0, \tag{21}$$

where we have set

$$A := \mathbf{P} \cdot \mathbf{X} - Ht. \tag{22}$$

Equation (21) becomes a strict conservation law (no source in the right-hand side) for a *scleronomic* (contrary of rheonomic) homogeneous system in the absence of body force; see eqn. (5.66) in [24]. The scalar defined in (22) is of utmost importance in physics, although completely ignored in continuum mechanics. To realize this importance and connect with wave mechanics, one has to remember that a *phase* for plane travelling waves is usually defined on the material manifold by

$$\varphi = \mathbf{K} \cdot \mathbf{X} - \omega t, \tag{23}$$

where $\mathbf{K}$ is a material wave vector and ω is a circular frequency. Then the analogy between (22) and (23) is made crystal-clear if we remember that in elementary wave mechanics (Max Planck and Louis de Broglie), we have for a particle $H = \hbar \omega$, $\mathbf{P} = \hbar \mathbf{K}$, –where $\hbar$ is Planck's reduced action quantum–, and thus $A = \hbar \varphi$. Accordingly, A, just like $\hbar$, is an *action*. From this we deduce (*cf.* [31, 32]) that the action (22) will play a prevailing role in wave studies in dynamic (bulky) materials.

3 General Case of Simple Materials

In the possible presence of dissipative processes (heat conduction, viscosity, plasticity, damage, creep, *etc.*) but still keeping the framework of simple materials (in which the stress is the unique "internal force"), we cannot use a variational formulation. Instead, we start with a postulate of standard balance laws and thermodynamic principles. These can be read of in books (*e.g.*, [10, 41]). To make a long story short, we give directly the local expression of these laws and principles at a regular material point. We have thus in the presence of a body force $\mathbf{f}_0$ per unit reference volume

Balance of mass:

$$\left. \frac{\partial \rho_0}{\partial t} \right|_X = 0, \tag{24}$$

Balance of linear (physical) momentum:

$$\left. \frac{\partial (\rho_0 \mathbf{v})}{\partial t} \right|_X - \operatorname{div}_R \mathbf{T} = \mathbf{f}_0, \tag{25}$$

First law of thermodynamics:

$$\left. \frac{\partial (K + E)}{\partial t} \right|_X - \nabla_R \cdot (\mathbf{T} \cdot \mathbf{v} - \mathbf{Q}) = \mathbf{f}_0 \cdot \mathbf{v}, \tag{26}$$

where $K = \rho_0 \mathbf{v}^2 / 2$ is the kinetic energy density, E is the *internal* energy per unit reference volume, and $\mathbf{Q}$ is the material heat flux. This is complemented by the

Second law of thermodynamics:

$$\left. \frac{\partial S}{\partial t} \right|_X + \nabla_R \cdot \mathbf{S} \geq 0, \qquad \mathbf{S} = \frac{\mathbf{Q}}{\theta} + \mathbf{K}, \tag{27}$$

where S is the entropy density, $\theta > 0$ is the absolute temperature ($\inf \theta = 0$), and $\mathbf{S}$ is the entropy flux. The "extra entropy flux" $\mathbf{K}$ vanishes in most cases. Sometimes the material time derivative of a field $A(\mathbf{X}, t)$ is alternately noted $\dot{A} \equiv dA/dt = \partial A/\partial t|_X$. We assume that the above-recalled local equations have indeed been deduced from global statements as usual in contemporary continuum thermomechanics, but this is not given here in order to save space.

Canonical conservation laws of momentum and energy

It is assumed that (20) holds good (symmetry of the Cauchy stress). It remains to transform (26) and (27)$_1$ to useful forms and to deduce the generalization of (8) in this general (nonvariational) case. First, taking the inner product of (25) with **v**, performing some simple manipulations, combining with (26), and setting $E = W + S\theta$ (so that W must be the Helmholtz free energy) and for $\mathbf{K} = \mathbf{0}$, we obtain a local energy equation in the form

$$\frac{d(S\theta)}{dt} + \nabla_R \cdot \mathbf{Q} = h^{int}, \qquad h^{int} := \mathbf{T} : \dot{\mathbf{F}} - \left.\frac{\partial W}{\partial t}\right|_X . \tag{28}$$

Although this is not evident, this equation is the *canonical equation of energy* associated with a canonical conservation law of momentum that will be established in strict parallel. For the moment we simply note that the quantity in the right-hand side of Eq. (28)$_1$ is akin to a heat source and the quantity $S\theta$ is an energy. Guided by the operation just effected on (25) and noting that $\mathbf{F}$ is a material space-like quantity directly associated with the time like quantity $\mathbf{v}$, we apply $\mathbf{F}$ to the right of (25) while introducing plus and minus the material gradient of the energy density $W = \overline{W}(.,.,..,\mathbf{X})$, where the first variables other than $\mathbf{X}$ are not specified. We arrive thus at the following expression of the (non)-conservation of canonical momentum:

$$\frac{d\mathbf{P}}{dt} - \operatorname{div}_R \mathbf{b} = \mathbf{f}^{int} + \mathbf{f}^{ext} + \mathbf{f}^{inh}, \tag{29}$$

in which we have defined the material *Eshelby stress* **b**, the material *inhomogeneity force* $\mathbf{f}^{inh}$, the material *external* (or body) force $\mathbf{f}^{ext}$, and the material *internal* force $\mathbf{f}^{int}$ by

$$\mathbf{b} = -(L_W \mathbf{1}_R + \mathbf{T} \cdot \mathbf{F}), \qquad L_W := K - W, \tag{30}$$

$$\mathbf{f}^{inh} := \left.\frac{\partial L_W}{\partial \mathbf{X}}\right|_{expl} \equiv \left.\frac{\partial L_W}{\partial \mathbf{X}}\right|_{fixed\ fields} = \frac{\mathbf{v}^2}{2}\nabla_R \rho_0 - \left.\frac{\partial \overline{W}}{\partial \mathbf{X}}\right|_{expl}, \tag{31}$$

$$\mathbf{f}^{ext} := -\mathbf{f}_0 \cdot \mathbf{F}, \qquad \mathbf{f}^{int} := \mathbf{T} : (\nabla_R \mathbf{F})^T - \nabla_R W|_{impl}, \tag{32}$$

where the subscript notations *expl* and *impl* mean, respectively, the material gradient keeping the fields fixed (and thus extracting the *explicit* dependence on $\mathbf{X}$), and taking the material gradient only through the fields present in the function. Equation (29) was in fact obtained through a *canonical projection* of (25) onto the material manifold. It is the *canonical* conservation law of momentum of continuum mechanics for *simple* materials. Equations (28)$_1$ and (29) are general laws of physics, but they are *not* independent of the original equations (25) and (26). They are canonical because they are expressed in the material framework; they are directly complementary of one another in the underlying four-dimensional Euclidean framework made transparent by comparing the formal expressions of h^{int} and $\mathbf{f}^{int}$. But note that the terms $\mathbf{f}^{inh}$ and $\mathbf{f}^{ext}$ have no analogues in Eq. (28)$_1$. The reason for this is that they are not associated with any internal dissipation which in fact is what Eq. (28)$_1$

essentially captures as already noted. Notice that L_W has the form of a density of Lagrangian although no variational principle has been introduced. Indeed, like in the rest of continuum mechanics, one should now proceed to establishing relevant *constitutive equations* on the requirement that the inequality $(27)_1$ transformed into the *Clausius–Duhem* (CL) form

$$-\left(\frac{\partial W}{\partial t}\bigg|_X + S\frac{\partial \theta}{\partial t}\bigg|_X\right) + \mathbf{T}:\dot{\mathbf{F}} - \mathbf{S}\cdot\nabla_R\theta \equiv h^{int} - S\frac{\partial \theta}{\partial t}\bigg|_X - \mathbf{S}\cdot\nabla_R\theta \geq 0 \qquad (33)$$

be respected, and in fact used as a constraint.

Equations (24), (25), (28), (29) and (33) form the complete system of "primary" local balance and conservation (or non-conservation) laws for so-called simple materials, "simple" being here understood in the sense of W. Noll. However, dissipative cases accounting for internal variables of state, and even their gradients, are not excluded as these do not modify, before hand, the general statements of balance laws (because they cannot be acted upon directly from the outside in the bulk or at the surface of the body). They will manifest themselves only through their appearance in the –as yet unspecified– energy density and in the residual dissipation.

Examples

A. Pure homogeneous elasticity.

In that case $\rho_0 = $ const., and $W = \overline{W}(\mathbf{F})$ only. We have $h^{int} \equiv 0$, $\mathbf{f}^{int} \equiv \mathbf{0}$ since $\mathbf{T} = \partial\overline{W}/\partial\mathbf{F}$, and also $\mathbf{f}^{inh} = \mathbf{0}$, $\mathbf{Q} \equiv \mathbf{0}$. Equations (29) and $(28)_1$ reduce to the following system ($\theta_0 = $ const.):

$$\frac{d\mathbf{P}}{dt} - \text{div}_R\,\mathbf{b} = \mathbf{0}, \qquad \theta_0\frac{dS}{dt} = 0. \qquad (34)$$

B. Inhomogeneous thermoelasticity of conductors.

In that case $\rho_0 = \overline{\rho}_0(\mathbf{X})$, and $W = \overline{W}(\mathbf{F}, \theta; \mathbf{X})$. Still with $\mathbf{K} \equiv \mathbf{0}$. We have the constitutive equations

$$\mathbf{T} = \frac{\partial\overline{W}}{\partial\mathbf{F}}, \qquad S = -\frac{\partial\overline{W}}{\partial\theta}, \qquad (35)$$

that follow from a standard exploitation of the CL- inequality (33). Accordingly, we obtain that $\mathbf{f}^{int} \equiv \mathbf{f}^{th}$ and $h^{int} \equiv h^{th} := S\dot{\theta}$, where $\mathbf{f}^{th} := S\nabla_R\theta$ is the material thermal force introduced by Bui [2] in small strains and Epstein and Maugin [25] in general, so that (29) and $(28)_1$ are replaced by the following canonical system of balance of momentum and energy:

$$\frac{d\mathbf{P}}{dt} - \text{div}_R\,\mathbf{b} = \mathbf{f}^{inh} + \mathbf{f}^{th}, \qquad \frac{d(S\theta)}{dt} + \nabla_R\cdot\mathbf{Q} = h^{th}, \qquad (36)$$

as found by Maugin [30].

C. Homogeneous dissipative solid material described by means of an internal variable of state.

Let α the internal variable of state whose tensorial nature is not specified. This may relate to damage, or anelasticity of some sort. Then W is specified as the general sufficiently regular function $W = \overline{W}(\mathbf{F}, \theta, \alpha)$. The equations of state are given by Gibbs' equation as

$$\mathbf{T} = \frac{\partial \overline{W}}{\partial \mathbf{F}}, \qquad S = -\frac{\partial \overline{W}}{\partial \theta}, \qquad A := \frac{\partial \overline{W}}{\partial \alpha}, \tag{37}$$

where A is the thermodynamical force (not to be mistaken for the action) associated with α. We then find that

$$\mathbf{f}^{int} = \mathbf{f}^{th} + \mathbf{f}^{intr}, \qquad h^{int} = h^{th} + h^{intr}, \tag{38}$$

where the thermal sources have already been defined and the "*intrinsic*" sources are given by

$$\mathbf{f}^{intr} := A \cdot (\nabla_R \alpha)^T, \qquad h^{intr} := A \dot{\alpha}, \tag{39}$$

so that we have the following obviously consistent (compare the two *r-h-s*) system of canonical conservation laws:

$$\frac{d\mathbf{P}}{dt} - \mathrm{div}_R \mathbf{b} = \mathbf{f}^{th} + \mathbf{f}^{intr}, \qquad \frac{d(S\theta)}{dt} + \nabla_R \cdot \mathbf{Q} = h^{th} + h^{intr}, \tag{40}$$

while the residual dissipation reads

$$\Phi = h^{intr} - \mathbf{S} \cdot \nabla_R \theta \geq 0, \qquad \mathbf{K} \equiv \mathbf{0}. \tag{41}$$

This case can easily be generalized (*cf.* [30]) to the case where it is assumed that the gradient of θ plays a determining role (*e.g.*, in plasticity with nonuniform strain states). Indeed, a possible *standard* model of materially homogenous elasto-anelasticity (plasticity or viscoplasticity) is the one deduced from the free energy density

$$W = \overline{W}(\mathbf{F}^e, \theta, \alpha = \{\beta, \mathbf{F}^p\}), \tag{42}$$

where the set of internal variables is composed of a scalar α accounting for work hardening and the anelastic deformation "gradient" $\mathbf{F}^p$, having assumed the validity of the multiplicative decomposition $\mathbf{F} = \mathbf{F}^e \mathbf{F}^p$, of which none of the factors is a true gradient. But if the gradient of α becomes so much relevant, then we should consider the energy density

$$W = \overline{W}(\mathbf{F}, \theta, \alpha, \nabla_R \alpha), \tag{43}$$

and keep the possibility that $\mathbf{K}$ be *not zero*. The equations of state are given by Gibbs' equation as

$$\mathbf{T} = \frac{\partial \overline{W}}{\partial \mathbf{F}}, \qquad S = -\frac{\partial \overline{W}}{\partial \theta}$$

$$A := -\frac{\partial \overline{W}}{\partial \alpha}, \qquad \mathbf{B} := -\frac{\partial \overline{W}}{\partial (\nabla_R \alpha)} \tag{44}$$

We further set

$$\widetilde{A} \equiv -\frac{\delta \overline{W}}{\delta \alpha} := -\left(\frac{\partial \overline{W}}{\partial \alpha} - \nabla_R \cdot \frac{\partial \overline{W}}{\partial (\nabla_R \alpha)} \right) = A - \nabla_R \cdot \mathbf{B},$$

$$\widetilde{\mathbf{S}} := \theta^{-1} \widetilde{\mathbf{Q}}, \qquad \widetilde{\mathbf{Q}} = \mathbf{Q} - \mathbf{B} \dot{\alpha}, \tag{45}$$

where we have selected $\mathbf{K} = -\theta^{-1} \mathbf{B} \dot{\alpha}$. We then let the reader check that Eq. (40) and (41) are now replaced by the following equations:

$$\frac{d\mathbf{P}}{dt} - \operatorname{div}_R \widetilde{\mathbf{b}} = \mathbf{f}^{th} + \widetilde{\mathbf{f}}^{intr}, \qquad \frac{d(S\theta)}{dt} + \nabla_R \cdot \widetilde{\mathbf{Q}} = h^{th} + \widetilde{h}^{intr}, \tag{46}$$

and

$$\Phi = \widetilde{h}^{intr} - \widetilde{\mathbf{S}} \cdot \nabla_R \theta \geq 0 \tag{47}$$

where we have introduced the new definitions (*cf.* [30])

$$\widetilde{\mathbf{b}} = -\left(L \mathbf{1}_R + \mathbf{T} \cdot \mathbf{F} - \mathbf{B} \cdot (\nabla_R \alpha)^T \right), \qquad \widetilde{\mathbf{f}}^{intr} := \widetilde{A} \nabla_R \alpha, \qquad h^{intr} := \widetilde{A} \dot{\alpha}. \tag{48}$$

This is in the spirit of the approach advocated by Maugin [23] where we can deviate from the classical relation between entropy and heat vectors. This happens in media with species diffusion; see [4]. Here we note that an alteration in the entropy flux definition goes along with a parallel alteration in the expression of the Eshelby stress tensor, thus reinforcing the space-like complementarity of Eq. (46).

Equations (34), or (34)₁ and the reduced form of (26), are identities at regular material points. But they provide the bases on which are built the expressions for the driving force and associated energy-release rate in the so-called "analytical" theory of brittle fracture (such as shown by Dascalu and Maugin [24]), yielding unequivocally the driving force and the known J-integral. But the more general expressions such as (36) and (40) have similarly been used to compute the driving force on dislocations and other defects in materially inhomogeneous bodies, in thermoelasticity, and also in elastoplasticity. Works along these lines demonstrating the essential role played by the (non) conservation of canonical momentum in the presence of field singularities are now numerous especially in numerical treatments. A selection of papers illustrating this remarkable success from different schools is: Haddi and Weichert [14], Müller and Maugin [35], Denzer *et al.* [5], Simha *et al.* [39], Fischer [12], Fischer *et al.* [13]. In parallel, the jump relations associated with equations (28)₁ and (29) have provided the main ingredients in the thermomechanical description of the irreversible progress of phase-transformation fronts and shock waves (see, *e.g.*, the book by Berezovski *et al.* [1], and papers by Maugin [26, 27]). Other material conservation laws (rotation and virial) play a fundamental role in studying appropriate defects (disclinations and void growth –see [25]– in relation with L and M path-independent integrals) as well as in certain problems of the strength of structures (*cf.* [17]). Another type of application concerns the characterization of a system as being homogeneous or not in time and space, for instance while dealing with "dynamics" materials.

D. Dynamic materials

Consider hyperelastic materials with matter density and energy density such as given by (19). This very specific choice makes that in 1D linear elasticity Eq. (25) takes the following form in the absence of body force (with an obvious notation):

$$\rho_0(x)\,u_{tt} - E(t)\,u_{xx} = 0. \tag{49}$$

This is a linear wave equation with space-time varying characteristic velocity. For this field equation for the displacement u, the associated (non) conservation of energy and canonical momentum read

$$\frac{\partial H}{\partial t} - \frac{\partial Q}{\partial x} = h := -\left.\frac{\partial L}{\partial t}\right|_{expl} \tag{50}$$

and

$$\frac{\partial P}{\partial t} - \frac{\partial b}{\partial x} = f := \left.\frac{\partial L}{\partial x}\right|_{expl}, \tag{51}$$

where

$$L = \frac{1}{2}\rho_0(x)\,(u_t)^2 - \frac{1}{2}E(t)\,(u_x)^2, \quad H = \frac{1}{2}\rho_0(x)\,(u_t)^2 + \frac{1}{2}E(t)\,(u_x)^2, \tag{52}$$

$$Q = -u_t\frac{\partial L}{\partial u_x} = E\,u_t\,u_x, \quad P = -u_x\frac{\partial L}{\partial u_t} = -\rho_0\,u_t\,u_x, \tag{53}$$

and

$$b = -\left(L - u_x\frac{\partial L}{\partial u_x}\right) = -H. \tag{54}$$

This final expression is a peculiarity of the 1D model with quadratic energy. Here, the special forms of h and f are immediately read off from (52):

$$h = \frac{(u_x)^2}{2}\frac{dE}{dt}, \quad f = \frac{(u_t)^2}{2}\frac{d\rho_0}{dx}. \tag{55}$$

Because of these two terms, the studied physical system is *inhomogeneous* in both space and time, although fully linear from the point of view of partial differential equations. But it presents original properties, in particular from the point of view of time (that we report in a different work; *cf.* [37]). These have also been studied in greater detail by Lurié [22]. This author has envisaged such doubly-periodic systems forming a checkerboard in space-time, and studied the long time behavior of travelling-wave solutions resulting in a kind of homogenisation. Indeed, with a superimposed tilde denoting his solutions, his equations (2.92)

([22], p. 44) for energy and canonical momentum corresponding to the zeroth-order asymptotic homogenized solution read

$$\frac{\partial \tilde{H}}{\partial t} - \frac{\partial \tilde{Q}}{\partial x} = 0, \qquad \frac{\partial \tilde{P}}{\partial t} - \frac{\partial \tilde{b}}{\partial x} = 0. \tag{56}$$

The second of these means that the system is now viewed as materially homogeneous (which is the usual purpose of homogenisation in space: replacing an inhomogeneous medium by an effective homogeneous one). The first of these of course means that the system is no longer dissipative or gaining energy. This is "homogenisation in time". The result is tantamount to saying that the looked for special effects imagined for these dynamic materials disappear altogether in the homogenisation procedure.

4   Nonsimple and Microstructured Materials

Equations (24), (25), (28), (29) and (33) require amendments and/or generalization for non-simple materials (*i.e.*, materials with higher order deformation gradients), for materials endowed with active internal degrees of freedom (such as in micropolar and micromorphic materials, according to Eringen's classification of media with microstructures), and media in interaction with, *e.g.*, electromagnetic fields. To the benefit of the readers, we content ourselves with citing the most representative works, *e.g.*, Maugin and Trimarco [34] for materials accounting for strain gradients in their energy density, Maugin [28] for polar materials endowed with a rigidly rotating microstructure, and to Trimarco and Maugin [42] for simple materials in interaction with electromagnetic fields. These, together with allied works by the same authors, demonstrate the usefulness of the material conservation laws.

5   The Appearance of Wave Action and Its Conservation

The Hamiltonian action A_H on which the formulation (1)–(8) was based reads

$$A_H = \int_t \int_{B_R} L\left(\frac{\partial \overline{\mathbf{x}}}{\partial \mathbf{X}} = \mathbf{F}, \frac{\partial \overline{\mathbf{x}}}{\partial t} = \mathbf{v}; \mathbf{X}, t\right) dV \, dt. \tag{57}$$

Let us consider Whitham's [45, 46] theory of the so-called averaged Lagrangian and the corresponding wave system. We remind the reader that the phase of a *plane* linear wave in a continuum is defined in the material description by

$$\varphi(\mathbf{X}, t) = \tilde{\varphi}(\mathbf{K}, \omega) = \mathbf{K} \cdot \mathbf{X} - \omega t, \tag{58}$$

where $\mathbf{K}$ is the material wave vector and ω is the associated circular frequency. But in the kinematic wave theory due mostly to Lighthill [21], Whitham [45, 46], and Hayes [15] a general phase function

$$\varphi = \overline{\varphi}(\mathbf{X}, t). \tag{59}$$

is introduced from which the material wave vector $\mathbf{K}$ and the frequency ω are defined by

$$\mathbf{K} = \frac{\partial \overline{\varphi}}{\partial \mathbf{X}} = \nabla_R \overline{\varphi}, \qquad \omega = -\frac{\partial \overline{\varphi}}{\partial t}, \tag{60}$$

Whence there follows at once the two equations (curl-free nature of $\mathbf{K}$, and *conservation* of wave vector)

$$\nabla_R \times \mathbf{K} = \mathbf{0} \tag{61}$$

and

$$\frac{\partial \mathbf{K}}{\partial t} + \nabla_R \omega = \mathbf{0}. \tag{62}$$

In particular, Eq. (60) are trivially satisfied for plane wave solutions for which the last of (58) holds true. For an *inhomogeneous rheonomic* linear behavior with dispersion we would have a dispersion relation $\omega = \Omega(\mathbf{K}; \mathbf{X}, t)$ while in an inhomogeneous rheonomic dispersive *nonlinear* material, the frequency will also depend on the amplitude. Let $\mathbf{a}$ the n-vector of R^n that characterizes this small slowly varying amplitude of a complex system (in general with several degrees of freedom). Thus, now,

$$\omega = \Omega(\mathbf{K}; \mathbf{X}, t; \mathbf{a}). \tag{63}$$

The relationship with the developments in Section 2 above follows from the remarkable considerations of Whitham on a so-called *averaged Lagrangian*. For a wave motion depending on the phase (59) and with all characteristic quantities varying slowly over space-time (derivatives of $\mathbf{a}$, ω, and $\mathbf{K}$ are small in an appropriate mathematical sense and can thus be neglected), Whitham proposes to replace the initial variational problem, say that based on a (57) for elasticity, by one pertaining to the averaged Lagrangian, *i.e.*,

$$\delta \int \widetilde{L} \, d\mathbf{X} \, dt = 0, \qquad \widetilde{L} = \frac{1}{2\pi} \int_0^{2\pi} L \, d\varphi, \tag{64}$$

with (compare to the integrand in (57))

$$\widetilde{L} = \widetilde{L}\left(\frac{\partial \overline{\varphi}}{\partial \mathbf{X}} = \mathbf{K}, \frac{\partial \overline{\varphi}}{\partial t} = -\omega, \mathbf{a}; \mathbf{X}, t \right), \tag{65}$$

where now the *fields* are the amplitude $\mathbf{a}$ and the phase φ. Accordingly, the associated equations replacing on one hand (1) and on the other hand (7) and (8) are the *field equations*

$$\frac{\partial \widetilde{L}}{\partial \mathbf{a}} = \mathbf{0}, \tag{66}$$

$$\frac{\partial \widetilde{S}}{\partial t} - \nabla_R \cdot \mathbf{W} = 0, \tag{67}$$

by direct variation and the two *conservation laws* (energy and canonical momentum)

$$\frac{\partial \widetilde{H}}{\partial t} - \nabla_R \cdot \widetilde{\mathbf{Q}} = \widetilde{h}, \qquad \frac{\partial \widetilde{\mathbf{P}}}{\partial t} - \text{div}_R \, \widetilde{\mathbf{b}} = \widetilde{\mathbf{f}}^{inh}, \tag{68}$$

by application of Noether's theorem for the translational invariance under t and $\mathbf{X}$, where we have set

$$\widetilde{S} := \frac{\partial \widetilde{L}}{\partial \omega}, \qquad \mathbf{W} := \frac{\partial \widetilde{L}}{\partial \mathbf{K}}, \tag{69}$$

$$\widetilde{H} = \omega \widetilde{S} - \widetilde{L}, \qquad \widetilde{\mathbf{Q}} = \omega \mathbf{W}, \qquad \widetilde{h} = -\frac{\partial \widetilde{L}}{\partial t}\bigg|_{expl}, \tag{70}$$

$$\widetilde{\mathbf{P}} = \widetilde{S}\mathbf{K}, \qquad \widetilde{\mathbf{b}} = -(\widetilde{L}\mathbf{1}_R - \mathbf{W} \otimes \mathbf{K}), \qquad \widetilde{\mathbf{f}}^{inh} = \frac{\partial \widetilde{L}}{\partial \mathbf{X}}\bigg|_{expl}. \tag{71}$$

Dimensionally, $\widetilde{S}$ is an *action* and may be called the **wave action**, while Eq. (67) may be referred to as a strict *conservation law* for the wave action in which $\mathbf{W}$ is the *action flux*. Here above the material co-vector $\widetilde{\mathbf{P}}$ may be called the **material wave momentum** (notice that its formula reminds us of the already mentioned quantum wave-mechanics relationship due to de Broglie: $\widetilde{\mathbf{P}} = \hbar \mathbf{K}$, where $\hbar$ is the reduced elementary quantum of action (Planck's constant)), and $\widetilde{\mathbf{b}}$, the associated flux, may be called the **material wave Eshelby stress**material wave Eshelby stress. This tensor is not symmetric unless $\mathbf{W}$ is proportional to $\mathbf{K}$. The wave action conservation equation (67) plays here the central role (equivalent to the balance of linear physical momentum (25)). Indeed, in the same way as Eq. (4) and (5) for elasticity can be deduced from (44) by right scalar multiplication by $\mathbf{v}$ and $\mathbf{F}$, respectively as shown above in Section 3, Eq. $(5.12)_{1-2}$ can be deduced from (67) by simple and tensorial multiplication, respectively, by ω and $\mathbf{K}$ on account of the functional dependency assumed for $\widetilde{L}$ (*cf.* [31, 32]). Equations (66) through (68) can be used in devising the equation that governs the slowly varying small amplitude of localized modulated waves (*e.g.*, bright and dark solitons) in mechanical nonlinear and dispersive elastic systems such as those mentioned in Section 4. An example of application is given in [31]. Of course, reasoning in $(\mathbf{K}, \omega)$ space is somewhat dual to that in space-time $(\mathbf{X}, t)$, the duality being understood in the sense of the phase function, hence of the action function according to the kinematic-wave theory. A final remark is in order: taking the inner product in material space of the first of (71) and combining with the product of the first of (70) by time t, we obtain the following remarkable result:

$$A = \widetilde{\mathbf{P}} \cdot \mathbf{X} - \widetilde{H}t = \widetilde{S}(\mathbf{K} \cdot \mathbf{X} - \omega t) + \widetilde{L}t. \tag{72}$$

For a *linear* elastic body in 1D (case of phonons) it is shown that ([31]; Eqs. (4.5)–(4.6))

$$\widetilde{L} = D_L(\omega, K)\mathbf{a}^2, \tag{73}$$

where $D_L(\omega,K) = 0$ happens to be the "linear" dispersion relation (as a matter of fact, the "field equation" associated with the amplitude; cf. Eq. (66)). Accordingly, in this case (72) reduces to

$$A = \widetilde{\mathbf{P}} \cdot \mathbf{X} - \widetilde{H}t = \widetilde{S}(\mathbf{K} \cdot \mathbf{X} - \omega t) = \widetilde{S}\varphi, \tag{74}$$

in agreement with the Planck–de Broglie formula of proportionality between action and phase. The more general formula (72) is still admissible as the right-hand side may be rewritten as (compare Eq. (44.3) in [20])

$$A = \widetilde{\mathbf{P}} \cdot \mathbf{X} - \widetilde{H}t = \widetilde{S}\mathbf{K} \cdot \mathbf{X} - \widetilde{H}t, \tag{75}$$

where the first contribution in the r-h-s may be considered to be the "reduced" action density of Landau and Lifshitz for the present formulation. Such notions can prove useful in some wave problems.

References

1. Berezovski, A., Engelbrecht, J., Maugin, G.A.: Numerical Simulation of Waves and Fronts. World Scientific, Singapore (2008)
2. Bui, H.D.: Mécanique de la rupture fragile. Masson, Paris (1978)
3. Dascalu, C., Maugin, G.A.: Forces matérielles et taux de restitution de l'énergie dans les corps élastiques homogènes avec défauts. C.R. Acad. Sci. Paris II-317, 1135–1140 (1993)
4. de Groot, S., Mazur, P.: Non-equilibrium thermodynamics. North-Holland, Amsterdam (1962)
5. Denzer, R., Liebe, T., Kuhl, E., Barth, F.J., Steinmann, P.: Material force method, Continuum damage and thermo-hyperelasticity. In: Steinmann, P., Maugin, G.A. (eds.) Mechanics of Material Forces, pp. 95–104. Springer, New York (2005)
6. Epstein, M., Maugin, G.A.: Thermoelastic material forces: definition and geometric aspects. C.R. Acad. Sci. Paris II-320, 63–68 (1995)
7. Epstein, M., Maugin, G.A.: Thermomechanics of Volumetric Growth in Uniform Bodies. International Journal of Plasticity 16, 951–978 (2000)
8. Epstein, M., Maugin, G.A.: Remark on the Universality of the Eshelby Stress. Math. and Mech. Solids (in press, 2009)
9. Eringen, A.C.: Theory of Micropolar Elasticity. In: Liebowitz, H. (ed.) Fracture, ch. 7, vol. II. Academic Press, New York (1968)
10. Eringen, A.C.: Mechanics of Continua. Revised and Augmented Edition, Krieger, Florida (1980); original edn. J. Wiley, New York (1967)
11. Eringen, A.C. (ed.): Continuum Physics, 4 vols. Academic Press, New York (1971-1976)
12. Fischer, F.D.: Design of damage resistant materials by configurational forces. Paper at ISDMM 2009, Trento, July 6-9 (2009)
13. Fischer, F.D., Predan, J., Kolednik, O., Simha, N.K.: Application of material forces to fracture of inhomogeneous materials: illustrative examples. Arch. Appl. Mech. 77, 95–112 (2007)
14. Haddi, A., Weichert, D.: On the Computation of the J-integral for Three-dimensional Geometries in Inhomogeneous Materials. Comput. Mat. Sci. 5, 143–150 (1995)

15. Hayes, W.D.: Conservation of action and modal wave action. Proc. Roy. Soc. Lond. A 320, 187–206 (1970)
16. Holland, P.R.: The quantum theory of motion. Cambridge University Press, U.K (1993)
17. Kienzler, R., Herrmann, G.: Mechanics in material space. Springer, Berlin (2000)
18. Kosevich, A.M.: The crystal lattice (phonons, solitons, dislocations). Wiley-VCH, Berlin (1999)
19. Lanczos, C.: Variational Principles in Mechanics. Toronto Univ. Press, Toronto (1962)
20. Landau, L.D., Lifshitz, E.M.: Mécanique (translation form the Russian). MIR, Moscow (1965)
21. Lighthill, M.J.: Contribution to the theory of waves in nonlinear dispersive systems. J. Inst. Maths. Applics. 1, 269–306 (1965)
22. Lurié, K.A.: An Introduction to the Mathematical Theory of Dynamic Materials. Springer, New York (2007)
23. Maugin, G.A.: Internal variables and dissipative structures. J. Non-Equilibr. Thermodynam. 15, 173–192 (1990)
24. Maugin, G.A.: Material inhomogeneities in elasticity. Chapman and Hall, London (1993)
25. Maugin, G.A.: Material forces: concepts and applications. ASME Appl. Mech. Rev. 48, 213–245 (1995)
26. Maugin, G.A.: Thermomechanics of inhomogeneous-heterogeneous systems: application to the irreversible progress of two- and three-dimensional defects. ARI 50, 41–56 (1997)
27. Maugin, G.A.: On Shock waves and Phase-transition Fronts in Continua. ARI 50, 141–150 (1998)
28. Maugin, G.A.: On the Structure of the Theory of Polar Elasticity. Phil. Trans. Roy. Soc. London A 356, 1367–1395 (1998)
29. Maugin, G.A.: Nonlinear waves in elastic crystals. Oxford University Press, U.K (1999)
30. Maugin, G.A.: On the Thermomechanics of Continuous Media with Diffusion and/or Weak Nonlocality. Arch. Appl. Mech. 75, 723–738 (2006)
31. Maugin, G.A.: Nonlinear Kinematic Wave Mechanics of Elastic Solids. Wave Motion 44(6), 472–481 (2007)
32. Maugin, G.A.: On Phase, Action and Canonical Conservation Laws in Kinematic-wave Theory. Fizika Nizkikh Temperatur (Ukraine; Issue dedicated to the late Kosevich, A.M.) 34(7), 721–724 (2008) (in Russian); Low Temperature Physics 34, 7 (2008) (in English)
33. Maugin, G.A.: On Inhomogeneity, Growth, Ageing and Dynamic Materials. J. Mech. Materials Structures (Issue in Memory of Herrmann, G.) 4, 731–741 (2009)
34. Maugin, G.A., Trimarco, C.: Pseudo-momentum and Material Forces in Nonlinear Elasticity: Variational Formulations and Application to Brittle Fracture. Acta Mechanica 94, 1–28 (1992)
35. Müller, R., Maugin, G.A.: On Material Forces and Finite Element Discretizations. Computational Mechanics 29(1), 52–60 (2002)
36. Noether, E.: Invariante Variations problem. Klg-Ges. Wiss. Nach. Göttingen, Math-Physik Kl.2, 235–257 (1918)
37. Rousseau, M., Maugin, G.A., Berezovski, M.: Prolegomena to studies on dynamical materials (submitted for publication, 2009)
38. Schrade, D., Müller, R., Gross, D., Utschig, T., Shur, V.Y., Lupascu, D.C.: Interaction of domain walls with defects in ferroelectric materials. Mechanics of Materials 39(2), 161–174 (2007)
39. Simha, N.K., Fischer, F.D., Shan, G.X., Chen, C.R., Kolednik, O.: J-integral and Crack driving force in elastic-plastic materials. J. Mech. Phys. Solids 56, 2876–2895 (2008)
40. Soper, D.E.: Classical Field Theory. J. Wiley, New York (1976)

41. Suhubi, E.S.: Thermoelasticity. In: Eringen, A.C. (ed.) Continuum Physics, vol. II. Academic Press, New York (1973)
42. Trimarco, C., Maugin, G.A.: Material Mechanics of Electromagnetic Bodies. In: Kienzler, R., Maugin, G.A. (eds.) Configurational Mechanics of Materials, Udine. Lecture Notes CISM, pp. 129–171. Springer, Wien (2001)
43. Truesdell, C.A., Noll, W.: Nonlinear Field Theories of Mechanics. In: Flügge, S. (ed.) Handbuch der Physik, Bd.III/3. Springer, Berlin (1965)
44. Truesdell, C.A., Toupin, R.A.: The Classical Theory of Fields. In: Flügge, S. (ed.) Handbuch der Physik, Bd. III/1. Springer, Berlin (1960)
45. Whitham, G.B.: A general approach to linear and nonlinear dispersive waves using a Lagrangian. J. Fluid Mech. 22, 273–283 (1965)
46. Whitham, G.B.: Linear and nonlinear waves. Interscience, J. Wiley, New York (1974)

Waves and Dislocations

Łukasz A. Turski

Abstract. I will review of works extending over almost 44 years on gauge field theory approach to the description of dislocations in continuous media with particular focus on properties of various waves propagating in these media (matter waves, acoustic waves, spin waves etc.).

Salutation

Almost forty four years ago, right after receiving my M.Sc. in physics from the Warsaw University I have joined Henryk Zorski research group at the Institute for Fundamental Technological Research (IPPT) in Warsaw. Already working for my thesis advisor at Hoża street[1], Ryszard Gajewski, was a scientific "cold shower". Our theoretical group at the University was heavily inclined to study the abstract subjects, like axiomatic field theory, so even the general relativity was considered as "applied" topic. Working for Ryszard I had to accept that the "numbers" count, and that I have to provide solutions to the equations I used. Few first months at Henryk group was much worse. It was like taking the raft ride on one of the Siberian rivers. People were talking on topics I never heard in my life, for example plasticity, column buckling, cracks propagation and **dislocations** *and* **disclinations**. *One day distinguished visitor from Carnegie–Mellon in Pittsburgh, Walter Noll, arrived[2] and, during his numerous talks, was writing on the blackboard expressions which*

Łukasz A. Turski
Center for Theoretical Physics, Polish Academy of Sciences and Department of Mathematics and Natural Sciences, Cardinal Wyszyński University. Warsaw, Poland
e-mail: laturski@cft.edu.pl

[1] Hoża 68 is the address of the hub of Polish theoretical and to the large extend also experimental physics. Here Polish physics was recreated from the ashes of the II World War.

[2] Later, I was frequently meeting Walter Noll when I was at Carnegie Mellon Physics Department in early seventies.

looked even more complicated (they were) than those I was accustomed on Hoża. Soon I learned that these talks dealt with applications of thermodynamics to the sophisticated problems in continuous media theory. I realized that there is one individual hanging around who was quite literate in those complicated issues and who, by himself, was giving similarly difficult for me to follow, seminars. He was a "peripatetic scientist" arriving now and then to the grim building on Świetokrzyska street in Warsaw from town Łódź. It took then less time to get from Łódź to Warszawa by train than now. After the seminar hours it was easy to locate that individual in the famous little "cafeteria", on the first floor of the Institute building, enjoying delicious sandwiches and sipping what used to be called coffee. This individual was, sure enough, Krzysztof Wilmański. I left IPPT right after my PhD thesis defense and Krzysztof continued his peripatetic life exploring various regions of the scientific world. Our life and research path did cross, however, very often during the elapsed time. I believe that during that time he had learned to accept that "numbers do count" and that not all complex equations make sense. I learned from him, that thermodynamics is truly fascinatiing subject and it pays to comprehend those strangely looking formulas, he and Noll, used to write.

Ad multos annos Krzysztof.

1 Introduction

Most of the modern textbooks of solid state physics, immediately after some introductory material, provide more or less extensive introduction to the lattice theory [1, 2]. We learn there what reciprocal lattice is, what are the point groups etc. All that refer to the perfect crystals, objects never produced by the Mother Nature, and perhaps for good reasons. Those who do not specialize in solid state physics or mechanics stop their education there. One has to devour almost 2/3 of the Ashcroft and Mermin book [1] to learn that what makes solids so interesting and useful are various defects–departures from the ideal structure and its perfect crystallographic beauty.

Sure enough I have heard very little of these defects before joining Zorski's group. I never really bother to learn before, that in addition to point defects, there are also extended lattice defects called dislocations, and right from the very beginning that was dumped on me in the form of the theory of dislocations. To make my life even more difficult to two of them at the same time. Henryk Zorski was just completing his theory od dislocations, based on the extensive applications of the physical field theory to the dynamics of dislocations defined, as I learned in spirit of Somiglina and Volterra [3], as surfaces of singularity in perfect elastic continuum [4]. The other theory, called continuous theory of dislocations, was developed by many researchers over the world, by Billby in England [6] (soon I have learned him on one of the famous conferences held at IPPT against all the odds of the communism time in Poland), Kondo in Japan [7], Ekkehart Kröner in Germany [8] and Marek Żórawski in Warsaw [9]. Although the Schouten's *Ricci Calculus* [10] was our favorite past time reading at Hoża I gave up on Kondo's papers very quickly.

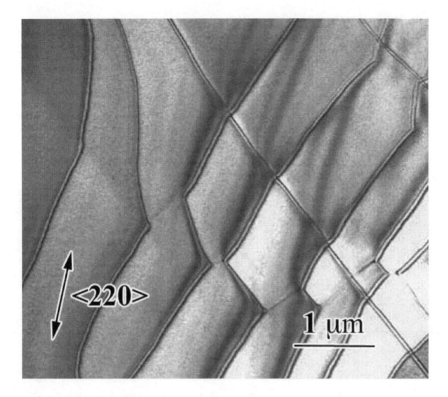

Fig. 1 The [110] dislocation in BISCO. From Ref.[12]

Kröner stuff and Żórawski book and papers were clearly written, and particularly Kröner formulation was very appealing for physicists. It was like doing general relativity with the chance of deriving "numbers" measurable in the laboratory by means of the X-ray analysis or neutron scattering done on real materials [11].

In early sixties one of the new concepts in theoretical physics, which clearly changed the way we describe nature, was the gauge field theory. The early papers by Utiyama [13], Kibble [14] and others become very popular and I attempted to use this approach to reformulation of the theory of elasticity. The resulting description turned out to be similar to the Kröner formulation of the continuum theory of dislocations. After I have written a paper on the use of the gauge field theory to formulate the theory of dislocation [15] I left that field and started to work on variety of other topics. I have returned to the theory of dislocations in mid nineties in a series of papers written jointly with Richard Bausch and Rudi Schmitz. These papers analyze the waves (including matter waves) interactions with the dislocations and some of these results will be reviewed in the following sections.

2 Elasticity Theory

The gauge theory approach to dislocations theory outlined in [15] is not the only one proposed. The other one has been developed by Edelen and collaborators [16].

The first formulation [15], relies on Euclidean symmetry of the elastic energy density in the undistorted reference state. Accordingly, the gauge group consists of *linear transformations of the Lagrange* coordinates, representing the positions of material points in the reference state, *at constant Euler coordinates* which represent the displaced positions in the strained material. The merit of the approach [15] is that it incorporates some important nonlinear couplings between dislocation-induced and externally generated distortions, including those due to sound waves with independently tunable amplitudes.

In the Edelen [16] formulation, a gauge group consists of *linear transformations of the Euler* coordinates at *fixed values of the Lagrange* coordinates. The reason of this second choice is that, independently of the symmetry of the material parameters, already the strain tensor of the reference system is invariant under such transformations. Only, if all types of distortion fields present are treated as perturbations of equal magnitude, both theories coincide.

In our recent publication [17] we have discussed the formulation from [15] in view of its application to scattering of a sound wave on edge and screw dislocation. Here we repeat salient features of that work.

The standard expression for the elastic energy density of the reference system reads,

$$e(x) = \frac{1}{2} \varepsilon_{ij}(x) c^{ijk\ell} \varepsilon_{k\ell}(x) \tag{1}$$

where, in a Cartesian frame,

$$c^{ijk\ell} = \lambda \delta^{ij} \delta^{k\ell} + \mu \left(\delta^{ik} \delta^{j\ell} + \delta^{i\ell} \delta^{jk} \right) \tag{2}$$

are the elastic constants of an isotropic medium with Lamé coefficients λ, μ, and where, in terms of Lagrange and Euler coordinates x^i, and X^A,

$$\varepsilon_{ij}(x) \equiv \frac{1}{2} \left\{ \left[\partial_i X^A(x) \right] \delta_{AB} \left[\partial_j X^B(x) \right] - \delta_{ij} \right\} \tag{3}$$

defines the strain tensor. The components of the Euler position vectors have been labelled by capitals A, B, in order to indicate that they behave as scalars under linear transformations of the Lagrange coordinates.

According to Kibble [14, 15], the symmetry of the distorted medium under local coordinate transformations can now be achieved by replacing the partial derivatives in the expression (3) by covariant derivatives $D_\alpha \equiv B^i_\alpha(x) \partial_i$ where in [15] the compensating fields $B^i_\alpha(x)$ should be identified with the defect–induced distortions, appearing in Kröner's theory of dislocations [8].

The result for the local elastic energy in presence of the defects reads

$$E(x) = \frac{1}{2} \varepsilon_{\alpha\beta}(x) c^{\alpha\beta\gamma\delta} \varepsilon_{\gamma\delta}(x) \qquad (4)$$

where $c^{\alpha\beta\gamma\delta}$ is numerically identical to $c^{ijk\ell}$, and

$$\varepsilon_{\alpha\beta}(x) \equiv \frac{1}{2} \left\{ \left[D_\alpha X^A(x) \right] \delta_{AB} \left[D_\beta X^B(x) \right] - \delta_{\alpha\beta} \right\} . \qquad (5)$$

The elastic energy density from Eq.(4) is the basic ingredient of the equilibrium theory of dislocations. Using it one can derive the (linear) theory of elasticity expressions for the Kröner's distortions $\mathbf{B}_\alpha = \mathbf{1} + \beta$ [8].

Indeed, the explicit expression for the equilibrium conditions reads, in the form of the strain tensor for frozen-in strain field due to dislocations $E_{\alpha\beta}(x) = \frac{1}{2}\left[B^i_\alpha \delta_{ij} B^j_\beta - \delta_{\alpha\beta} \right]$, and the elastic energy $U(x) \equiv \frac{1}{2} E_{\alpha\beta}(x) c^{\alpha\beta\gamma\delta} E_{\gamma\delta}(x)$,

$$D_\alpha \, \partial U / \partial B^i_\alpha = 0. \qquad (6)$$

For example, for a single screw dislocation along the z-axis of the laboratory frame we have $\mathbf{b} = (0,0,b)$, and the distortion field β^α_i has the only nonzero components $\beta^3_1(x) = -(b/4\pi) \partial_2 \ln[(x^1)^2 + (x^2)^2]$, and $\beta^3_2(x) = (b/4\pi) \partial_1 \ln[(x^1)^2 + (x^2)^2]$. Similar explicit expressions are known for other types of line defects [8, 18].

Following [17] we will now use this in order to describe sound propagation in the distorted medium. To do so, we introduce in (5) a time dependence of the Euler coordinate $X^A(x,t)$ and a displacement field $u_i(x,t)$ by writing $X^A(x,t) = \delta^A_i x^i + \delta^{iA} u_i(x,t)$. The total strain is now given by

$$\varepsilon_{\alpha\beta}(x,t) = E_{\alpha\beta}(x) + \frac{1}{2} \left[B^i_\alpha(x) D_\beta + B^i_\beta(x) D_\alpha \right] u_i(x,t)$$
$$+ \frac{1}{2} [D_\alpha u_i(x,t)] \delta^{ij} [D_\beta u_j(x,t)] . \qquad (7)$$

Expanding (7) to linear order in the quantities $\partial_j u_i$ and $\beta^i_\alpha \equiv B^i_\alpha - \delta^i_\alpha$ one recovers the strain tensor, appearing in the Kadic-Edelen approach [16]. The neglected higher-order terms, however, turn out to contribute in an essential way to the process of sound propagation.

The resulting wave equation for the sound modes, derived in [17] has a form analogous to the Lame equations from linear elasticity with two important changes. First the space derivative are replaced by the covariant ones, and second, the constant elastic coefficients from Eq.(2) are repalced by the space dependent effective elastic coefficients depending on the distortion field $\mathbf{B}_\alpha$. We have then

$$\rho \, \delta^{ij} \partial^2_t u_j(x,t) = D_\alpha C^{\alpha i \beta j}(x) D_\beta u_j(x,t) \qquad (8)$$

where ρ is a constant mass parameter, and

$$C^{\alpha i \beta j}(x) \equiv B^i_\gamma(x) \, c^{\alpha \gamma \beta \delta} \, B^j_\delta(x) + c^{\alpha \beta \gamma \delta} E_{\gamma \delta}(x) \, \delta^{ij} . \tag{9}$$

3   Sound Waves and Dislocations

Eq.(8) is valid for arbitrary arrangements and types of dislocations in isotropic materials. Whereas the topological nature of the defects is essentially hidden in the covariant derivatives, the defect-induced changes of the elastic properties are contained in the coefficients (9). It should be noted that in practical applications the core regions of the defects in general require separate considerations [18]. As pointed out by Kosevich [18], the change of the elastic coefficients in the core region of dislocation can, in case of the edge dislocation, result in presence of the localized modes of elastic vibrations. These kind effects is not included into our formulation. We can, however study possibility of the appearance of the localized modes next to the edge dislocation by studying the solutions of Eq.(8) for a single straight edge dislocation localized along z- direction of the laboratory frame with the Burgers vector $\mathbf{b} = (b,0,0)$. Writing $u_i(\mathbf{x},t) = v_i(x,y)\exp[\imath(kz - \omega t)]$ and working out equation fort the envelope $v_i(x,y)$ to the first order in b/r and κ/k, where κ is the inverse size of the localized state, we found[17] the Schrödinger like equation for the v_z

$$\left[\partial^2 + (\omega/c_\parallel)^2 - k^2 - V(x,y) \right] v_z(x,y) = 0 \tag{10}$$

where $\partial^2 = \partial_x^2 + \partial_y^2$, $c_\parallel \equiv \sqrt{(\lambda + 2\mu)/\rho}$, $c_\perp \equiv \sqrt{\mu/\rho}$, and the "potential" $V(x,y) = (\lambda k^2/\rho c_\parallel^2)\delta^{ab}E_{ab}$, which is purely attractive and, therefore, leads to the existence of bound phonon states [19].

The analysis of Eq.(8) for a screw dislocation leads to quite different results. Assuming that dislocation is, again, put along the z axis and with the Burgers vector $\mathbf{b} = (0,0,b)$, using explicit expressions for the Kröner distortions β_b^a, we found in [17], following the cylindrical symmetry Ansatz from [20] $v_z(r,\phi) = \chi(r)\exp\{\imath[m - kb/(2\pi)]\phi\}$

$$\left[\partial_r^2 + \frac{1}{r}\partial_r + \left(\frac{\omega}{c_\parallel}\right)^2 - k^2 - V(r) \right] \chi(r) = 0 \tag{11}$$

where the potential reads

$$V(r) \equiv \left[\frac{c_\perp^4}{(c_\parallel^2 - c_\perp^2)c_\parallel^2}\left(\frac{kb}{2\pi}\right)^2 + 2\frac{c_\parallel^2 + c_\perp^2}{c_\parallel^2}\left(\frac{kb}{2\pi}\right)m + m^2 \right]\frac{1}{r^2} . \tag{12}$$

Since this potential is repulsive for all known materials, we conclude that in our model localized vibrational states do not exist in the strain field of a screw dislocation.

We shall conclude discussion of application of the gauge theory description of the sound waves (phonons) interactions with the dislocations by analysis of the phonons scattering off the screw dislocations. In analysis of this phenomenon it is essential to keep full covariant derivatives appearing tin the wave equation (8) for the defect-generated topology of space is felt at arbitrary distance from the dislocation line and the ratio κ/k does not serve as the small parameter for the perturbative expansion any longer.

In terms of the full covariant derivatives

$$\mathbf{D} = [D_1, D_2, D_3] = [\partial_1 + \imath\partial_1\Phi(x^1, x^2), \partial_2 + \imath\partial_2\Phi(x^1, x^2), \imath k] \qquad (13)$$
$$= \exp(-i\Phi)[\partial_1, \partial_2, \imath k]\exp(i\Phi),$$

where the $\Phi = (kb/2\pi)\arctan(x^2/x^1)$ the wave equation (8) for the envelope $\mathbf{v}$ assumes the form:

$$-\rho\,\omega^2\mathbf{v} = (\lambda + \mu)\mathbf{D}(\mathbf{D}\cdot\mathbf{v}) + \mu\mathbf{D}^2\mathbf{v}. \qquad (14)$$

Decomposing equation (14) into longitudinal and transverse parts by means of generalized transformation [17] involving the integral operator D^{-2}

$$\mathbf{v} = \mathbf{D}\left(\mathbf{D}\cdot\mathbf{D}^{-2}\mathbf{v}\right) - \mathbf{D}\times\left(\mathbf{D}\times\mathbf{D}^{-2}\mathbf{v}\right) \equiv \mathbf{v}_{\parallel} + \mathbf{v}_{\perp} \qquad (15)$$

we obtain two independent wave equations for longitudinal $\mathbf{v}_{\parallel}$ and transverse $\mathbf{v}_{\perp}$ modes

$$-\omega^2\mathbf{v}_a = c_a^2\mathbf{D}^2\mathbf{v}_a \;,\; a = \parallel, \perp. \qquad (16)$$

Assuming the incoming wave with the wave vector $[-q_a, 0, k]$ of the form $\mathbf{v}_a^{in} = \mathbf{e}_a\exp(-\imath q_a x^1)\exp(-\imath\Phi(x^1, 0))$, where $\mathbf{e}_a$ are the polarization vectors, observing that in our model the polarization of the wave is unaffected by the scattering, we found the scattered sound modes of the form

$$\mathbf{v}_a^{sc} = \mathbf{e}_a\frac{e^{iq_a r}}{(2\pi q_a r)^{1/2}}\sin(kb/2)\frac{e^{i\phi/2}}{\cos(\phi/2)}. \qquad (17)$$

This solutions resemble the solution for the electron wave function scattering off the magnetic flux line in the Aharonov-Bohm effect [21]. Our analysis here is quite similar to that proposed originally by Kawamura [22] in his description of the electron scattering on dislocation lines and discussed by Bausch, Schmitz and myself in [23, 24].

The above outlined results of the sound waves interaction with the dislocations, can be generalized for other topological defects which can be brought within the realm of the gauge field theory formulation of defects as proposed in [15]. This kind of defects theory formulation can also be derived from the more "microscopic" analysis presented in [25]. That procedure leads to the Schrödinger equation for the wave function describing the particle progagating in the crystal containing dislocation, which is identical to that obtained from the gauge theory approach. In the following section I shall describe that kind of the analysis as applied to the spin wave excitations in the classical Heisenberg model of the ferromagnet containing dislocations.

4 Spin Waves and Dislocations

In our earlier publication [25] we have shown systematic procedure to derive the
Schrödinger equation from them "microscopic" lattice model, in that case tight binding one. That procedure relies on generalization of the expansion procedure leading
from the lattice (discrete) to the continuum description of the medium properties,
which consists of formal $\mathcal{O}(a^2)$ expansion in the bare lattice constant and $\mathcal{O}(\beta^2(\mathbf{x}))$
expansion in the Kröner distortions β describing topological defect distribution
in the lattice. Here we show how that procedure, equivalent to the gauge theory
approach, works for the spin waves description for classical Heisenberg model. We
shall formulate here the general theory and show exact results for the spin wave
interaction with a frozen in distortion due to a screw dislocation [26].

The fundamental equation in the theory of the magnetizm, the Landau-Lifshitz
equation for spin vector field $\mathbf{S}(\mathbf{x})$,

$$\partial_t \mathbf{S}(\mathbf{x}) = Ja^2 \mathbf{S}(\mathbf{x}) \times \nabla^2 \mathbf{S}(\mathbf{x}) \tag{18}$$

after linearization around static and constant "magnetization" $\mathbf{S}_0$ describes spin
waves with dispersion relation $\omega(\mathbf{q}) = Ja^2 S_0 \mathbf{q}^2$. Eq.(18) follows from the continuum approximation ($\mathcal{O}(a^2)$; $a \to 0$ expansion) of the equations of motion for classical spins described by the classical Heisenberg ferromagnet Hamiltonian

$$\mathcal{H} = \frac{1}{2} \sum_{\mathbf{n},\mathbf{a}(\mathbf{n})} J(\mathbf{n}, \mathbf{n} + \mathbf{a}(\mathbf{n})) \mathbf{S}(\mathbf{n}) \cdot \mathbf{S}(\mathbf{n} + \mathbf{a}(\mathbf{n})) \tag{19}$$

where the vector $\mathbf{n}$ denotes a simple cubic lattice site position, $\{\mathbf{a}_\alpha(\mathbf{n})\}$; $\alpha = 1 \ldots d$
is the set of the lattice vectors and $\mathbf{S}(\mathbf{n})$ the spin vector on that site. $J(\mathbf{n}, \mathbf{n} + \mathbf{a}(\mathbf{n}))$ denotes the nearest-neighbor exchange interaction. In the undistorted lattice, in some
coordinate system, $a_\alpha^i(\mathbf{n}) = a\delta_\alpha^i$, where a is the bare lattice constant. In real materials lattice is not perfect and, as discussed in Section(2), might contain topological
defects–dislocations. The main goal of this Section is to find the generalization of
Eq.(18) for such a lattice and analyze the consequences of the changes in it on propagation of the spin waves.

We proceed, following [25, 26], by observing that the Kröner's lattice distortions
$\beta_j^i(\mathbf{n}) = B_\alpha^i(\mathbf{n})\delta_j^\alpha - \delta_j^i$, [8], change the global equivalence of the set of the lattice
vectors $\{\mathbf{a}_\alpha(\mathbf{n})\}$. To carry out the sum in Eq.(19) we use the discussed in Ref[25]
procedure of decomposing that set into $\{\mathbf{a}_\alpha^+(\mathbf{n}), \mathbf{a}_\alpha^-(\mathbf{n})\}$, where $\mathbf{a}_\alpha^-(\mathbf{n})$ are essentially
opposite to $\mathbf{a}_\alpha^+(\mathbf{n})$ (in the sense of the distorted lattice geometry). To the second order
in $\mathbf{B}_\alpha$ we have [25]:

$$\mathbf{a}_\alpha^\pm(\mathbf{n}) = \pm a \mathbf{B}_\alpha(\mathbf{n}) + \frac{a^2}{2} \left(\mathbf{B}_\alpha(\mathbf{n}) \cdot \frac{\partial}{\partial \mathbf{n}} \right) \mathbf{B}_\alpha(\mathbf{n}), \tag{20}$$

and subsequently up to $\mathcal{O}(a^2, \mathbf{B}_\alpha)$

$$S(n+a_\alpha^\pm(n)) = S(n) \pm a\left(B_\alpha(n)\cdot\frac{\partial}{\partial n}\right)S(n)$$

$$+\frac{a^2}{2}\left(B_\alpha(n)\cdot\frac{\partial}{\partial n}\right)^2 S(n). \tag{21}$$

In the limit $a\to 0$ the matrix fields B_α and its inverse $B_\alpha^i B_j^\alpha = \delta_j^i$, provide the continuum description of the distorted crystal by a Riemann-Cartan manifold with the metric tensor $g_{ij} = \delta_{\alpha\beta}B_i^\alpha B_j^\beta$ and affine connection $\Gamma_{ij}{}^k = B_\alpha^k \partial_i B_j^\alpha$, [8, 15, 25, 27]. Note, that this manifold is flat (its Riemann tensor $R_{ijk}{}^\ell = 0$), but its torsion $T_{ij}{}^\ell = \Gamma_{ij}{}^\ell - \Gamma_{ji}{}^\ell$ does not vanish and it provides the measure of the dislocation density. Following the discussion in [15, 25] we introduce the covariant derivatives ∇_i with respect to the connection $\Gamma_{ij}{}^k$ and the related derivatives $\nabla_I^T = \nabla_i + 2T_{ik}{}^k$. With g^{ij} being the inverse of g_{ij} and $g = \det g_{ij}$ we obtain the $\mathcal{O}(a^2,\beta^2)$ Heisenberg Hamiltonian (19) in the covariant form

$$\mathcal{H} = \frac{Ja^2}{2}\int d^d x\sqrt{g}g^{ij}S(x)\cdot\nabla_i\nabla_j S(x), \tag{22}$$

In (22) we have retained only those terms in the Hamiltonian (19) expansion which correspond to the exchange interaction J independent of the lattice position. I emphasizes that the Hamiltonian (22) describes purely topological in nature interaction between the dislocation and magnetization $\propto S(x)$ via the differential geometric structure of the effective Riemann-Cartan manifold describing the medium with defects.

In a real crystal one should also include two additional couplings between the magnetic degrees of freedom and the crystalline lattice. The first one is due to the localized change in the exchange integral $J(n)$ caused by the presence of the dislocation core, and should play the role similar to the dislocation core modification of the sound waves interactions with dislocations mentioned in Section(3) [18]. The second one is caused by the magnetostriction, that is dependence of J on the lattice vibrations–phonon's described by the field $u(n)$. some discussion of that point has been offered in [26]. I shall alos add that present description of the spin waves interactions with dislocations differ essentially from that one discussed by Kutchko and collaborators [28, 29]. The origins of this difference has been also discussed in [26].

Using the spin Poisson Brackets $\{S^a(x), S^b(y)\} = (\delta(x-y)/\sqrt{g})\,\varepsilon^{abc}S^c(x)$ we found, from (22), that the covariant form of the Landau-Lifshitz equation of motion (18) reads:

$$\partial_t S(x) = Ja^2 S(x)\times\hat{\mathcal{B}}S(x). \tag{23}$$

Here $\hat{\mathcal{B}}$ denotes the covariant wave operator

$$\hat{\mathcal{B}} = g^{ij}\left[\nabla_i^T\nabla_j + \nabla_i^T T_{jk}{}^k\right]. \tag{24}$$

featuring the derivatives ∇_i^T and the torsion tensor T_{ij}^k.

Linearizing Eq.(23) around a constant is space stationary solution $\mathbf{S} = S_0 e_z$ we obtain the covariant equation for spin waves in the ferromagnet containing defects which density is described by the Riemann-Cartan manifold torsion tensor $T_{ij}{}^k$. Writing $S(x,t) = S_0 + \delta S(x,t)$ and introducing for convenience complex amplitude $S_0 \Psi(x,t) = (\delta S_x(x,t) + \iota \delta S_y(x,t))$ we obtain the Schrödinger like equation for Ψ, namely

$$\iota \partial_t \Psi(x,t) = -\frac{1}{2\mu} \hat{\mathscr{B}} \Psi(x,t) \qquad (25)$$

where $\mu = 1/2Ja^2 S_0$. Eq.(25) is the spin wave analogue of the Kawamura equation [22, 25, 27] which, again, leads to the Aharonov-Bohm [21] like solutions for the waves interacting with defects.

We shall now present solution of Eq.(25) for a single screw dislocation along z-axis with the Burgers vector $b = be_z$. Assuming in the cylindrical coordinates $\Psi(r,\phi,z,t) = \chi(r,\phi) \exp(ikz + i\omega t)$ we found that the envelope $\chi(r,\phi)$ obeys the Aharonv-Bohm equation [21]

$$\left[\frac{\partial^2}{\partial r^2} + \frac{1}{r}\frac{\partial}{\partial r} + \frac{1}{r^2}\left(\frac{\partial}{\partial \phi} + \iota\alpha \right)^2 - q^2 \right] \chi(r,\phi) = 0, \qquad (26)$$

where $\alpha = kb/2\pi$ and $q^2 = k^2 + 2\mu\omega$. Using the asymptotic $r \to \infty$ solution for that equation [21, 22, 25] we found

$$\begin{pmatrix} \delta S_x \\ \delta S_y \end{pmatrix} = S_0 \begin{pmatrix} \cos \\ -\sin \end{pmatrix} (\alpha\phi + qr\cos\phi)$$

$$+ \frac{\sin(\pi\alpha)}{\cos(\phi/2)\sqrt{2\pi qr}} \begin{pmatrix} \cos \\ \sin \end{pmatrix} (qr + \alpha(\phi - \pi)/2|\alpha| - \pi/4) \qquad (27)$$

In the absence of the dislocation the (pseudo)flux $\alpha = 0$ and we recover standard spin wave solution of Eq.(18). The first tem on the rhs of Eq.(27) shows the helical structure of the incomming spin wave due to global distortion of the lattice and the second describes scattering phase shift due to the presence of dislocation. In Fig(2) we have shown *Mathematica 7* generated density plot for the δS_y component of the solution (27). The δS_x component exhibits identical structure with trivial phase shift found from (27).

As seen from the above analysis the spin waves propagating on a lattice containing screw dislocation exhibits, similarly to the sound waves the Aharonov-Bohm like oscillations. It seem that such an oscillations are prevailing in all wave phenomena in condensed matter where the medium carrying waves can be modeled as the manifold with torsion sharing topological similarity with the simple case discussed in the pioneering work [21]. In the following section we briefly show that this is also the case for the scattering of the matter waves from the dislocation.

δSy component of the Spin Wave

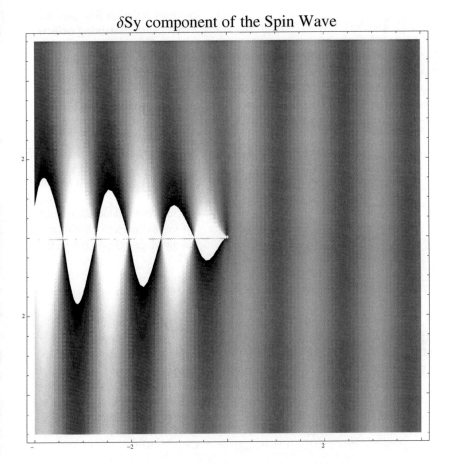

Fig. 2 The y-component of the δS spin wave shown using *Mathematica 7* density plot. From Ref.[26]

5   Matter Waves and Dislocations

In this section I shall briefly present results from [25]. The starting point in our analysis there was the tight binding model for a single electron propagating on a ideal crystalline lattice which Hamiltonian reads:

$$\mathcal{H} = -\frac{1}{2}\sum_{\mathbf{n},\mathbf{m}} \phi^{\dagger}(\mathbf{n})\,(t(\mathbf{n},\mathbf{m}) + t(\mathbf{n},\mathbf{m})^{*})\,\phi(\mathbf{n}) \tag{28}$$

where, as in Section(4) the $\mathbf{n}$ labels the lattice sites and $\phi(\mathbf{n})$ and $\phi(\mathbf{n})^{\dagger}$ are the electron annihilation and creation operators, respectively. The transition amplitudes $t(\mathbf{n},\mathbf{m})$ are defined in terms of the single particle lattice Hamiltonian and depend on the electron–lattice ions interaction potential $V(\mathbf{x} - \mathbf{n})$. For sake of convenience it was assumed in [25] that $t(\mathbf{n},\mathbf{m})$ describe only the nearest-neighbor transitions on the lattice and, therefore, can be written as $t(\mathbf{n},\mathbf{m}) = \sum_{\mathbf{a(n)}} t(\mathbf{a(n)})\delta_{\mathbf{n}+\mathbf{a(n)}}$, where as previously $\mathbf{a(n)}$ are the lattice vectors.

The procedure outlined in Section(4), with field operators ϕ replacing the spin vectors in eq.(21), but taking care for the changes in the transition amplitude $t(\mathbf{a(n)})$ implied by the expansion in (20) leads, in the $\mathcal{O}(a^2,\beta^2)$, to the continuum version of the electron Hamiltonian consisting of two parts: covariant (analogous to Eq.(22) and the noncovariant following from explict space dependence of the transition amplitudes T.

$$\mathcal{H}_{cov} = -\frac{\hbar^2}{2m}\int d^d x\sqrt{g}\phi^\dagger(x)\hat{\mathcal{B}}\phi(x), \tag{29}$$

and

$$\mathcal{H}_{noncov} = -\frac{\hbar^2}{2ma^2}\int d^d x\sqrt{g}\left(\sum_\alpha \frac{1}{g_{ij}B_\alpha^i B_\alpha^j} - 1\right)\left[2\phi^\dagger\phi + a^2\phi^\dagger\nabla^2\phi\right], \tag{30}$$

where ∇^2 is the usual Euclidean Laplacian and $\hat{\mathcal{B}}$ is the covariant wave operator (24)

Using Hamiltonians (29) and (30) ona can analyze variety of problems reseultin from the topological in nature interaction between electorns and dislocations. Here I shall just quote results of the analysis how the matter wave (electron) scatters off the screw dislocation placed, as previously, along the z-axis. Using the explicit expressions for the Kröner's distortions and writing the wave function for the spinless electron $\psi(x,t)$ as $\psi(\mathbf{r},z,t) = \chi(\mathbf{r})\exp(\imath kz)\exp(-\imath Et/\hbar)$ we found in [30] the Schrödinger equation for the envelope $\chi(\mathbf{r}) \equiv \chi(r,\phi)$ of the form:

$$\left[\frac{1}{r}\partial_r r\partial_r + \frac{1}{r^2}(\partial_\phi + \imath\alpha)^2 - \frac{\beta^2}{r^2} + q^2\right]\chi(r,\phi) = 0, \tag{31}$$

where $\alpha = (kb)/2\pi$, $\beta^2 = [2-(ka)^2](b/2\pi a)^2$, and $E = (\hbar^2/2m)(q^2+k^2)$. Here α plays the role of the vector potential in the conventional Aharonov-Bohm equation while β^2 (which in the continuum limit $ka < 1$ is always positive) measures the strength of the repulsive scalar potential due to the change in the local value of the transition amplitude $t(\mathbf{a(n)})$.

Equation (31) can be solved exactly [30]. In the asymptotic region one finds

$$\psi \to \exp[-\imath(qr\cos(\phi)+\alpha\phi - kz)] + f(\phi)\frac{\exp(\imath qr+kz)}{\sqrt{2\pi\imath qr}} \tag{32}$$

where the scattering amplitude $f(\phi)$ is given in terms of the infinite sum of exponentials [30]. It is the angular dependence of the $f(\phi)$ which distinguishes the scattering cross section $\sigma(\phi) \propto |f(\phi)|^2$ for the screw dislocation from the Aharonov-Bohm expression. Indeed the cross section here is skew-symmetric, an obvious fact for

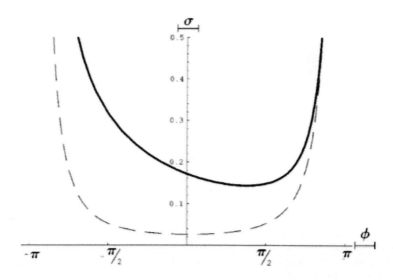

Fig. 3 Scattering cross section for a single screw dislocation[30] for $ka = qa = 0.7$ and $b = a$. The dashed line represents the usual Aharonov-Bohm cross section obtained, for screw dislocation, by Kawamura[21, 22]

anybody who ever had driven a car into the ramp of the parking garage. Fig.(3) shows the scattering cross section obtained in [30].

This form of the scattering cross section should be the one used in the analysis of the rest resistivity of the metals due to dislocations. The fact that space dependence of the transition amplitude leads to the skew-symmetry of σ implies that also for the other scattering of waves, discussed in this paper, namely for sound waves and spin waves, we should expect imprtant changes when the "scalar" potentials due to dislocation core (for sound modes) and local changes in the exchange coupling J (in case of the ferromagnet) will be included. Some of these issues are now being investigated. I should also add that the gauge theory formulation of the dislocation theory allows also for novel discussion of the issue of the scalar diffusion in the field of dislocations [31]. Here the results are quite interesting and one of them is the possibilty of observing more than one dimensional Sinay diffusion in the case of the diffusion in the field of disclinations. As discussed in [31] this seems to be, unfortunately, rather academic example, due to the unphysical features of this particular type of disclinations in solids.

Acknowledgment

This work was supported in part by the contribution from the LFPPI network and Polish MNiSz grant N202 042 32/1171.

References

1. Ashcroft, N.W., Mermin, N.D.: Solid State Physics. Holt, Reinehart (1976)
2. Ibach, H., Lüth, H.: Solid State Physics: An Introduction to Principles of Materials Science. Springer, Heidelberg (1995)
3. For moder interoduction to the theory of defects developed along that line see: Kléman M.: Points, LInes and Walls. J. Wiley, New York (1983)
4. Zorski, H.: Theory of discrete defects. Archives of Mechanics 3 (1966)
5. Zorski, H.: Statistical theory of dislocations. Int. J. of Solids and Structures 4 (1968)
6. Bilby, B.A., Bullough, R., Smith, E.: Proc. Roy. Soc. London A 231, 263–273 (1955)
7. Kondo, K.: RAAG Memoirs of the unifying study of the basic problems in physics and engeneering science by means of geometry, vol. 1. Gakujutsu Bunken Fukyu-Kay, Tokio (1952); vol. 2. Gakujutsu Bunken Fukyu-Kay, Tokio (1955)
8. For the lucid presentation of Ekkehart Kröner contributions see: Kröner, E.: Physics of Defects (Les Houches 1980), Balian, R., Klman, N.J., Poirier, P. (eds.). North-Holland, Amsterdam (1981)
9. Zórawski, M.: Théorie Mathématique des Dislocations. Dunod, Paris (1967)
10. Schouten, J.A.: Ricci-Calculus: An Introduction to Tensor Analysis and its Geometrical Applications. Springer, Heidelberg (1954)
11. Fleck, N.A., Muller, G.M., Ashby, M.F., Hutchinson, J.W.: Acta metal. matter. 42, 475 (1994)
12. This unpublished electron microscope picture was taken by Professor Marek Niewczas from the McMaster University. I thank Professor Niewczas for permission to use it prior to publication
13. Utiyama, R.: Phys. Rev. 101, 1597 (1956)
14. Kibble, T.: Journ. of Math. Phys. 2, 212 (1961)
15. Turski, L.A.: Bull. Polon. Acad. Sci. serie IV. 14, 289 (1966)
16. Kadic, A., Edelen, D.G.B.: A Gauge Theory of Dislocations and Disclinations. Springer, Heidelberg (1983)
17. Turski, L.A., Bausch, R., Schmitz, R.: J. Phys. Condens. Matter. 19, 096211 (2007)
18. Kosevich, A.M.: The crystal lattice. Wiley-VCH, Berlin (1999); Russian edition: Naukova Dumka, Kharkov (1981)
19. Lifshitz, I.M., Pushkarov, K.I.: JETP Lett. 11, 310 (1970)
20. Kosevich, A.M.: Sov. J. Low Temp. Phys. 4, 429 (1978)
21. Aharonov, Y., Bohm, D.: Phys. Rev. 115, 485 (1959)
22. Kawamura, K.: Z. Physik 29, 101 (1978)
23. Bausch, R., Schmitz, R., Turski, Ł.A.: Phys. Rev. Lett. 80, 2257 (1998)
24. Bausch, R., Schmitz, R., Turski, Ł.A.: Phys. Rev. B 59, 13491 (1999)
25. Bausch, R., Schmitz, R., Turski, Ł.A.: Ann. Phys. (Leipzig) 8, 181 (1999)
26. Turski, L.A., Mińkowski, M.: J. Phys.: Condens. Matter 21, 376001 (2009)
27. Bausch, R., Schmitz, R., Turski, Ł.A.: Zeit. für Physik B97, 171 (1995)
28. Gorobets, Y.I., Kutchko, A.N., Reshetnyak, S.A.: The Physics of Metals and Metallography 83, 344 (1997)
29. Kutchko, A.N.: Metallofiz. Noveishie Tekhnol (in Russian) 27, 1001 (2005)
30. Bausch, R., Schmitz, R., Turski, Ł.A.: Phys. Rev. B59, 13491 (1999)
31. Bausch, R., Schmitz, R., Turski, Ł.A.: Phys. Rev. Lett. 73, 2382 (1994)

Part VI

PHASE TRANSITIONS

Liquid-Solid Phase Transitions in a Deformable Container

Pavel Krejčí, Elisabetta Rocca, and Jürgen Sprekels

Abstract. We propose a model for water freezing in an elastic container, taking into account differences in the specific volume, specific heat and speed of sound in the solid and liquid phases. In particular, we discuss the influence of gravity on the equilibria of the system.

Introduction

Water is a substance with extremely peculiar physical properties. A nice survey of the challenges in modeling water behavior can be found on the web page [22]. Being aware of the obstacles, we try to develop some mathematical models related to freezing of water in a container. In [11] and [12], we have proposed an approach to model the occurrence of high stresses due to the difference between the specific volumes of the solid and of the liquid phase, assuming first that the speed of sound and the specific heat are the same in solid and in liquid. We have proved there the existence and uniqueness of global solutions, as well as the convergence of the solutions to equilibria. In reality, the specific heat in water is about the double, while

Pavel Krejčí
Institute of Mathematics, Academy of Sciences of the Czech Republic, Žitná 25,
11567 Praha 1, Czech Republic
e-mail: krejci@math.cas.cz

Elisabetta Rocca
Dipartimento di Matematica, Università di Milano, Via Saldini 50, 20133 Milano, Italy
e-mail: elisabetta.rocca@unimi.it

Jürgen Sprekels
Weierstrass Institute for Applied Analysis and Stochastics (WIAS), Mohrenstr. 39,
10117 Berlin, Germany
e-mail: sprekels@wias-berlin.de

the speed of sound in water is less than one half of the one in ice. The main goal of this contribution is to include this dependence into the model. We discuss here the modeling issues and investigate in detail the equilibria. For containers of reasonable shape and reasonable height (a few kilometers at most), filled with water in a uniform gravity field, we obtain a unique equilibrium, which is either pure solid, or pure liquid, or a solid layer above a liquid layer separated by a horizontal surface, in dependence on the surrounding temperature. New mathematical and modeling challenges arise and it is not our aim here to solve the problem completely. In particular, the proof of well-posedness of the resulting nonlinear evolution system will be the subject of a subsequent paper.

There is an abundant classical literature on phase transition processes, see e.g. the monographs [2], [4], [20] and the references therein. It seems, however, that only few publications take into account different mass densities/specific volumes of the phases. In [5], the authors proposed to interpret a phase transition process in terms of a balance equation for macroscopic motions, and to include the possibility of voids. Well-posedness of an initial-boundary value problem associated with the resulting PDE system is proved there and the case of two different densities ρ_1 and ρ_2 for the two substances undergoing phase transitions has been pursued in [6].

Let us also mention the papers [16] and [17] dealing with macroscopic stresses in phase transitions models, where the different properties of the viscous (liquid) and elastic (solid) phases are taken into account and the coexisting viscous and elastic properties of the system are given a distinguished role, [13] and [14], which pertains to nonlinear thermoviscoplasticity, and [3] where another coupled system for temperature, displacement, and phase parameter has been derived in order to model the full thermomechanical behavior of shape memory alloys. First mathematical results were published in [3], while a long list of references for further developments can be found in the monographs [4] and [20].

The main advantage of our approach is that we deal exclusively with physically measurable quantities. All parameters have a clear physical meaning. The derivation is carried out under the assumption that the displacements are small. This enables us to state the system in Lagrangian coordinates. The main difference with respect to the Eulerian framework e.g. in [6] is that in Lagrangian coordinates, the mass conservation law means that the mass density is constant and does not depend on the phase, while the specific volumes of the liquid and solid phases are possibly different. For simplicity, we still assume that viscosity and thermal expansion coefficient do not depend on the phase, the evolution is slow, and the shear viscosity, shear stresses, and inertia effects are negligible.

In Section 1, we describe the model, and the balance equations (energy balance, quasistatic momentum balance, and a phase dynamics equation) are derived in Section 2. Questions of thermodynamic consistency are discussed in Section 3, and in Section 4 we state and prove Theorem 1 on existence and uniqueness of equilibrium configurations in the limit case of rigid boundary. The elastic case can be treated in a similar way, just the computations are slightly more involved.

1   The Model

As reference state, we consider a liquid substance contained in a bounded connected container $\Omega \subset \mathbb{R}^3$ with boundary of class $C^{1,1}$. The state variables are the absolute temperature $\theta > 0$, the displacement $\mathbf{u} \in \mathbb{R}^3$, and the phase variable $\chi \in [0,1]$. The value $\chi = 0$ means solid, $\chi = 1$ means liquid, $\chi \in (0,1)$ is a mixture of the two.

We make the following modeling hypotheses.

(A1) The displacements are small. Therefore, we state the problem in *Lagrangian coordinates*, in which mass conservation is equivalent to the condition of a constant mass density $\rho_0 > 0$.

(A2) The substance is isotropic and compressible; the speed of sound and the specific heat may depend on the phase χ.

(A3) The evolution is slow, and we neglect shear viscosity and inertia effects.

(A4) We neglect shear stresses.

In agreement with **(A1)**, we define the strain ε as an element of the space $\mathbb{T}^{3\times3}_{\mathrm{sym}}$ of symmetric tensors by the formula

$$\varepsilon = \nabla_s \mathbf{u} := \frac{1}{2}(\nabla \mathbf{u} + (\nabla \mathbf{u})^T). \tag{1}$$

Let $\delta \in \mathbb{T}^{3\times3}_{\mathrm{sym}}$ denote the Kronecker tensor. By **(A4)**, the elasticity matrix $\mathbf{A}$ has the form

$$\mathbf{A}\varepsilon = \lambda(\chi)(\varepsilon : \delta)\,\delta, \tag{2}$$

where " $:$ " is the canonical scalar product in $\mathbb{T}^{3\times3}_{\mathrm{sym}}$, and $\lambda(\chi) > 0$ is the Lamé constant (or *bulk elasticity modulus*), which may depend of χ by virtue of **(A2)**.

We model the situation where the specific volume V_i of the solid phase is larger than the specific volume V_w of the liquid phase. In a homogeneous substance, the speed of sound v_0 is related to the bulk elasticity modulus λ through the formula $v_0 = \sqrt{\lambda/\rho_0}$. Here, in agreement with the Lagrange description, the speeds of sound v_w in water and v_i in ice are related to the corresponding elasticity moduli λ_w, λ_i through the formulas $\lambda_w = v_w^2/V_w$, $\lambda_i = v_i^2/V_i$. For the moment, we do not specify any particular interpolation $\lambda(\chi)$ between λ_i and λ_w for $\chi \in (0,1)$. This will only be done in Section 4 together with a motivation for the corresponding choice.

Considering the liquid phase as the reference state, we introduce the dimensionless phase expansion coefficient $\alpha = (V_i - V_w)/V_w > 0$, and we define the phase expansion strain $\tilde{\varepsilon}$ by

$$\tilde{\varepsilon}(\chi) = \frac{\alpha}{3}(1-\chi)\delta. \tag{3}$$

The stress tensor σ is decomposed into the sum $\sigma^v + \sigma^e$ of the viscous component σ^v and elastic component σ^e, which are assumed in the form

$$\sigma^v = v(\varepsilon_t : \delta)\delta \tag{4}$$

$$\sigma^e = (\lambda(\chi)(\varepsilon : \delta - \alpha(1-\chi)) - \beta(\theta - \theta_c))\,\delta, \tag{5}$$

where $v > 0$ is the volume viscosity coefficient and β is the thermal expansion coefficient, which are both assumed constant.

Our main concern is to define the free energy properly. We proceed formally, assuming that the absolute temperature remains positive. This will have to be proved in a subsequent analysis. The process is governed by the following three physical principles:

$$-\operatorname{div}\sigma = \mathbf{f}_{vol} \qquad \text{(mechanical equilibrium)} \qquad (6)$$

$$\rho_0 e_t + \operatorname{div}\mathbf{q} = \sigma : \varepsilon_t \qquad \text{(energy balance)} \qquad (7)$$

$$\rho_0 s_t + \operatorname{div}\frac{\mathbf{q}}{\theta} \geq 0 \qquad \text{(entropy inequality)} \qquad (8)$$

where $\mathbf{f}_{vol}$ is a given volume force density (the gravity force)

$$\mathbf{f}_{vol} = -\rho_0 g\,\delta_3 , \qquad (9)$$

with standard gravity g and vector $\delta_3 = (0,0,1)$, e is the specific internal energy, s is the specific entropy, and $\mathbf{q}$ is the heat flux vector that we assume for simplicity in the form

$$\mathbf{q} = -\kappa(\chi)\nabla\theta \qquad (10)$$

with a heat conductivity $\kappa(\chi) > 0$ depending possibly on χ.

We assume the specific heat $c_V(\chi,\theta)$ in the form

$$c_V(\chi,\theta) = c_0(\chi)c_1(\theta). \qquad (11)$$

This is still a rough simplification, and further generalizations are desirable. According to [9, Chapter VI] or [15, Section 5], the purely caloric parts e_{cal} and s_{cal} of the specific internal energy and specific entropy are given by the formulas $e_{cal}(\chi,\theta) = c_0(\chi)e_1(\theta)$, $s_{cal}(\chi,\theta) = c_0(\chi)s_1(\theta)$, with

$$e_1(\theta) = \int_0^\theta c_1(\tau)\,d\tau, \quad s_1(\theta) = \int_0^\theta \frac{c_1(\tau)}{\tau}\,d\tau. \qquad (12)$$

By virtue of (7)–(8), the specific free energy $f = e - \theta s$ satisfies the conditions $\sigma^e = \rho_0\partial_\varepsilon f$, $s = -\partial_\theta f$. With a prescribed constant latent heat L_0 and freezing point at standard atmospheric pressure $\theta_c > 0$, the specific free energy f necessarily has the form

$$f = c_0(\chi)f_1(\theta) + \frac{\lambda(\chi)}{2\rho_0}((\varepsilon - \tilde\varepsilon(\chi)):\delta)^2 \qquad (13)$$

$$-\frac{\beta}{\rho_0}(\theta - \theta_c)\varepsilon : \delta + L_0\chi\left(1 - \frac{\theta}{\theta_c}\right) + \tilde f(\chi),$$

where

$$f_1(\theta) = e_1(\theta) - \theta s_1(\theta) = \int_0^\theta c_1(\tau)\left(1 - \frac{\theta}{\tau}\right) d\tau,$$

and $\tilde{f}$ is a arbitrary function of χ (integration "constant" with respect to θ and ε). We choose $\tilde{f}$ so as to ensure that the values of χ remain in the interval $[0,1]$, and that the phase transition under standard pressure takes place at temperature θ_c. More specifically, we set

$$\tilde{f}(\chi) = L_0 I(\chi) - c_0(\chi) f_1(\theta_c).$$

where I is the indicator function of the interval $[0,1]$. Below in (38)–(40), we come back to the principles of thermodynamics.

For specific entropy s and specific internal energy e we obtain

$$s = -\partial_\theta f = c_0(\chi) s_1(\theta) + \frac{\beta}{\rho_0}\varepsilon : \delta + \frac{L_0}{\theta_c}\chi, \qquad (14)$$

$$e = c_0(\chi)(e_1(\theta) - f_1(\theta_c)) + \frac{\lambda(\chi)}{2\rho_0}(\varepsilon : \delta - \alpha(1-\chi))^2$$

$$+ \frac{\beta}{\rho_0}\theta_c\varepsilon : \delta + L_0(\chi + I(\chi)). \qquad (15)$$

2   Balance Equations

As another formal consequence of the entropy balance (8), we have the inequality $\chi_t \partial_\chi f \leq 0$ for every process. This will certainly be satisfied if we assume that $-\chi_t$ is proportional to $\partial_\chi f$ with proportionality constant (relaxation time) $\gamma_0 > 0$. It determines how fast the system reaches an equilibrium. We thus consider the evolution system

$$-\operatorname{div}\sigma = \mathbf{f}_{vol}, \qquad (16)$$

$$\rho_0 e_t + \operatorname{div}\mathbf{q} = \sigma : \varepsilon_t, \qquad (17)$$

$$-\gamma_0\chi_t \in \partial_\chi f, \qquad (18)$$

where ∂_χ is the partial Clarke subdifferential with respect to χ. The scalar quantity

$$p := -\nu\varepsilon_t : \delta - \lambda(\chi)(\varepsilon : \delta - \alpha(1-\chi)) + \beta(\theta - \theta_c) \qquad (19)$$

is the *pressure* and the stress has the form $\sigma = -p\delta$. The equilibrium equation (16) can be rewritten in the form $\nabla p = \mathbf{f}_{vol}$, hence

$$p(x,t) = P(t) - \rho_0 g x_3, \qquad (20)$$

where P is a function of time only, which is to be determined. Recall that in the reference state $\varepsilon : \delta = \varepsilon_t : \delta = 0$, $\chi = 1$, and at standard pressure P_{stand}, the freezing temperature is θ_c. We thus see from (19) that $P(t)$ is in fact the deviation from the standard pressure. We assume also the external pressure in the form $P_{ext} = P_{stand} + p_0$ with a constant deviation p_0. The normal force acting on the boundary is $(P(t) - \rho_0 g x_3 - p_0)\mathbf{n}$, where $\mathbf{n}$ denotes the unit outward normal vector. We assume an elastic response of the boundary, and a heat transfer proportional to the inner and outer temperature difference. On $\partial\Omega$, we thus prescribe boundary conditions for $\mathbf{u}$ and θ in the form

$$(P(t) - \rho_0 g x_3 - p_0)\mathbf{n} = \mathbf{k}(x)\mathbf{u}, \tag{21}$$

$$\mathbf{q} \cdot \mathbf{n} = h(x)(\theta - \theta_\Gamma) \tag{22}$$

with a given symmetric positive definite matrix $\mathbf{k}$ (elasticity of the boundary), a positive function h (heat transfer coefficient), and a constant $\theta_\Gamma > 0$ (external temperature). This enables us to find an explicit relation between $\mathrm{div}\,\mathbf{u}$ and P. Indeed, on $\partial\Omega$ we have by (21) that $\mathbf{u} \cdot \mathbf{n} = (P(t) - \rho_0 g x_3 - p_0)\mathbf{k}^{-1}(x)\mathbf{n}(x) \cdot \mathbf{n}(x)$. Assuming that $\mathbf{k}^{-1}\mathbf{n} \cdot \mathbf{n}$ belongs to $L^1(\partial\Omega)$, we set

$$\frac{1}{K_\Gamma} = \int_{\partial\Omega} \mathbf{k}^{-1}(x)\mathbf{n}(x) \cdot \mathbf{n}(x)\,\mathrm{d}\sigma(x), \quad m_\Gamma = K_\Gamma \int_{\partial\Omega} \mathbf{k}^{-1}(x)\mathbf{n}(x) \cdot \mathbf{n}(x) x_3\,\mathrm{d}\sigma(x), \tag{23}$$

and obtain by Gauss' Theorem that

$$U_\Omega(t) := \int_\Omega \mathrm{div}\,\mathbf{u}(x,t)\,\mathrm{d}x = \frac{1}{K_\Gamma}(P(t) - \rho_0 g m_\Gamma - p_0). \tag{24}$$

Under the small strain hypothesis, the function $\mathrm{div}\,\mathbf{u}$ describes the local relative volume increment. Hence, Eq. (24) establishes a linear relation between the total relative volume increment $U_\Omega(t)$ and the relative pressure $P(t) - p_0$. We have $\varepsilon : \delta = \mathrm{div}\,\mathbf{u}$, and thus the mechanical equilibrium equation (20), due to (19) and (24), reads

$$\nu\mathrm{div}\,\mathbf{u}_t + \lambda(\chi)(\mathrm{div}\,\mathbf{u} - \alpha(1-\chi)) - \beta(\theta - \theta_c) + \rho_0 g(m_\Gamma - x_3) = -p_0 - K_\Gamma U_\Omega(t). \tag{25}$$

As a consequence of (10), (13), and (15), the energy balance and the phase relaxation equation in (17)–(18) have the form

$$\rho_0 c_0(\chi) e_1(\theta)_t - \mathrm{div}\,(\kappa(\chi)\nabla\theta) + \rho_0 c_0'(\chi)\chi_t(e_1(\theta) - f_1(\theta))$$

$$= \nu(\mathrm{div}\,\mathbf{u}_t)^2 - \beta\theta\mathrm{div}\,\mathbf{u}_t + \rho_0 \gamma_0 \chi_t^2 - \rho_0 L_0 \frac{\theta}{\theta_c}\chi_t, \tag{26}$$

$$-\rho_0 \gamma_0 \chi_t - \frac{\lambda'(\chi)}{2}(\mathrm{div}\,\mathbf{u} - \alpha(1-\chi))^2 - \alpha\lambda(\chi)(\mathrm{div}\,\mathbf{u} - \alpha(1-\chi))$$

$$\in \rho_0 c_1'(\chi)(f_1(\theta) - f_1(\theta_c)) + \rho_0 L_0 \left(1 - \frac{\theta}{\theta_c}\right) + \partial I(\chi). \tag{27}$$

Note that mathematically, the subdifferential $\partial I(\chi)$ is the same as $\rho_0 L_0 \partial I(\chi)$. For simplicity, we now set

$$c(\chi) := \rho_0 c_0(\chi), \quad \gamma := \rho_0 \gamma_0, \quad L := \rho_0 L_0. \tag{28}$$

The system now reduces to the following three scalar equations – one PDE and two "ODEs", for three unknown functions θ, χ, and $U = \operatorname{div} \mathbf{u}$.

$$c(\chi) e_1(\theta)_t - \operatorname{div}(\kappa(\chi)\nabla\theta) = c'(\chi)\chi_t(f_1(\theta) - e_1(\theta))$$

$$+ \nu U_t^2 - \beta \theta U_t + \gamma \chi_t^2 - L\frac{\theta}{\theta_c}\chi_t, \tag{29}$$

$$\nu U_t + \lambda(\chi)(U - \alpha(1 - \chi)) - \beta(\theta - \theta_c) = \rho_0 g(x_3 - m_\Gamma) - p_0 - K_\Gamma U_\Omega(t), \tag{30}$$

$$-\gamma\chi_t - \frac{\lambda'(\chi)}{2}(U - \alpha(1 - \chi))^2 - \alpha\lambda(\chi)(U - \alpha(1 - \chi))$$

$$\in c'(\chi)(f_1(\theta) - f_1(\theta_c)) + L\left(1 - \frac{\theta}{\theta_c}\right) + \partial I(\chi) \tag{31}$$

with $U_\Omega(t) = \int_\Omega U(x,t)\,dx$, and with boundary condition (22), (10). To find the vector function $\mathbf{u}$, we first define Φ as a solution to the Poisson equation $\Delta\Phi = U$ with the Neumann boundary condition $\nabla\Phi \cdot \mathbf{n} = (K_\Gamma U_\Omega(t) + \rho_0 g(m_\Gamma - x_3))\mathbf{k}^{-1}(x)\mathbf{n}(x) \cdot \mathbf{n}(x)$. With this Φ, we find $\tilde{\mathbf{u}}$ as a solution to the problem

$$\operatorname{div}\tilde{\mathbf{u}} = 0 \quad \text{in } \Omega \times \infty, \tag{32}$$

$$\left.\begin{array}{r}\tilde{\mathbf{u}} \cdot \mathbf{n} = 0 \\ (\tilde{\mathbf{u}} + \nabla\Phi - (K_\Gamma U_\Omega + \rho_0 g(m_\Gamma - x_3))\mathbf{k}^{-1}\mathbf{n}) \times \mathbf{n} = 0\end{array}\right\} \quad \text{on } \partial\Omega \times (0, \infty), \tag{33}$$

and set $\mathbf{u} = \tilde{\mathbf{u}} + \nabla\Phi$. Then $\mathbf{u}$ satisfies a.e. in Ω the equation $\operatorname{div}\mathbf{u} = U$, together with the boundary condition (21), that is, $\mathbf{u} = (K_\Gamma U_\Omega + \rho_0 g(m_\Gamma - x_3))\mathbf{k}^{-1}\mathbf{n}$ on $\partial\Omega$.

For the solution to (32)–(33), we refer to [8, Lemma 2.2] which states that for each $\mathbf{g} \in H^{1/2}(\partial\Omega)^3$ satisfying $\int_{\partial\Omega} \mathbf{g} \cdot \mathbf{n}\,d\sigma(x) = 0$ there exists a function $\tilde{\mathbf{u}} \in H^1(\Omega)^3$, unique up to an additive function $\mathbf{v}$ from the set V of divergence-free $H^1(\Omega)$ functions vanishing on $\partial\Omega$, such that $\operatorname{div}\tilde{\mathbf{u}} = 0$ in Ω, $\tilde{\mathbf{u}} = \mathbf{g}$ on $\partial\Omega$. In terms of the system (32)–(33), it suffices to set $\mathbf{g} = ((\nabla\Phi - (K_\Gamma U_\Omega + \rho_0 g(m_\Gamma - x_3))\mathbf{k}^{-1}\mathbf{n}) \times \mathbf{n}) \times \mathbf{n}$ and use the identity $(\mathbf{b} \times \mathbf{n}) \times \mathbf{n} = (\mathbf{b} \cdot \mathbf{n})\mathbf{n} - \mathbf{b}$ for every vector $\mathbf{b}$. Moreover, the estimate

$$\inf_{\mathbf{v} \in V} \|\tilde{\mathbf{u}} + \mathbf{v}\|_{H^1(\Omega)} \le C\|\mathbf{g}\|_{H^{1/2}(\partial\Omega)} \le \tilde{C}\|\Phi\|_{H^2(\Omega)} \tag{34}$$

holds with some constants $C, \tilde{C}$. The required regularity is available here by virtue of the assumption that Ω is of class $C^{1,1}$, provided $\mathbf{k}^{-1}$ belongs to $H^{1/2}(\partial\Omega)$. Note that a weaker formulation of problem (32)–(33) can be found in [1, Section 4].

Due to our hypotheses (A3), (A4), we thus lose any control on possible volume preserving turbulences $\mathbf{v} \in V$. This, however, has no influence on the system

(29)–(31), which is the subject of our interest here. Inequality (34) shows that if U is small in agreement with hypothesis **(A1)**, then also $\mathbf{v}$ can be chosen in such a way that hypothesis **(A1)**, interpreted in terms of H^1, is not violated.

3 Energy and Entropy

In terms of the new variables θ, U, χ, the energy e and entropy s can be written as

$$e = c_0(\chi)(e_1(\theta) - f_1(\theta_c)) + \frac{\lambda(\chi)}{2\rho_0}(U - \alpha(1 - \chi))^2$$

$$+ \frac{\beta}{\rho_0}\theta_c U + L_0(\chi + I(\chi)), \tag{35}$$

$$s = c_0(\chi)s_1(\theta) + \frac{L_0}{\theta_c}\chi + \frac{\beta}{\rho_0}U. \tag{36}$$

The energy functional has to be supplemented with the boundary energy term

$$E_\Gamma(t) = \frac{K_\Gamma}{2}\left(U_\Omega(t) + \frac{p_0 + \rho_0 g m_\Gamma}{K_\Gamma}\right)^2, \tag{37}$$

as well as with the gravity potential $-\rho_0 g x_3 U$. The energy and entropy balance equations now read

$$\frac{\mathrm{d}}{\mathrm{d}t}\left(\int_\Omega \rho_0(e(x,t) - g x_3 U)\,\mathrm{d}x + E_\Gamma(t)\right) = \int_{\partial\Omega} h(x)(\theta_\Gamma - \theta)\,\mathrm{d}\sigma(x), \tag{38}$$

$$\rho_0 s_t + \mathrm{div}\frac{\mathbf{q}}{\theta} = \frac{\kappa(\chi)|\nabla\theta|^2}{\theta^2} + \frac{\gamma}{\theta}\chi_t^2 + \frac{\nu}{\theta}U_t^2 \geq 0, \tag{39}$$

$$\frac{\mathrm{d}}{\mathrm{d}t}\int_\Omega \rho_0 s(x,t)\,\mathrm{d}x = \int_{\partial\Omega}\frac{h(x)}{\theta}(\theta_\Gamma - \theta)\,\mathrm{d}\sigma(x) \tag{40}$$

$$+ \int_\Omega\left(\frac{\kappa(\chi)|\nabla\theta|^2}{\theta^2} + \frac{\gamma}{\theta}\chi_t^2 + \frac{\nu}{\theta}U_t^2\right)\mathrm{d}x.$$

The entropy balance (39) says that the entropy production on the right hand side is nonnegative in agreement with the second principle of thermodynamics. The system is not closed, and the energy supply or the energy loss through the boundary is given by the right hand side of (38).

We prescribe the initial conditions

$$\theta(x,0) = \theta^0(x) \tag{41}$$

$$U(x,0) = U^0(x) \tag{42}$$

$$\chi(x,0) = \chi^0(x) \tag{43}$$

for $x \in \Omega$, and compute from (35)–(36) the corresponding initial values e^0, E_{Γ}^0, and s^0 for specific energy, boundary energy, and entropy, respectively. Let $E^0 = \int_{\Omega} \rho_0 e^0 \, dx$, $S^0 = \int_{\Omega} \rho_0 s^0 \, dx$ denote the total initial energy and entropy, respectively. From the energy end entropy balance equations (38), (40), we derive the following crucial (formal for the moment) balance equation for the "extended" energy $\rho_0(e - \theta_{\Gamma} s)$:

$$\int_{\Omega} \left(c(\chi)(e_1(\theta) - f_1(\theta_c)) + \frac{\lambda(\chi)}{2}(U - \alpha(1 - \chi))^2 \right)(x,t) \, dx$$

$$+ \int_{\Omega} (\beta \theta_c U + L\chi - \rho_0 g x_3 U)(x,t) \, dx$$

$$+ \frac{K_{\Gamma}}{2} \left(U_{\Omega}(t) + \frac{p_0 + \rho_0 g \, m_{\Gamma}}{K_{\Gamma}} \right)^2$$

$$+ \theta_{\Gamma} \int_0^t \int_{\Omega} \left(\frac{\kappa(\chi)|\nabla \theta|^2}{\theta^2} + \frac{\gamma}{\theta}\chi_t^2 + \frac{\nu}{\theta}U_t^2 \right)(x,\tau) \, dx \, d\tau$$

$$+ \int_0^t \int_{\partial \Omega} \frac{h(x)}{\theta}(\theta_{\Gamma} - \theta)^2(x,\tau) \, d\sigma(x) \, d\tau$$

$$= E^0 + E_{\Gamma}^0 - \theta_{\Gamma} S^0 + \theta_{\Gamma} \int_{\Omega} \left(c(\chi)s_1(\theta) + \frac{L}{\theta_c}\chi + \beta U \right)(x,t) \, dx. \quad (44)$$

We assume that both $c(\chi)$ and $\lambda(\chi)$ are bounded from above and from below by positive constants. The growth of $s_1(\theta)$ is dominated by $e_1(\theta)$ as a consequence of the inequality

$$\frac{s_1(\theta) - s_1(\theta^*)}{e_1(\theta) - e_1(\theta^*)} \leq \frac{1}{\theta^*} \qquad \forall \theta > \theta^* > 0.$$

Hence, there exists a constant $C > 0$ independent of t such that for all $t > 0$ we have

$$\int_{\Omega} (e_1(\theta) + U^2)(x,t) \, dx + \int_0^t \int_{\Omega} \left(\frac{|\nabla \theta|^2}{\theta^2} + \frac{\chi_t^2}{\theta} + \frac{U_t^2}{\theta} \right)(x,\tau) \, dx \, d\tau \quad (45)$$

$$+ \int_0^t \int_{\partial \Omega} \frac{h(x)}{\theta}(\theta_{\Gamma} - \theta)^2(x,\tau) \, d\sigma(x) \, d\tau \leq C.$$

4 Equilibria

It follows from (22) and (29) that the only possible equilibrium temperature is $\theta = \theta_{\Gamma}$, and the equilibrium configurations $U_{\infty}, \chi_{\infty}$ for U, χ satisfy for a.e. $x \in \Omega$ the equations

$$\lambda(\chi_{\infty}(x))(U_{\infty}(x) - \alpha(1 - \chi_{\infty}(x))) = \beta(\theta_{\Gamma} - \theta_c)$$

$$+ \rho_0 g(x_3 - m_{\Gamma}) - p_0 - K_{\Gamma} \int_{\Omega} U_{\infty}(x') \, dx', \quad (46)$$

$$L\left(\frac{\theta_\Gamma}{\theta_c} - 1\right) + c'(\chi_\infty(x))(f_1(\theta_c) - f_1(\theta_\Gamma))$$

$$-\frac{1}{2}\lambda'(\chi_\infty(x))(U_\infty(x) - \alpha(1 - \chi_\infty(x)))^2$$

$$-\alpha\lambda(\chi_\infty(x))(U_\infty(x) - \alpha(1 - \chi_\infty(x))) \in \partial I(\chi_\infty(x)), \quad (47)$$

as a consequence of (30), (31). We now eliminate U_∞ from the above equations. To simplify the formulas, we introduce the notation

$$\left. \begin{array}{ll} S := \int_\Omega (1 - \chi_\infty(x'))\,dx', & U_\Omega := \int_\Omega U_\infty(x')\,dx', \\[2mm] \Lambda := \int_\Omega \frac{dx'}{\lambda(\chi_\infty(x'))}, & m_\lambda := \frac{1}{\Lambda}\int_\Omega \frac{x'_3}{\lambda(\chi_\infty(x'))}\,dx'. \end{array} \right\} \quad (48)$$

We see that S is the total solid content, and U_Ω is the total volume increment. We now divide (46) by $\lambda(\chi_\infty(x))$ and integrate over Ω. This yields

$$(1 + K_\Gamma \Lambda)U_\Omega = \alpha S + \Lambda(\beta(\theta_\Gamma - \theta_c) - p_0 + \rho_0 g(m_\lambda - m_\Gamma)).$$

This enables us to replace U_Ω on the right hand side of (46) and to obtain

$$\lambda(\chi_\infty(x))(U_\infty(x) - \alpha(1 - \chi_\infty(x)))$$

$$= \frac{\beta(\theta_\Gamma - \theta_c) - p_0 - \alpha K_\Gamma S}{1 + K_\Gamma \Lambda} + \rho_0 g(x_3 - m^*), \quad (49)$$

where m^* is a convex combination of m_Γ and m_λ, given by

$$m^* = \frac{1}{1 + K_\Gamma \Lambda}m_\Gamma + \frac{K_\Gamma \Lambda}{1 + K_\Gamma \Lambda}m_\lambda. \quad (50)$$

Eq. (47) can thus be rewritten as

$$L\left(\frac{\theta_\Gamma}{\theta_c} - 1\right) + c'(\chi_\infty(x))(f_1(\theta_c) - f_1(\theta_\Gamma))$$

$$-\frac{\lambda'(\chi_\infty(x))}{2\lambda^2(\chi_\infty(x))}\left(\frac{\beta(\theta_\Gamma - \theta_c) - p_0 - \alpha K_\Gamma S}{1 + K_\Gamma \Lambda} + \rho_0 g(x_3 - m^*)\right)^2$$

$$-\alpha\left(\frac{\beta(\theta_\Gamma - \theta_c) - p_0 - \alpha K_\Gamma S}{1 + K_\Gamma \Lambda} + \rho_0 g(x_3 - m^*)\right) \in \partial I(\chi_\infty(x)). \quad (51)$$

Approximate values of the physical constants are listed in Table 1, see [7, 18, 19, 22].

In order to draw some conclusions about the solutions to (51), we eliminate the χ-dependence and non-monotonicities in θ_Γ on the left hand side of (51) by choosing the following nonlinearities:

Table 1 Physical constants for water

Specific volume of water	$V_w = 1/\rho_0$	10^{-3}	m^3/kg
Specific volume of ice	V_i	$1.09 \cdot 10^{-3}$	m^3/kg
Speed of sound in water	v_w	$1.5 \cdot 10^3$	m/s
Speed of sound in ice	v_i	$3.12 \cdot 10^3$	m/s
Elasticity modulus of water	$\lambda_w = v_w^2/V_w$	$2.25 \cdot 10^9$	$Pa = J/m^3 = kg/m s^2$
Elasticity modulus of ice	$\lambda_i = v_i^2/V_i$	$9 \cdot 10^9$	$Pa = J/m^3 = kg/m s^2$
Specific heat of water	c_w	$4.2 \cdot 10^3$	$J/kgK = m^2/s^2K$
Specific heat of ice	c_i	$2.1 \cdot 10^3$	$J/kgK = m^2/s^2K$
Latent heat	L_0	$3.34 \cdot 10^5$	$J/kg = m^2/s^2$
Thermal expansion coefficient	β	$4.5 \cdot 10^5$	$J/m^3K = kg/m s^2K$
Melting temperature at standard pressure	θ_c	273	K
Standard atmospheric pressure	p_0	10^5	$Pa = J/m^3 = kg/m s^2$
Phase expansion coefficient	$\alpha = (V_i - V_w)/V_w$	0.09	
Gravity constant	g	9.8	m/s^2

$$\lambda(\chi) = \left(\frac{1}{\lambda_i} + \left(\frac{1}{\lambda_w} - \frac{1}{\lambda_i} \right) \chi \right)^{-1}, \tag{52}$$

$$c(\chi) = \frac{c_i}{V_i} + \left(\frac{c_w}{V_w} - \frac{c_i}{V_i} \right) \chi, \tag{53}$$

$$c_1(\theta) = \left(\frac{\theta}{\theta_c} \right)^{\xi}, \tag{54}$$

with a constant $\xi > 0$. The function f_1 is, consequently,

$$f_1(\theta) = -\frac{1}{\xi(1+\xi)} \frac{\theta^{1+\xi}}{\theta_c^{\xi}}. \tag{55}$$

This is again a very rough approximation. In reality, for temperatures near zero Kelvin, the exponent ξ should be 3 according to the Einstein-Debye law, while for large temperatures, it should vanish. Our choice is motivated by the effort to keep the number of parameters as low as possible.

Assuming (52)–(54), we write (51) in explicit form

$$L\left(\frac{\theta_\Gamma}{\theta_c} - 1 \right) + \frac{\theta_c}{\xi(1+\xi)} \left(\frac{c_w}{V_w} - \frac{c_i}{V_i} \right) \left(\left(\frac{\theta_\Gamma}{\theta_c} \right)^{1+\xi} - 1 \right)$$

$$+ \frac{1}{2} \left(\frac{1}{\lambda_w} - \frac{1}{\lambda_i} \right) \left(\frac{\beta(\theta_\Gamma - \theta_c) - p_0 - \alpha K_\Gamma S}{1 + K_\Gamma \Lambda} + \rho_0 g(x_3 - m^*) \right)^2$$

$$- \alpha \left(\frac{\beta(\theta_\Gamma - \theta_c) - p_0 - \alpha K_\Gamma S}{1 + K_\Gamma \Lambda} + \rho_0 g(x_3 - m^*) \right) \in \partial I(\chi_\infty(x)), \tag{56}$$

with

$$\Lambda = \frac{1}{\lambda_w}|\Omega| - \left(\frac{1}{\lambda_w} - \frac{1}{\lambda_i}\right) S. \tag{57}$$

To estimate an appropriate value of ξ, let us neglect the gravity forces (which are indeed very small as we shall see) and assume the rigid limit $K_\Gamma \to \infty$. We have

$$R := \frac{S}{\lambda_i \Lambda} = \frac{\frac{1}{\lambda_i}\frac{S}{|\Omega|}}{\frac{1}{\lambda_w} - \left(\frac{1}{\lambda_w} - \frac{1}{\lambda_i}\right)\frac{S}{|\Omega|}} \in [0,1]. \tag{58}$$

Eq. (56) then reads in dimensionless form

$$\frac{L}{\alpha^2 \lambda_i}\left(\frac{\theta_\Gamma}{\theta_c} - 1\right) + \frac{1}{\xi(1+\xi)}\frac{\theta_c}{\alpha^2 \lambda_i}\left(\frac{c_w}{V_w} - \frac{c_i}{V_i}\right)\left(\left(\frac{\theta_\Gamma}{\theta_c}\right)^{1+\xi} - 1\right)$$

$$+ \frac{1}{2}\left(\frac{\lambda_i}{\lambda_w} - 1\right)R^2 + R \in \partial I(\chi_\infty(x)). \tag{59}$$

For $\theta_\Gamma \geq \theta_c$, the left hand side of (59) is nonnegative, hence necessarily $\chi_\infty(x) = 1$ for (almost) all $x \in \Omega$ and $S = R = 0$. Because of the pressure increase due to solidification, the liquid phase persists also for temperatures below θ_c. We only obtain pure ice $\chi_\infty = 0$ if the left hand side of (59) with $R = 1$ is nonpositive, that is, if $\theta_\Gamma \leq y\theta_c$, where $y \in (0,1)$ is the solution (if it exists) to the equation

$$C_1(y-1) + \frac{C_2}{\xi(1+\xi)}(y^{1+\xi} - 1) + C_3 = 0, \tag{60}$$

with dimensionless constants

$$C_1 = \frac{L}{\alpha^2 \lambda_i}, \quad C_2 = \frac{\theta_c}{\alpha^2 \lambda_i}\left(\frac{c_w}{V_w} - \frac{c_i}{V_i}\right), \quad C_3 = \frac{1}{2}\left(\frac{\lambda_i}{\lambda_w} - 1\right) + 1.$$

For the values of the constants in Table 1, we obtain

$$C_1 \approx 4.58, \quad C_2 \approx 8.5, \quad C_3 \approx 2.5, \tag{61}$$

hence the solution $y = y(\xi)$ to (60) exists for all $\xi > 0$, and we easily compute the limits $\lim_{\xi \to 0+} y(\xi) = 1$, $\lim_{\xi \to +\infty} y(\xi) = 1 - C_3/C_1$. Assume that we know the full solidification temperature θ_s, and that

$$(1 - C_3/C_1)\theta_c < \theta_s < \theta_c. \tag{62}$$

Then we identify the value of ξ from the equation $y(\xi) = \theta_s/\theta_c$, that is,

$$\varphi(\xi) := C_2\left(\frac{\theta_s}{\theta_c}\right)^{1+\xi} + \left(C_3 + C_1\left(\frac{\theta_s}{\theta_c} - 1\right)\right)\xi(1+\xi) = 1. \tag{63}$$

The function φ is convex in $(0,\infty)$, $\varphi(0) < 1$, $\varphi(+\infty) = +\infty$. Eq. (63) thus determines the desired value of ξ uniquely.

Still in the rigid limit $K_\Gamma \to \infty$, consider now the gravity effects in Eq. (56). Then, by (48) and (50), we have $m^* = m_\lambda \in (a,b)$, and the counterpart of Eq. (59) reads

$$(C_3 - 1)(R - \eta(x_3 - m_\lambda))^2 + (R - \eta(x_3 - m_\lambda)) - C_4(\theta_\Gamma) \in \partial I(\chi_\infty(x)), \qquad (64)$$

where

$$C_4(\theta_\Gamma) := C_1 \left(1 - \frac{\theta_\Gamma}{\theta_c}\right) + \frac{C_2}{\xi(1+\xi)} \left(1 - \left(\frac{\theta_\Gamma}{\theta_c}\right)^{1+\xi}\right),$$

C_1, C_2, C_3 are as above, and

$$\eta = \frac{\rho_0 g}{\alpha \lambda_i} \approx 1.2 \cdot 10^{-5} \, [m^{-1}]. \qquad (65)$$

The left hand side of (64) is a function of x_3 only. Let the interval (a,b) be the projection of Ω onto the x_3-axis, that is,

$$x_3 \in (a,b) \Leftrightarrow \exists (x_1,x_2) \in \mathbb{R}^2 : (x_1,x_2,x_3) \in \Omega.$$

We prove the following result.

Theorem 1. *Let the height $b - a$ of the container satisfy the inequality*

$$2\eta(b-a)(C_3 - 1) < 1. \qquad (66)$$

Then Eq. (64) admits a solution $\chi_\infty \in L^\infty(\Omega)$. Moreover, there exist temperatures $\theta_w > \theta_c > \theta_i > 0$ such that $\chi_\infty \equiv 1$ if $\theta_\Gamma \geq \theta_w$, $\chi_\infty \equiv 0$ if $\theta_\Gamma \leq \theta_i$, and for $\theta_\Gamma \in (\theta_i, \theta_w)$ there exists $z \in (a,b)$ such that $\chi_\infty(x) = 1$ for $x_3 < z$, $\chi_\infty(x) = 0$ for $x_3 > z$.

Condition (66) is not too restrictive. With the values in (61) and (65), the maximal admissible height is almost 30 km. The solution may not be unique if the shape of Ω is very irregular. If Ω is a straight vertical cylinder $\Omega = \Omega_{2D} \times (a,b)$, for example, where $\Omega_{2D} \subset \mathbb{R}^2$ is fixed, the proof below shows that the solution is unique.

The interval (θ_c, θ_w) of "overheated ice temperatures" is very narrow, of the size of $\eta(b-a)$, and corresponds to the low pressure ice layer on the top of the container.

Proof. The left hand side of (64) is always nonnegative if $4C_4(\theta_\Gamma)(C_3 - 1) + 1 \leq 0$, that is, if θ_Γ is above a certain temperature slightly bigger than θ_c. In this case, $\chi_\infty(x) = 1$ for all $x \in \Omega$ independently of the height $b - a$. Assume now

$$4C_4(\theta_\Gamma)(C_3 - 1) + 1 > 0.$$

Then the left hand side of (64) is positive if and only if

$$\eta(x_3 - m_\lambda) < R + \frac{1}{2(C_3 - 1)} \left(1 - \sqrt{4C_4(\theta_\Gamma)(C_3 - 1) + 1}\right) \qquad (67)$$

or

$$\eta(x_3 - m_\lambda) > R + \frac{1}{2(C_3 - 1)}\left(1 + \sqrt{4C_4(\theta_\Gamma)(C_3 - 1) + 1}\right). \qquad (68)$$

Condition (68) is in contradiction with the assumption (66), hence the exists at most one

$$z = m_\lambda + \frac{1}{\eta}\left(R + \frac{1}{2(C_3 - 1)}\left(1 - \sqrt{4C_4(\theta_\Gamma)(C_3 - 1) + 1}\right)\right) \in (a, b)$$

such that the left hand side of (64) is positive for $x_3 < z$ and negative for $x_3 > z$. By definition of the subdifferential of the indicator function on the right hand side of (64), we then have $\chi_\infty(x) = 1$ for $x_3 < z$, $\chi_\infty(x) = 0$ for $x_3 > z$, as expected. It remains to determine z. Assume first that such a z exists. Then both $R = R(z)$ and $m_\lambda = m_\lambda(z)$ are functions of z. We denote

$$\Omega(z) = \{x = (x_1, x_2) \in \mathbb{R}^2 : (x_1, x_2, z) \in \Omega\}.$$

The set $\Omega(z)$ is empty for $z \geq b$ and for $z \leq a$. Let $\omega(z)$ be the 2D Lebesgue measure of $\Omega(z)$. Then, by (58), we have

$$R(z) = \frac{\frac{1}{\lambda_i}\int_z^b \omega(s)\,ds}{\frac{1}{\lambda_w}\int_a^z \omega(s)\,ds + \frac{1}{\lambda_i}\int_z^b \omega(s)\,ds},$$

and by (48),

$$m_\lambda(z) = \frac{\frac{1}{\lambda_w}\int_a^z s\omega(s)\,ds + \frac{1}{\lambda_i}\int_z^b s\omega(s)\,ds}{\frac{1}{\lambda_w}\int_a^z \omega(s)\,ds + \frac{1}{\lambda_i}\int_z^b \omega(s)\,ds}.$$

The dependence of z on θ_Γ is given by the equation

$$z - m_\lambda(z) - \frac{1}{\eta}R(z) = \frac{1}{\eta}\left(\frac{1}{2(C_3 - 1)}(1 - \sqrt{4C_4(\theta_\Gamma)(C_3 - 1) + 1})\right). \qquad (69)$$

The left hand side of (69) is a continuous function of z, which is negative for $z = a$ and positive for $z = b$, and the statement of Theorem 1 easily follows. For a straight cylinder $\Omega = \Omega_{2D} \times (a, b)$, where $\Omega_{2D} \subset \mathbb{R}^2$ is fixed, the left hand side of (69) is an increasing function of z, hence the solution is unique.   □

Remark 1. We can interpret Eqs. (46)–(47) in another way. On the interface x_3 between water and ice, the left hand side of (47) vanishes, and (46) has the form

$$\lambda(\chi_\infty(x))(U_\infty(x) - \alpha(1 - \chi_\infty(x))) = \beta(\theta_\Gamma - \theta_c) - P_\infty, \qquad (70)$$

where $P_\infty = p_0 + K_\Gamma U_\Omega + \rho_0 g(m_\Gamma - x_3)$ is the equilibrium pressure in agreement with (19). Hence, (47) can be reformulated in terms of P_∞ as

$$L\left(\frac{\theta_\Gamma}{\theta_c} - 1\right) + \left(\frac{c_w}{V_w} - \frac{c_i}{V_i}\right)(f_1(\theta_c) - f_1(\theta_\Gamma))$$

$$+ \frac{1}{2}\left(\frac{1}{\lambda_w} - \frac{1}{\lambda_i}\right)(\beta(\theta_\Gamma - \theta_c) - P_\infty)^2 - \alpha(\beta(\theta_\Gamma - \theta_c) - P_\infty) = 0. \quad (71)$$

This would be the Clausius-Clapeyron relation in the sense of [21, Equation (288)] if $c_w/V_w = c_i/V_i$ and $\lambda_i = \lambda_w$, namely

$$\frac{P_\infty}{\theta_\Gamma - \theta_c} = \frac{L_\beta}{\theta_c(V_w - V_i)},$$

where $L_\beta = L_0 - (\alpha\beta\theta_c)/\rho_0$ is the modified latent heat. In our case, the modified latent heat contains additional terms related to the differences in elasticity moduli and in specific heat capacities.

Remark 2. Note that in the fully solidified rigid limit case, the equilibrium pressure is very high, namely (up to negligible contributions due to gravity and thermal expansion) $P_\infty \approx \alpha\lambda_i \approx 0.81\,GPa$.

Conclusion

A model is proposed for describing the dynamics of freezing/melting of water in an elastic container, taking into account the differences in specific volume, specific heat, and speed of sound in water and in ice. The process is described by one parabolic PDE, one integrodifferential ODE, and one differential inclusion for three unknown functions – the absolute temperature, relative volume increment, and liquid fraction. A study of the equilibria in the rigid limit is carried out in detail.

References

1. Amrouche, C., Girault, V.: Decomposition of vector spaces and application to the Stokes problem in arbitrary dimension. Czechoslovak Math. J. 44 (119), 109–140 (1994)
2. Brokate, M., Sprekels, J.: Hysteresis and Phase Transitions. Appl. Math. Sci. 121 (1996)
3. Colli, P., Frémond, M., Visintin, A.: Thermo-mechanical evolution of shape memory alloys. Quart. Appl. Math. 48, 31–47 (1990)
4. Frémond, M.: Non-smooth thermo-mechanics. Springer, Berlin (2002)
5. Frémond, M., Rocca, E.: Well-posedness of a phase transition model with the possibility of voids. Math. Models Methods Appl. Sci. 16, 559–586 (2006)
6. Frémond, M., Rocca, E.: Solid liquid phase changes with different densities. Q. Appl. Math. 66, 609–632 (2008)
7. Gerthsen, C.: Physics, 9th edn. Springer, Heidelberg (1966)
8. Girault, V., Raviart, P.-A.: Finite Element Methods for Navier-Stokes Equations. Springer, Berlin (1986)
9. Joos, G.: Lehrbuch der theoretischen Physik. Akademische Verlagsgesellschaft, Leipzig (1939) (in German)

10. Krejčí, P., Rocca, E., Sprekels, J.: A nonlocal phase-field model with nonconstant specific heat. Interfaces and Free Boundaries 9, 285–306 (2007)
11. Krejčí, P., Rocca, E., Sprekels, J.: A bottle in a freezer. To appear in SIAM J. Math. Anal.
12. Krejčí, P., Rocca, E., Sprekels, J.: Phase separation in a gravity field. To appear in DCDS-S
13. Krejčí, P., Sprekels, J., Stefanelli, U.: Phase-field models with hysteresis in one-dimensional thermoviscoplasticity. SIAM J. Math. Anal. 34, 409–434 (2002)
14. Krejčí, P., Sprekels, J., Stefanelli, U.: One-dimensional thermo-visco-plastic processes with hysteresis and phase transitions. Adv. Math. Sci. Appl. 13, 695–712 (2003)
15. Madelung, E.: Die mathematischen Hilfsmittel des Physikers, 6th edn. Springer, Heidelberg (1957) (in German)
16. Rocca, E., Rossi, R.: A nonlinear degenerating PDE system modelling phase transitions in thermoviscoelastic materials. J. Differential Equations 245, 3327–3375 (2008)
17. Rocca, E., Rossi, R.: Global existence of strong solutions to the one-dimensional full model for phase transitions in thermoviscoelastic materials. Appl. Math. 53, 485–520 (2008)
18. Schulson, E.M.: The Structure and Mechanical Behavior of Ice. JOM 51, 21–27 (1999)
19. Vogt, C., Laihem, K., Wiebusch, C.: Speed of sound in bubble-free ice. J. Acoust. Soc. Am. 124, 3613–3618 (2008)
20. Visintin, A.: Models of Phase Transitions. In: Progress in Nonlinear Differential Equations and their Applications, vol. 28. Birkhäuser, Basel (1996)
21. Yildirim Erbil, H.: Surface Chemistry of Solid and Liquid Interfaces. Blackwell Publishing, John Wiley & Sons (2006)
22. http://www1.lsbu.ac.uk/water/anmlies.html

Composite Beams with Embedded Shape Memory Alloy

Mieczysław Kuczma

Abstract. We are concerned with the bending problem of composite beams with embedded shape memory alloy in the form of fibres or strips. As a special case, the proposed formulation covers the case of a monolithic beam made of shape memory alloy. Shape memory alloys (SMAs) are materials that may undergo a temperature- or stress-induced martensitic phase transformation, which results in the shape memory effect and the pseudoelastic (superelastic) behaviour. The unique behaviour of SMAs opens new possibilities in design of adaptive structures. Herein we have formulated the quasi-static bending problem for a SMA beam in the form of a evolution variational inequality. After the finite element approximation, the variational inequality is solved incrementally as a sequence of linear complementarity problems. Finally, the predictions of the proposed model, including hysteresis loops, are illustrated with many results of numerical simulations for SMA beams.

1 Introduction

Shape memory alloys (SMAs) constitute a new class of functional, smart materials that have found many technological applications and offer innovative solutions in design of adaptive structures and systems [18, 3, 17, 8]. A typical example of SMAs is Nickel-Titanium that exhibits the shape memory effect and the property of pseudoelasticity (superelasticity), i.e. is capable of recovering large strains during mechanical loading-unloading cycles conducted at a constant temperature. This unique material response is attributed to a martensitic phase transformation, which is a first-order reversible transformation from a high temperature phase with greater symmetry, called *austenite*, to a low temperature phase with lower symmetry, called

Mieczysław Kuczma
University of Zielona Góra, Institute of Building Engineering, ul. Z. Szafrana 1,
65-516 Zielona Góra, Poland
e-mail: m.kuczma@ib.uz.zgora.pl

martensite, [12]. The martensitic phase transformation can be induced by stresses, changes of temperature, or a magnetic field.

In this contribution we formulate and numerically solve the quasi-static, isothermal bending problem of composite beams reinforced with SMA wires. As a special case, our formulation covers the case of a monolithic beam made of shape memory material. A characteristic feature of SMA behaviour are hysteretic loops manifested in different quantities describing the martensitic phase transformation process. The flexural characteristics are important in design of SMA devices constructed from strips, tubes, beams, plates and shells. Modelling of SMAs is an interdisciplinary research activity. Various aspects and models of SMAs are studied in [1, 4, 10, 14, 2, 19, 9, 11, 6, 5, 20, 16], where further references are included. The bending problem of SMA beams was investigated in [15, 13], for example.

We have formulated the bending problem for a beam with shape memory alloy in the form of an evolution variational inequality which, after the finite element approximation, is solved incrementally as a sequence of linear complementarity problems. The obtained results of numerical simulations for SMA beams present the distribution of volume fractions of martensites and the characteristic hysteresis loops of pseudoelasticity.

2 Constitutive Relations for a Shape Memory Alloy

In this section we present constitutive relations for shape memory alloys in the range of pseudoelasticity. The presented model accounts for the characteristic hysteresis loops in both tension and compression states as illustrated in a typical one-dimensional stress-strain diagram in Fig. 2. It should be noted that a shape memory alloy is itself a kind of composite in which concentrations (volume fractions) of its constituents are not known in advance, but they evolve as a result of the deformation process. In a general case, the martensite may exist in many variants [12]. In the one-dimensional case of the bending problem considered here, we can treat the A/M-mixture as if it were composed of austenite A and only two 'averaged' variants of martensite M_1 and M_2. The variant M_1 corresponds to the phase transformation strain η_1 measured in a tensile test, while the variant M_2 corresponds to the phase transformation strain η_2 measured in a compression test, see e.g. [15]. Furthermore, we can assume that the eigenstrain of austenite $\eta_3 = 0$. Let c_1 and c_2 denote a volume fraction of martensite M_1 and M_2, respectively, and $c_3 = 1 - c_1 - c_2$ be the volume fraction of austenite. By definition, the volume fractions satisfy the condition $0 \leq c_i \leq 1$ for $i = 1, 2, 3$. In the studied case of two variants of martensite M_1 or M_2, which can be induced by positive stresses or negative stresses, respectively, at a given material point we can have only one variant active, i.e. $c_1 \cdot c_2 = 0$ must hold.

We define the averaged specific free energy of the A/M-mixture by the formula

$$\widetilde{W} = \sum_{i=1}^{3} c_i W_i + W_{\text{mix}} + I_{[0,1]}(c_1 + c_2) \tag{1}$$

where W_i $(i = 1,2,3)$ is a free energy of the different, particular phase (or variant), W_{mix} denotes an energy of mixing and $I_{[0,1]}(\cdot)$ stands for the indicator function of the interval $[0,1]$ that is a formal imposition of the constraint on volume fractions c_1 and c_2.

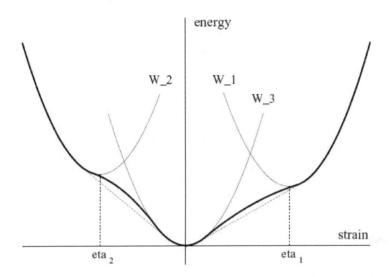

Fig. 1 Quasiconvexified three-well energy function (bold line). The deshed line corresponds to convexification

When adopting the quadratic energy functions for each phase $(i = 1,2,3)$,

$$W_i = \frac{1}{2}E\,(\varepsilon_i^e)^2 + \varpi_i(\theta), \qquad W_{mix} = \sum_{i=1}^{2} \frac{1}{2}B_i\,(1-c_i)c_i \qquad (2)$$

in which $\varepsilon_i^e = \varepsilon - c_i\eta_i$, is an elastic part of the total strain ε and $\varpi_i(\theta)$ is an energy in a stress-free state, we finally obtain the following expression for the quasiconvexified free energy of the A/M-mixture, see Fig. 1,

$$\widetilde{W}(\varepsilon,\mathbf{c}) = \frac{1}{2}E\left(\varepsilon - \sum_{i=1}^{2}c_i\eta_i\right)^2 + \sum_{i=1}^{3}c_i\varpi_i + \sum_{i=1}^{2}\frac{1}{2}B_i(1-c_i)c_i + \partial I_{[0,1]}(c_1+c_2) \quad (3)$$

In formulae (2) and (3), E is the Young modulus of austenite and martensite, $B_i, \eta_i (i = 1,2)$ are material parameters which can be calculated from the stress-strain diagram of pseudoelasticity as shown in Fig. 2, see [10, 6]. By θ we denote temperature, which is assumed to be constant in our considerations herein, and $\partial I_{[0,1]}(\cdot)$ is a subdifferential of the indicator function.

Fig. 2 Hysteretic stress-
strain diagram of pseudo-
elasticity

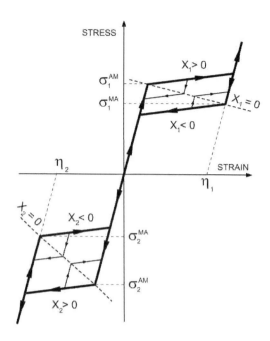

In order to account for the dissipative effects in the martensitic phase transformation process whose manifestation are the hysteresis loops in Fig. 2, we make use of
the rate of dissipation inequality

$$D = \sigma\dot{\varepsilon} - \dot{\widetilde{W}} - s\dot{\theta} \geq 0 \tag{4}$$

In the light of the formulae for the stress

$$\sigma \equiv \partial\widetilde{W}/\partial\varepsilon = E(\varepsilon - c_1\eta_1 - c_2\eta_2) \tag{5}$$

and entropy

$$s \equiv -\partial\widetilde{W}/\partial\theta \tag{6}$$

equation (4) can be reduced to

$$D = X_1\dot{c}_1 + X_2\dot{c}_2 \geq 0 \tag{7}$$

In eqn. (7) X_i is a thermodynamically conjugate variable to volume fraction c_i

$$X_i \equiv -\partial\widetilde{W}/\partial c_i, \quad i = 1, 2 \tag{8}$$

which is the driving force of the reversible phase transformation A $\rightarrow$ M$_i$,

$$X_i(\varepsilon, c_i) = E(\varepsilon - c_i\eta_i)\eta_i - (\varpi_i - \varpi_3) - 0.5\, B_i(1 - 2c_i) - R_i, \quad R_i \in \partial I_{[0,1]}(c_i) \tag{9}$$

The driving forces X_i are positive for the forward transformation from austenite to martensite $A \rightarrow M_i$ ($i = 1, 2$), negative for the reverse transformation $M_i \rightarrow A$, and equal to zero at the characteristic diagonals in Fig. 2.

In reference to the hysteretic behaviour shown in Fig. 2, the evolutions of volume fractions c_i, $i = 1, 2$, are supposed to be governed by the following phase transformation rule (PTR)

$$
\begin{array}{llll}
\text{if} & X_i(\varepsilon, c_i) = \kappa_{3 \rightarrow i}(c_i) & \text{then} & \dot{c}_i \geq 0 \\
\text{if} & X_i(\varepsilon, c_i) = \kappa_{i \rightarrow 3}(c_i) & \text{then} & \dot{c}_i \leq 0 \qquad (10) \\
\text{if} & \kappa_{i \rightarrow 3}(c_i) < X_i(\varepsilon, c_i) < \kappa_{3 \rightarrow i}(c_i) & \text{then} & \dot{c}_i = 0
\end{array}
$$

where $\kappa_{3 \rightarrow i} \geq 0$, and $\kappa_{i \rightarrow 3} \leq 0$ are threshold functions, which are presumed in the form $\kappa_{3 \rightarrow i}(c_i) = L_i c_i$, $\kappa_{i \rightarrow 3}(c_i) = L_i(c_i - 1)$, with L_i being material parameters. The forward or reverse phase transformation commences and continues only then when the driving force reaches and remains equal to the current value of the threshold function. In the interior of the hysteresis loops the material response is elastic.

In the sequel we will use the presented relations in defining the bending problem of composite beams with SMA wires or layers, or made entirely of such a phase-transforming material.

3 Bending of a SMA Composite Beam

In this section we have formulated the quasi-static bending problem for a composite beam enforced with SMA wires or strips. The starting point is the classical beam theory, complemented with a layerwise approach in which the cross-section of the beam is divided into a number of layers, N_l. In a particular case, the proposed layerwise model covers also the instance of monolithic shape memory alloy beams.

Let $w_0 = w_0(x, t)$ denote the deflection of beam's axis, wherein x is the abscissa, $0 \leq x \leq l_b$, and t is a time-like parameter. Making use of the Bernoulli kinematical hypothesis, we can express the normal strain ε as

$$
\varepsilon = \varepsilon(x, z, t) = \varepsilon_0(t) - \frac{\partial^2 w_0}{\partial x^2}(z - z_0(t)) = \varepsilon_0 - w_0'' z \qquad (11)
$$

where z is the through-the-thickness coordinate, ε_0 is a normal strain at the axis located at $z = z_0$. In the sequel, for simplicity, we suppose that the properties of the cross-section of the beam are symmetrical with respect to the axis at $z_0 = 0$ and presume $\varepsilon_0 = 0$.

For a typical kth layer, whose location in the cross-section is determined by the z-coordinates z_k, z_{k+1} and hight $h_k = z_{k+1} - z_k$, let $E_m^{(k)}, v_m^{(k)}$ and $E_f^{(k)}, v_f^{(k)}$ denote the Young modulus and volume fraction of the matrix and the fibres, respectively. Further, let $E_{mf}^{(k)} = v_m^{(k)} E_m^{(k)} + v_f^{(k)} E_f^{(k)}$ be the effective modulus of the composite material in kth layer, and let $b = b(z)$ indicate the width of the beam cross-section.

By $\mathscr{I}_f$ we designate the set of indices of the beam layers which are reinforced with SMA fibres.

Having defined the normal stress in the matrix by

$$\sigma(x,z,t) = E_m^{(k)}\varepsilon(x,z,t), \qquad \text{with } z \in (z_k, z_{k+1}) \tag{12}$$

and that in the SMA fibres by

$$\sigma(x,z,t,\mathbf{c}^{(k)}) = E_f^{(k)}\left(\varepsilon(x,z,t) - \eta_1^{(k)}c_1^{(k)}(x,z,t) - \eta_2^{(k)}c_2^{(k)}(x,z,t)\right), \quad z \in (z_k, z_{k+1}) \tag{13}$$

we suppose that the normal stress in the kth layer with SMA fibres is given by

$$\sigma(x,z,t,\mathbf{c}^{(k)}) = E_{mf}^{(k)}\varepsilon(x,z,t) - v_f^{(k)}E_f^{(k)}\left(\eta_1^{(k)}c_1^{(k)}(x,z,t) + \eta_2^{(k)}c_2^{(k)}(x,z,t)\right) \tag{14}$$

with $z \in (z_k, z_{k+1})$. In the above formula (14), $c_1^{(k)}$ and $c_2^{(k)}$ are the volume fractions of martensites M_1 and M_2, respectively, in the fibres of kth layer. It should be noted that values of quantities $c_1^{(k)}$ and $c_2^{(k)}$ are not known in advance. By contrast, the evolutions of $c_1^{(k)}$ and $c_2^{(k)}$ constitute additional unknowns of the problem. Collecting them together, we shall denote $\mathbf{c}^{(k)} \equiv \left(c_1^{(k)}, c_2^{(k)}\right)$, and for all N_f layers with SMA fibres we have $\mathbf{c} \equiv \left(\mathbf{c}^{(1)}, \mathbf{c}^{(2)}, \dots, \mathbf{c}^{(N_f)}\right)$. In the one-dimensional bending problem under consideration, one of $c_1^{(k)}$ and $c_2^{(k)}$ equals zero, i.e. $c_1^{(k)}(x,z,t)\,c_2^{(k)}(x,z,t) = 0$ at all material points (x,z) and all time levels t. It is worth to note that this situation is different from that considered in [7] where many variants of martensite may appear at a material point as a result of relaxation of the original problem.

The use of both $c_1^{(k)}$ and $c_2^{(k)}$ in eqn (14) allows us to consistently formulate the boundary value problem for the SMA beam as presented below. The considered problem is nonlinear since at a given point (x,z,t) it is not known in advance: (i) if the martensitic phase transformation process is active; if the answer is 'yes', then (ii) if it is induced by tension and then accompanied by occurrence of martensite M_1, or (iii) if it is induced by compression and then accompanied by occurrence of martensite M_2. The answers to all these questions can be obtained by solving the corresponding, evolutional boundary value problem as a whole.

The equilibrium condition of the beam subjected to load $f(x,t)$ can be written as the variational equation (equation of virtual work)

$$a(w_0, \mathbf{c}; v) = F(t,v) \qquad \forall v \in H_0^2(0, l_b) \tag{15}$$

wherein the linear and bilinear forms are defined by

$$F(t,v) = \int_0^{l_b} f(x,t)v\,dx \tag{16}$$

$$a(w_0, \mathbf{c}; v) = \int_0^{l_b} v'' \left(\sum_{k=1}^{N_l} \int_{z_k}^{z_{k+1}} E_{mf}^{(k)} z^2 b \, dz \right) w_0'' \, dx + \tag{17}$$

$$\int_0^{l_b} v'' \left(\sum_{k \in \mathscr{I}_f} \int_{z_k}^{z_{k+1}} v_f^{(k)} E_f^{(k)} \eta_1^{(k)} c_1^{(k)} \right) z \, b \, dz \, dx + \tag{18}$$

$$\int_0^{l_b} v'' \left(\sum_{k \in \mathscr{I}_f} \int_{z_k}^{z_{k+1}} v_f^{(k)} E_f^{(k)} \eta_2^{(k)} c_2^{(k)} \right) z \, b \, dz \, dx \tag{19}$$

In order to express in weak form the PTR (10) of SMA wires in kth layer, we define first the phase transformation functions for forward and reverse martensitic phase transformations as follows

$$\Phi_{i,+}^{(k)} = v_f^{(k)} \int_{z_k}^{z_{k+1}} \left(\kappa_{i,+}^{(k)}(c_i^{(k)}) - X_i(w_0, c_i^{(k)}) \right) b \, dz \geq 0$$
$$\Phi_{i,-}^{(k)} = v_f^{(k)} \int_{z_k}^{z_{k+1}} \left(X_i(w_0, c_i^{(k)}) - \kappa_{i,-}^{(k)}(c_i^{(k)}) \right) b \, dz \geq 0 \tag{20}$$

with

$$\kappa_{i,+}^{(k)}(c_i^{(k)}) \equiv \max\{L_i^{(k)} c_i^{(k)}, 0\}$$
$$\kappa_{i,-}^{(k)}(c_i^{(k)}) \equiv \min\{L_i^{(k)}(c_i^{(k)} - 1), 0\} \tag{21}$$

and forces X_i expressed in terms of w_0 via (9) and (11).

Let us introduce the notation

$$\mathbf{T}^{(k)} = \mathrm{col}\,(\Phi_{1,+}^{(k)}, \Phi_{2,+}^{(k)}, \Phi_{1,-}^{(k)}, \Phi_{2,-}^{(k)}) \tag{22}$$

$$\mathbf{u}^{(k)} = \mathrm{col}\,(\dot{c}_{1,+}^{(k)}, \dot{c}_{2,+}^{(k)}, \dot{c}_{1,-}^{(k)}, \dot{c}_{2,-}^{(k)}) \tag{23}$$

where

$$\dot{c}_{i,+}^{(k)} \equiv \max\{\dot{c}_i^{(k)}, 0\} \qquad \dot{c}_{i,-}^{(k)} \equiv \min\{\dot{c}_i^{(k)}, 0\} \tag{24}$$

are positive and negative rates of $c_i^{(k)}$.

Now we can impose the conditions (10) in the form of the variational inequality for each kth layer with SMA fibres,

$$\left\langle \mathbf{T}^{(k)}(w_0, \mathbf{c}^{(k)}; \mathbf{u}^{(k)}), \mathbf{s} - \mathbf{u}^{(k)} \right\rangle \geq 0 \quad \forall \mathbf{s} \in C, \qquad k \in \mathscr{I}_f \tag{25}$$

where

$$\langle \mathbf{s}, \mathbf{v} \rangle = \int_0^{l_b} \mathbf{s}(x) \cdot \mathbf{v}(x) \, dx$$

and C is a positive cone in $(L_2(0, l_b))^4$.

The equation (15) and inequalities (25) completely describe the evolution of the beam in bending, automatically accounting for the progress of the phase transformation process under study. After the finite element approximation of (15) and (25), we can solve the resulting system of matrix equations and inequalities for finite increments as a series of linear complementarity problems.

4 Incremental Problem

We shall solve the system (15) and (25) in an incremental manner. Let us select a number of instants t_n, $n = 0, 1, \ldots$, on a time-interval of interest, and for a typical time-step $t_{n-1} \to t_n$ define a finite increment of the displacement field w_0 and volume fractions $\mathbf{c}^{(k)}$ according to the scheme

$$(\bullet)_n = (\bullet)_{n-1} + \Delta(\bullet)_n$$

With respect to space variables x and z, we will approximate the considered problem by making use of the finite element method (FEM). Let $\phi_i = \phi_i(x)$, $(i = 1, \ldots, DOF_w)$ stand for basis functions of finite element approximation of the beam displacement w_0. Further, we denote by $\psi_j = \psi_j(x)$, $(j = 1, \ldots, DOF_{c_x})$, and $\chi_k = \chi_k(z)$, $(k = 1, \ldots, DOF_{c_z})$ finite element basis functions for volume fractions along the beam and through its thickness, respectively. Having applied the finite element approximation to the conditions (15) and (25), we can reduce them to the following linear complementarity problem that is solved for finite increments of nodal values (parameters) of displacement field $\Delta \mathbf{w}_n$ and for positive $\Delta \mathbf{c}_n^+$ and negative $\Delta \mathbf{c}_n^-$ increments of nodal values of volume fraction fields $\Delta \mathbf{c}_n = \mathrm{col}\left(\Delta \mathbf{c}_n^{(k)}, \ldots, \Delta \mathbf{c}_n^{(k+l)}, \ldots, \Delta \mathbf{c}_n^{(N_f)}\right)$, where $N_f = \dim \mathscr{I}_f$.

The incremental problem for conditions (15) and (25) can be formulated as the following linear complementarity problem:

At time t_n find a vector $\mathbf{x}_n$ such that

$$\mathbf{D}\mathbf{x}_n + \mathbf{y}_n = \mathbf{b}_{n,n-1}$$
$$\mathbf{x}_n' \geq 0, \qquad \mathbf{y}_n^1 = 0, \qquad \mathbf{y}_n \geq 0, \qquad \mathbf{x}_n \cdot \mathbf{y}_n = 0 \tag{26}$$

The matrix $\mathbf{D}$, vector of unknowns $\mathbf{x}_n$ and vector of data $\mathbf{b}_{n,n-1}$ in (26) have the following structure

$$\mathbf{D} = \begin{bmatrix} -\mathbf{K} & -\mathbf{G} & \mathbf{G} & & \\ -\mathbf{G} & -\mathbf{H} & \mathbf{H} & -\mathbf{I} & \\ \mathbf{G} & \mathbf{H} & -\mathbf{H} & & -\mathbf{I} \\ & \mathbf{I} & & & \\ & & \mathbf{I} & & \end{bmatrix}, \quad \mathbf{x}_n = \begin{Bmatrix} \Delta \mathbf{w}_n \\ \Delta \mathbf{c}_n^+ \\ \Delta \mathbf{c}_n^- \\ \mathbf{r}_n^1 \\ \mathbf{r}_n^0 \end{Bmatrix}, \quad \mathbf{b}_{n,n-1} = \begin{Bmatrix} \mathbf{f}_{n,n-1}^w \\ \mathbf{b}_{n-1}^+ \\ \mathbf{b}_{n-1}^- \\ \mathbf{1} - \mathbf{c}_{n-1} \\ \mathbf{c}_{n-1} \end{Bmatrix}$$

The stiffness matrix $\mathbf{K}$ is generated by the first term of bilinear form a defined in (17), in FEM terms its entries K_{ij} depend on the assumed basis (shape) functions ϕ_i. The coupling matrices $\pm\mathbf{G}$ are defined by (18) and (19) and depend on all the basis functions ϕ_i, ψ_j and χ_k, whereas the matrix $\mathbf{H}$ results from the expression (20) and depends on basis functions ψ_j and χ_k. The constraint of non-negativity is not imposed on the displacement nodal values, $\Delta\mathbf{w}_n$, so $\mathbf{x}'_n = \mathrm{col}(\Delta\mathbf{c}_n^+, \Delta\mathbf{c}_n^-, \mathbf{r}_n^1, \mathbf{r}_n^0)$. The vectors $\mathbf{r}_n^1, \mathbf{r}_n^0$ play a role of Lagrange multipliers of the constraints imposed on increments of volume fractions ($\mathbf{I}$ is an identity matrix), implied by the subdifferential ∂I in formula (9).

For solving the system (26) we have devised a computer program in Fortran 95, based on a direct method for this mixed (or nested) linear complementarity problem. We have applied piecewise polynomials of 3rd order for ϕ_i, piecewise linear functions for ψ_j and piecewise constant functions for χ_k.

5   Numerical Examples

In order to illustrate the response of the shape memory alloy model and to verify the proposed formulation we have carried out a number of numerical simulations on test examples. The presented results pertain to two monolithic SMA beams and an epoxy resin beam reinforced with two SMA strips symmetrically disposed across the cross-section of the beam. For all the three beams a rectangular cross-section of dimensions 2.5×10 mm was assumed. The monolithic beams were dived into twenty layers of equal thickness, whereas the composite beam was divided into five layers whose location is defined by z_k coordinates: -5.0, -4.75, -3,75, 3.75, 4.75, 5.0 mm. Layers 2 and 4 are SMA strips of thickness 1 mm, i.e. $\mathscr{I}_f = \{2,4\}$. The material data used correspond to the CuZuAl shape memory alloy (cf. [10, 6]): $E_f^{(k)} = 12.30$ GPa, $B_1^{(k)} = B_2^{(k)} = 1.30$ MPa, $L_1^{(k)} = L_2^{(k)} = 1.01B^{(k)} = 1.313$ MPa, $\eta_1^{(k)} = -\eta_2^{(k)} = 0.0686$, $(\varpi_i^{(k)} - \varpi_3) = 3.756$ MJ/m^3, $i = 1,2$, and for the epoxy resin: $E_m^{(k)} = 5.00$ GPa.

Example 1
First, let us consider a one-span beam of length $l_b = 150$ mm, fixed at both ends and subjected to half-span variable loading ($e = l_b/2$) as shown in Fig. 3. The history of the concentrated force $P = P(t)$ is displayed in Fig. 4.

Fig. 7 illustrates the distribution of volume fractions of martensites M_1 and M_2 along the beam and through its thickness. As expected the martensitic phase transformation takes place in regions of greatest stress, i.e. at the supports and the application point of force. These regions of highly localized deformation form the so-called superelastic (pseudoplastic) hinges which allow for considerable rotations and, resulting from these, deflections of the beam, see Fig. 6. In contrast to the usual plastic hinges, the superelastic hinges are reversible, provided that the stresses in the material will not exceed some critical strength.

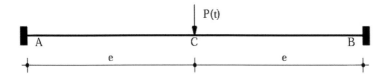

Fig. 3 Clamped beam loaded with concentrated force $P = P(t)$ applied at its mid span point C ($e = l_b/2 = 0.075$ m)

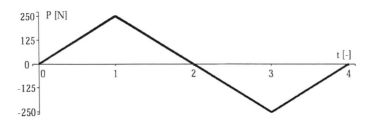

Fig. 4 History of concentrated force $P = P(t)$

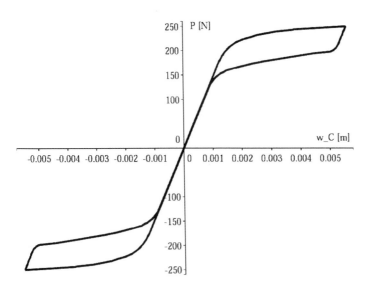

Fig. 5 Histeresis loops in displacement w_C at the centre of beam in Fig. 3 induced by variable half-span loading $P = P(t)$ with history in Fig. 4

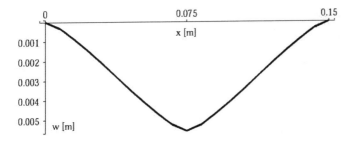

Fig. 6 Displacement $w(x,t)$ of the beam in Fig. 3 by half-span point load $P = 250$ N at $t = 1$ in Fig. 4

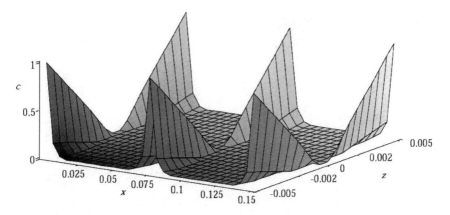

Fig. 7 Volume fraction distribution of two variants of martensites, $c \in \{c_1, c_2\}$, in the beam of Fig. 3 as a function of abscissa $x \in [0, 0.15]$ and through-the-thickness coordinate $z \in [-0.005, 0.005)]$ at time $t = 1$ in Fig. 4

Example 2

Example 2 illustrates the influence of SMA strips on a hysteretic response of a composite beam. The geometrical data and loading conditions of the composite beam are same as that of Example 1. The difference is that instead of monolithic SMA beam of Example 1, in Example 2 the beam is made of an epoxy resin and two SMA strips located in cross-section at $z \in (-4.75, -3.75) \cup (3.75, 4.75)$. Fig. 5 shows the obtained hysteresis loops in the deflection of centre point, w_C, and the concentrated load $P(t)$ defined in Fig. 4.

Example 3

We have analyzed the two-span beam shown in 9 loaded by two variable point forces with history defined in Fig. 10. As can be seen in figs. 11 – 13, for two time levels, $t = 1$ and $t = 2$, the superelastic hinges may occur and disappear, which is directly related to the distribution of martensitic variants.

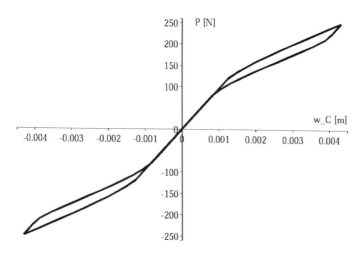

Fig. 8 Histeresis loops in displacement w_C as in Fig. 5 when the beam is made of epoxy resin and two SMA strips instead of the monolithic SMA cross-section

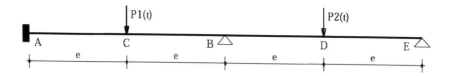

Fig. 9 Two-span beam subjected to concentrated forces $P_1 = P_1(t)$ and $P_2 = P_2(t)$ with their history shown in Fig. 10

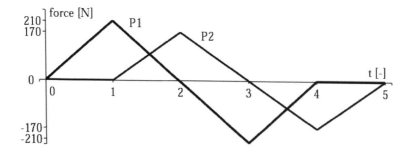

Fig. 10 History of concentrated forces $P_1 = P_1(t)$ and $P_2 = P_2(t)$

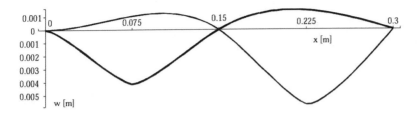

Fig. 11 Displacement $w(x,t)$ of the two-span beam at $t = 1$ ($P_1 = 210$ N, $P_2 = 0$) – thick line, and at $t = 2$ ($P_1 = 0$, $P_2 = 170$ N) – thin line, see Fig. 10

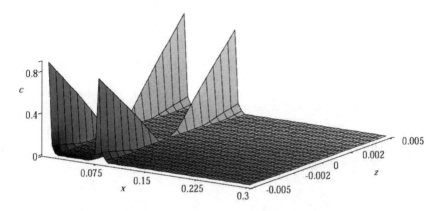

Fig. 12 Volume fraction distribution of two variants of martensites, $c \in \{c_1, c_2\}$, in the two-span beam in Fig. 9 as a function of abscissa $x \in [0, 0.30]$ and through-the-thickness coordinate $z \in [-0.005, 0.005]$ at time $t = 1$ in Fig. 10

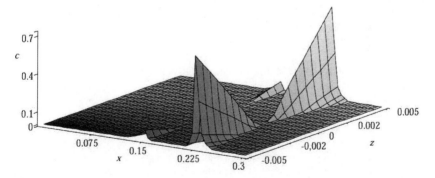

Fig. 13 Volume fraction distribution of two variants of martensites, $c \in \{c_1, c_2\}$, in the beam in Fig. 9 as a function of $x \in [0, 0.30]$ and $z \in [-0.005, 0.005]$ at time $t = 2$ in Fig. 10

6 Conclusion

The presented theoretical model and devised computer program are capable of modelling the hysteretic behaviour of SMA beams. The stress-induced martensitic phase transformation in a SMA beam results in occurrence of reversible superelastic hinges that allow for large reversible deflections of the SMA beam.

References

1. Ball, J., James, R.: Fine phase mixtures as minimizers of energy. Arch. Rational Mech. Anal. 100, 13–52 (1987)
2. Bhattacharya, K.: Comparison of the geometrically nonlinear and linear theories of martensitic transformation. Continuum Mech. and Themodyn. 5, 205–242 (1993)
3. Ghandi, M.V., Thompson, B.S.: Smart Materials and Structures. Chapman & Hall, London (1992)
4. Kohn, R.: The relaxation of a double-well energy. Continuum Mech. Thermodyn. 3, 193–236 (1991)
5. Kuczma, M.S.: Application of variational inequalities in the mechanics of plastic flow and martensitic phase transformations. Politechnika Poznańska, Poznań (1999), http://www.uz.zgora.pl/~mkuczma/hab352.pdf
6. Kuczma, M.S., Mielke, A.: Influence of hardening and inhomogeneity on internal loops in pseudoelasticity. ZAMM 80, 291–306 (1999)
7. Kuczma, M.S.: Modelling of hysteresis in multiphase martensitic materials (in Polish). Transactions from the mechanics of structures and materials, Politechnika Krakowska 302, 161–169 (2004)
8. Lagoudas, D. (ed.): Shape Memory Alloys. In: Modeling and Engineering Applications. Springer, New York (2008)
9. Levitas, V.I., Stein, E., Lengnick, M.: On a unified approach to the description of phase transition and strain localization. Arch. Appl. Mech. 66, 242–254 (1996)
10. Müller, I., Xu, H.: On the pseudoelastic hysteresis. Acta metall. mater. 39, 263–271 (1991)
11. Nowacki, W.K. (ed.): Foundations of the thermodynamics of shape memory materials (in Polish). In: IPPT PAN, Warszawa (1996)
12. Otsuka, K., Wayman, C.M.: Shape Memory Materials. Cambridge Univ. Press, Cambridge (1998)
13. Purohit, P.K., Bhattacharya, K.: On beams made of a phase-transforming material. Int. J. Solids and Structures 39, 3907–3929 (2002)
14. Raniecki, B., Lexcellent, C., Tanaka, K.: Thermodynamic models of pseudoelastic behaviour of shape memory alloys. Arch. Mech. 44, 261–284 (1992)
15. Raniecki, B., Rejzner, J., Lexcellent, C.: Anatomization of hysteresis loops in pure bending of ideal pseudoelastic SMA beams. Int. J. Mech. Sci. 43, 1339–1368 (2001)
16. Roubiček, T.: Models of microstructure evolution in shape memory alloys. In: Ponte Castañeda, P., Telega, J.J., Gambin, B. (eds.) Nonlinear Homogenization and its Applications to Composite, Polycrystals and Smart Materials, pp. 269–304. Kluwer, Dordrecht (2004)
17. Smith, R.C.: Smart Material Systems. SIAM, Philadelphia (2005)

18. Stöckel, D., Hornbogen, E., Ritter, F., Tautzenberger, P.: Legierungen mit Formgedächtnis. expert-Verlag, Ehningen bei Böblingen (1988)
19. Wilmanski, K.: Model of stress-strain hysteresis loops in shape memory alloys. Int. J. Engng Sci. 31(8), 13–52 (1993)
20. Żak, A., Cartmell, M.P., Ostachowicz, W., Wiercigroch, M.: One-dimensional SMA models for use with reinforced composite structures. Smart Materials and Structures 12, 338–346 (2003)

Microstructures in the $Ti_{50}Ni_{50-x}Pd_x$ Alloys' Cubic-to-Orthorhombic Phase Transformation: A Proposed Energy Landscape

Mario Pitteri

Abstract. Recent experimental results related to the cubic-to-orthorhombic phase transformation in the family of $Ti_{50}Ni_{50-x}Pd_x$ shape memory alloys show that the remarkable reduction of the hysteresis when the Pd content approaches 11% is accompanied by a noticeable change in the microstructure present in the martensitic phase; in particular, of the types of twins and of their arrangement. As a preliminary step towards modelling such a behavior, I consider the simplest model of Landau-type for this transformation and in a specific but not too special case compare the theoretical results with the experimental findings.

1 Introduction

Recently published experimental results related to the cubic-to-orthorhombic phase transformation in the family of $Ti_{50}Ni_{50-x}Pd_x$ shape memory alloys show that the remarkable reduction of the hysteresis when the Pd content approaches 11% is accompanied by a noticeable change in the microstructure present in the martensitic phase –see [5]. In particular, the types of twins and their arrangement change. As a preliminary step towards modelling such a behavior, we propose as simple a model as possible for a Landau-type polynomial free energy and analyze its outcome regarding the equilibria as temperature is varied across the critical temperature. Since the Landau potential is of sixth order, some of the stability analysis has to be done numerically for a choice of coefficients in the energy which guarantees the stability of the martensitic phase. In this case we compare the theoretical results with the experimental findings. These certainly remain true while the coefficients vary in some subset of $\mathbb{R}^7$ which is open but certainly not dense.

All the numerical calculations and plots are performed with Mathematica [1].

Mario Pitteri
DMMMSA–University of Padova, via Trieste 63, 35121 Padova, Italy
e-mail: pitteri@dmsa.unipd.it

2 Preliminaries

We use standard notations of continuum mechanics, as introduced in [14], for instance. The basic descriptor of deformation is the *deformation gradient* $\mathbf{F}$. Like any invertible euclidean tensor, it satisfies the *polar decomposition* theorem: $\mathbf{F}$ can be uniquely written in the form

$$\mathbf{F} = \mathbf{RU}, \tag{1}$$

where $\mathbf{U}$ is the *right stretch*, a symmetric, positive-definite tensor, and $\mathbf{R}$ is the *rotation*, an orthogonal tensor. The symmetric, positive-definite *right Cauchy–Green* tensor $\mathbf{C}$ is defined as usual by

$$\mathbf{C} := \mathbf{F}^t\mathbf{F} = \mathbf{U}^2, \tag{2}$$

where $\mathbf{U}$ is the right stretch in the polar decomposition (1) of $\mathbf{F}$, and ^{t} denotes transposition. A useful measure of deformation is the *Lagrangian strain tensor* $\mathbf{E}$:

$$\mathbf{E} = \frac{1}{2}(\mathbf{C} - \mathbf{1}) \quad \text{or} \quad \mathbf{C} = \mathbf{1} + 2\mathbf{E}. \tag{3}$$

As is known, for small deformations the Lagrangian strain tensor reduces to the infinitesimal strain tensor, usually denoted by ε, which is the symmetric part of the gradient of the displacement $\mathbf{u}$:

$$\varepsilon := \frac{1}{2}(\nabla\mathbf{u} + (\nabla\mathbf{u})^t). \tag{4}$$

In this same approximation

$$\mathbf{U} = \mathbf{1} + \varepsilon. \tag{5}$$

In this paper we consider homogeneous equilibrium states for a solid immersed in a heat-bath which, for simplicity, is characterized by its temperature θ_0 alone, to be regarded as control parameter, the hydrostatic load being considered negligible. Then the relevant thermodynamic potential is the Helmoltz free energy density, per unit reference volume, evaluated at the environmental temperature –henceforth indicated by just θ– and the stable, or metastable, equilibria are the local minimizers of the free energy with respect to deformation. Therefore we look for minimizers of the free energy density function

$$\phi = \tilde{\phi}(\mathbf{F}, \theta) \tag{6}$$

with respect to $\mathbf{F}$ at any temperature the heat-bath is allowed to take.

The free energy density must satisfy two invariance requirements: objectivity, or frame indifference, and material symmetry. Together, these imply

$$\phi = \check{\phi}(\mathbf{C}, \theta) \quad \text{and} \quad \check{\phi}(\mathbf{C}, \theta) = \check{\phi}(\mathbf{H}^t\mathbf{CH}, \theta) \quad \forall\,\mathbf{H} \in G, \quad \text{or} \tag{7}$$

$$\phi = \hat{\phi}(\mathbf{E}, \theta) \quad \text{and} \quad \hat{\phi}(\mathbf{E}, \theta) = \hat{\phi}(\mathbf{H}^t\mathbf{EH}, \theta) \quad \forall\,\mathbf{H} \in G; \tag{8}$$

here G, the material symmetry group, consists of unimodular tensors and is a material property. In this paper G will be one of the classical crystal holohedries, hence all tensors $\mathbf{H}$ will be orthogonal.

We assume henceforth $\breve{\phi}$ to be bounded below and an arbitrary additive constant to have been chosen in such a way that $\breve{\phi} \geq 0$.

We are interested in the properties of $\breve{\phi}$ near any one minimizing point, $\bar{\mathbf{C}}$ say, at the temperature $\bar{\theta}$. As is usual in this case, we change reference configuration and take the minimizer $\bar{\mathbf{C}}$ as the reference configuration. For simplicity we keep the same symbol $\breve{\phi}(\mathbf{C}, \theta)$ for the new constitutive function. Now, by assumption, $\breve{\phi}(\mathbf{C}, \theta)$ has a minimum at $(\mathbf{1}, \bar{\theta})$. In addition, we restrict the domain of $\breve{\phi}$ to a neighborhood N containing only one minimizer (that is, $\mathbf{1}$);[1] then one has, by hypothesis,

$$\breve{\phi}(\mathbf{C}, \bar{\theta}) > \breve{\phi}(\mathbf{1}, \bar{\theta}) \text{ for all } \mathbf{C} \in N \backslash \{\mathbf{1}\}. \tag{9}$$

Considering $\breve{\phi}(\mathbf{C}, \theta)$ as above, and assuming that it admits at least piecewise continuous second derivatives with respect to $\mathbf{C}$, we introduce the classical fourth-order elasticity tensor – also called 'the tensor of the elasticities':

$$\mathsf{C} = \frac{\partial^2 \breve{\phi}}{\partial \mathbf{C} \partial \mathbf{C}}, \quad \mathsf{C} : \mathbf{V} \mapsto \mathsf{C}[\mathbf{V}], \ \mathbf{V} \in Sym. \tag{10}$$

We indicate by L the tensor of the elasticities at the minimizer (natural state):

$$\mathsf{L} = \mathsf{C}(\mathbf{1}, \bar{\theta}). \tag{11}$$

The invariance (7) of $\breve{\phi}$ implies that the tensor L of the elasticities of a cubic crystalline solid in a natural state in which the reference axes are oriented along the cubic edges satisfies the following invariance condition:

$$\mathbf{Q}^t \mathsf{L}[\mathbf{V}] \mathbf{Q} = \mathsf{L}[\mathbf{Q}^t \mathbf{V} \mathbf{Q}] \text{ for all } \mathbf{V} \in Sym \text{ and } \mathbf{Q} \in \mathscr{C}_{\mathbf{ijk}}; \tag{12}$$

here, if $\mathbf{R}_{\mathbf{n}}^\alpha$ denotes the rotation by the angle α about the direction of $\mathbf{n}$,

$$\begin{aligned} \mathscr{C}_{\mathbf{ijk}} = \{ &\mathbf{1}, \mathbf{R}_{\mathbf{i}}^\pi, \mathbf{R}_{\mathbf{j}}^\pi, \mathbf{R}_{\mathbf{k}}^\pi, \mathbf{R}_{\mathbf{i}}^{\frac{\pi}{2}}, \mathbf{R}_{\mathbf{i}}^{\frac{3\pi}{2}}, \mathbf{R}_{\mathbf{j}}^{\frac{\pi}{2}}, \mathbf{R}_{\mathbf{j}}^{\frac{3\pi}{2}}, \mathbf{R}_{\mathbf{k}}^{\frac{\pi}{2}}, \mathbf{R}_{\mathbf{k}}^{\frac{3\pi}{2}}, \mathbf{R}_{\mathbf{i-j}}^\pi, \mathbf{R}_{\mathbf{i+j}}^\pi, \\ &\mathbf{R}_{\mathbf{i-k}}^\pi, \mathbf{R}_{\mathbf{i+k}}^\pi, \mathbf{R}_{\mathbf{j-k}}^\pi, \mathbf{R}_{\mathbf{j+k}}^\pi, \mathbf{R}_{\mathbf{i+j+k}}^{\frac{2\pi}{3}}, \mathbf{R}_{\mathbf{i+j+k}}^{\frac{4\pi}{3}}, \\ &\mathbf{R}_{\mathbf{i+j-k}}^{\frac{2\pi}{3}}, \mathbf{R}_{\mathbf{i+j-k}}^{\frac{4\pi}{3}}, \mathbf{R}_{\mathbf{i-j+k}}^{\frac{2\pi}{3}}, \mathbf{R}_{\mathbf{i-j+k}}^{\frac{4\pi}{3}} \mathbf{R}_{\mathbf{i-j-k}}^{\frac{2\pi}{3}}, \mathbf{R}_{\mathbf{i-j-k}}^{\frac{4\pi}{3}} \} \end{aligned} \tag{13}$$

is the subgroup of rotational symmetries of the cubic reference. This is sufficient because the central inversion satisfies (12) trivially.

The first-derivatives test for a minimum of $\breve{\phi}$ provides the equilibrium conditions

$$\frac{\partial \breve{\phi}}{\partial \mathbf{C}} = \mathbf{0}, \tag{14}$$

---

[1] In general, N is chosen to be a *weak-transformation neighborhood*. Definitions and properties are given in [11] and are not recalled here.

and the second-derivatives test requires the elasticity tensor C in (10) to be positive semi-definite at any equilibrium $(\mathbf{C}, \theta)$:

$$\mathbf{V} \cdot \mathsf{C}(\mathbf{C}, \theta)[\mathbf{V}] \geq 0 \quad \text{for all} \quad \mathbf{V} \in Sym. \tag{15}$$

In particular, if the minimizer at some temperature $\bar{\theta}$ is chosen as the reference configuration, then (15) becomes

$$\mathbf{V} \cdot \mathsf{L}[\mathbf{V}] \geq 0 \quad \text{for all} \quad \mathbf{V} \in Sym. \tag{16}$$

The analysis of bifurcation below rests on certain assumptions of 'genericity' on the constitutive equations, which are discussed in detail in [11]. Here we limit ourselves to indicating the consequences of genericity where needed.

First, assume $(\mathbf{1}, \bar{\theta})$ to solve the equilibrium conditions (14) and the tensor L to be non-singular. Then, by the implicit function theorem, there is a unique curve of equilibria $\mathbf{C} = \mathbf{C}(\theta)$ defined in a neighborhood of $\bar{\theta}$ and such that $\mathbf{C}(\bar{\theta}) = \mathbf{1}$. One can show that all the configurations on this curve have the same geometric symmetry as the reference itself; in this paper, cubic. Therefore, a phase transformation with change of symmetry can occur at $(\mathbf{1}, \bar{\theta})$ only if at least one of the eigenvalues of L vanishes. The invariance (7) of the free energy implies, by successive differentiation, identities among derivatives of $\check{\phi}$; in particular, certain eigenvalues of L are forced to be equal. Genericity then requires that no relation among derivatives except the ones implied by symmetry should hold; in particular, all the eigenspaces of L must be irreducible invariant under the action

$$\mathbf{V} \mapsto \mathbf{Q}^t \mathbf{V} \mathbf{Q} \quad \forall \mathbf{Q} \in G, \tag{17}$$

all the eigenvalues which are not forced to be equal by simmetry should be regarded as different, and exactly one of them should be zero.

In our case $G = \mathscr{C}_{\mathbf{ijk}}$ and – see [11] – there is a unique decomposition of Sym into irreducible invariant subspaces under the action of $\mathscr{C}_{\mathbf{ijk}}$. It consists of a one-dimensional space $\mathscr{V}_1 = \langle \mathbf{V}_1 \rangle$, a two-dimensional one, $\mathscr{V}_2 = \langle \mathbf{V}_2, \mathbf{V}_3 \rangle$, and a three-dimensional one, $\mathscr{V}_3 = \langle \mathbf{V}_4, \mathbf{V}_5, \mathbf{V}_6 \rangle$, with generators

$$\mathbf{V}_1 = \frac{1}{\sqrt{3}} \begin{pmatrix} 1 & 0 & 0 \\ 0 & 1 & 0 \\ 0 & 0 & 1 \end{pmatrix}, \ \mathbf{V}_2 = \frac{1}{\sqrt{2}} \begin{pmatrix} 1 & 0 & 0 \\ 0 & -1 & 0 \\ 0 & 0 & 0 \end{pmatrix}, \ \mathbf{V}_3 = \frac{1}{\sqrt{6}} \begin{pmatrix} 1 & 0 & 0 \\ 0 & 1 & 0 \\ 0 & 0 & -2 \end{pmatrix},$$

$$\mathbf{V}_4 = \frac{1}{\sqrt{2}} \begin{pmatrix} 0 & 0 & 0 \\ 0 & 0 & 1 \\ 0 & 1 & 0 \end{pmatrix}, \ \mathbf{V}_5 = \frac{1}{\sqrt{2}} \begin{pmatrix} 0 & 0 & 1 \\ 0 & 0 & 0 \\ 1 & 0 & 0 \end{pmatrix}, \ \mathbf{V}_6 = \frac{1}{\sqrt{2}} \begin{pmatrix} 0 & 1 & 0 \\ 1 & 0 & 0 \\ 0 & 0 & 0 \end{pmatrix}. \tag{18}$$

The typical $\mathbf{V} \in Sym$ can be represented in the othonormal basis $\mathbf{V}_1, \ldots, \mathbf{V}_6$; in particular we have, in a neighborhood of $\mathbf{C} = \mathbf{1}$,

$$\mathbf{C} = \mathbf{1} + 2\mathbf{E}, \quad \mathbf{E} = \sum_{i=1}^{6} y_1 \mathbf{V}_i. \tag{19}$$

Therefore the free energy can be thought of as a function of the y_i and θ:

$$\phi = \Phi(y_1,\ldots,y_6,\theta). \tag{20}$$

The explicit relation between the components of $\mathbf{E}$ and the y_i will be useful:

$$y_1 = \frac{1}{\sqrt{3}}\sum_{r=1}^{3} E_{rr}, \quad y_2 = \frac{1}{\sqrt{2}}(E_{11}-E_{22}), \quad y_3 = \frac{1}{\sqrt{6}}(E_{11}+E_{22}-2E_{33}),$$

$$y_4 = \sqrt{2}E_{23}, \quad y_5 = \sqrt{2}E_{13}, \quad y_6 = \sqrt{2}E_{12}, \tag{21}$$

The basis chosen in $\mathcal{V}_2$ is related to our intent to describe by the condition $y_2 = 0$ the subspace of $\mathcal{V}_2$ consisting of tetragonal configurations whose four-fold axis is along the z coordinate axis; this is conventionally assumed by both material scientists and elasticians.

The action of $\mathcal{C}_{\mathbf{ijk}}$ on Sym reduces to independent orthogonal actions on the subspaces $\mathcal{V}_r, r = 1,2,3$, and the matrix representation of anyone of its elements in the orthonormal basis chosen is a block matrix with orthogonal blocks of respective dimensions 1, 2 and 3. The action on $\mathcal{V}_1$ is trivial. The one on $\mathcal{V}_2$ is the group of symmetries of the equilateral triangle, with generators

$$\begin{pmatrix} -1 & 0 \\ 0 & 1 \end{pmatrix} \text{ representing } \mathbf{R}_{\mathbf{k}}^{\frac{\pi}{2}} \quad \text{and} \quad r(\frac{2\pi}{3}) \text{ representing } \mathbf{R}_{\mathbf{i+j+k}}^{\frac{4\pi}{3}}; \tag{22}$$

here $r(\alpha)$ is the counterclockwise 2×2 rotation by the angle α. The action on $\mathcal{V}_3$ is the group of symmetries of a regular tetrahedron, generated by

$$-R_1(\frac{\pi}{2}) \text{ representing } \mathbf{R}_{\mathbf{k}}^{\frac{\pi}{2}} \quad \text{and} \quad -R_2(\frac{\pi}{2}) \text{ representing } \mathbf{R}_{\mathbf{j}}^{\frac{\pi}{2}}; \tag{23}$$

here $R_i(\frac{\pi}{2})$ is the rotation by $\frac{\pi}{2}$ about the i-th axis in $\mathbb{R}^3$.

For the Ti$_{50}$Ni$_{50-x}$Pd$_x$ family, the phase transformation is from B$_2$ (a variant of cubic) to B$_{19}$ (a variant of orthorhombic), and the analysis in [11] implies that the kernel of L has to be $\mathcal{V}_3$.

3   The Twinning Condition

An exhaustive analysis of the twins in a weak-transformation neighborhood of a cubic reference is given in [11], where details can be found. For the Ti$_{50}$Ni$_{50-x}$Pd$_x$ family the martensitic twins are in the orthorhombic 'mixed axes' variant structure, with representative Cauchy-Green transformation strain tensor –see [11] §§5.1.3.2 and 9.2.3.2–

$$\mathbf{C} = \begin{pmatrix} C_{11} & C_{12} & 0 \\ C_{12} & C_{11} & 0 \\ 0 & 0 & C_{33} \end{pmatrix}. \tag{24}$$

In this case there are exactly two equivalence classes of twins, one of compound twins and one of pairs of conjugate Type-1 and Type-2 twins. Representative twinning elements, in particular the shear amplitude s, are given in [11], Table 9.3 for either class. In terms of the Cauchy-Green tensor components the shear amplitudes for the compound twins and for both conjugate Type-1 and Type-2 twins are

$$s_1^2 = \frac{4C_{12}^2}{C_{11}^2 - C_{12}^2} \quad \text{and} \quad s_2^2 = \frac{C_{11}^3 - C_{11}C_{12}^2 - 2C_{11}^2 C_{33} + 3C_{12}^2 C_{33} + C_{11}C_{33}^2}{C_{33}(C_{11}^2 - C_{12}^2)}, \quad (25)$$

respectively. By using the data in [5] we can test the simple-minded prejudice that smaller shear amplitude may be a favorable condition for twinning, actually present in the earliest theoretical attempts to model mechanical twinning; for instance by Vladimirskii (1947) cited in [9]. Another reasoning of a more dynamic character in favor of the small-shear criterion may be that small shear be associated to a smaller energy barrier to be overcome in a dynamic process whose starting and ending configurations belong to distinct energy wells.

In fact, for all the reported Pd concentrations $s_1 < s_2$, so compound twins seem to be always favorite by the small-shear criterion, contrary to the experimental evidence showing them to occur only rarely and when the Pd concentration is about 11%. We know by now the conditions for various microstructures in the martensite to be compatible with austenite; for a review and extensive references see [8] or [3]. For instance a single variant of martensite is only compatible with austenite if and only if the middle eigenvalue in the transformation stretch is 1 –see [2]– which condition is best approximated in the Pd_{11} alloy. Specific results are given

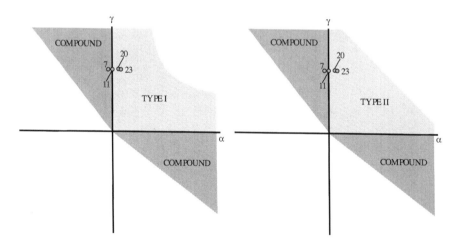

Fig. 1 Compatibility of compound or Type-1 and Type-2 twin laminates with austenite in the $Ti_{50}Ni_{50-x}Pd_x$ alloys' cubic-to-orthorhombic phase transformation. The dots represent the alloy for different Pd concentrations, which are correspondingly indicated. α, β (not represented here) and γ are the principal stretches, that is, the eigenvalues of the transformation stretch $\mathbf{U}$.

in [7] for cubic-to-orthorhombic transformations and the related martensitic twin laminates compatible with austenite. These results, in particular figures 1 and 2 for $0.9 < \beta < 1$, are used by Delville [4] to check that at most the Pd$_7$ alloy can have compound twin laminates compatible with austenite, Pd$_{11}$ is on the border line, and for the other examined compositions only laminates involving Type-1 and Type-2 twins are allowed. A sketch of Delville's picture is presented in figure 1, above. Since Pd$_{11}$ almost exactly satisfies the compatibility of a single variant of martensite with austenite, one would expect compound twins to be dominant in this alloy. While some isolate compound twins are indeed observed, more common are triangular patterns, probably of self-accommodation, in which the interfaces are of Type-1 or Type-2 twins. Below we infer from the energy landscape a tentative explanation for this behavior.

4   A Basic Free Energy Density

By a known theorem [13] the free energy, which is supposed to be smooth and $\mathscr{C}_{ijk}$-invariant, is a function of certain polynomial $\mathscr{C}_{ijk}$-invariants. Here we further assume it to be a polynomial in the variables y_i introduced in (19). There, the y_i represent the deformation from the reference configuration, which has been chosen to be the cubic configuration at the critical temperature. This is sufficient for the (local) analysis of the bifurcation near the critical point, but we are interested in stable orthorhombic configurations which cannot a priori be considered to be close to the critical one. Thus, in agreement with a standard interpretation in the Landau theory, the free energy introduced below measures, for each configuration and each temperature, the excess free energy from the one at the cubic equilibrium at that same temperature. Since below the critical temperature the cubic phase is unstable, hence unobservable, its dependence on temperature has to be somehow extrapolated. This is not always trivial, see [12], or also [11], §7.1.1.

For our purposes a sixth-order polynomial is needed, and we restrict the attention to the following invariants, in the notation of [10]:

$$I_1 = y_1, \quad I_2 = y_2^2 + y_3^2, \quad I_7 = y_3(y_4^2 + y_5^2 - 2y_6^2) + \sqrt{3}y_2(y_4^2 - y_5^2),$$
$$I_4 = y_4^2 + y_5^2 + y_6^2, \quad I_5 = y_4 y_5 y_6, \quad I_6 = y_4^4 + y_5^4 + y_6^4. \tag{26}$$

We also propose the following free energy function for the cubic-to-orthorhombic Ti$_{50}$Ni$_{50-x}$Pd$_x$ phase transformation as the simplest one for obtaining the transformation strains given by experiment:

$$\Phi = a_1 I_1^2 + a_2 I_2 + a_3 I_1 I_4 + a_4 I_7 + b I_4 + c I_5 + f I_4^3. \tag{27}$$

This potential always admit the (trivial) cubic equilibrium $y_1 = \ldots = y_6 = 0$, which is stable for $b > 0$ and unstable for $b < 0$. We are interested in the nontrivial equilibria. As is customary in a Landau-type approach to phase transformations, we assume all the coefficients in (27) to be constant except b, for which the following is assumed, θ_c denoting the critical temperature and α being a suitable positive constant:

$$b = \alpha(\theta - \theta_c). \tag{28}$$

Furthermore, the constants a_1, a_2, f are assumed to be strictly positive. Then the vanishing of the partial derivatives of Φ with respect to y_1, y_2 and y_3, which is part of the equilibrium equations, is equivalent to the conditions

$$y_1 = -\frac{a_3}{2a_1} I_4, \quad y_2 = -\frac{a_4}{2a_2}\sqrt{3}(y_4^2 - y_5^2), \quad y_3 = -\frac{a_4}{2a_2}(y_4^2 + y_5^2 - 2y_6^2). \tag{29}$$

By substituting these values for y_1, y_2 and y_3 in the expression (27) of Φ we obtain the *reduced* or *Landau* potential:

$$\tilde{\Phi} = b I_4 + c I_5 + d I_4^2 + e I_6 + f I_4^3, \tag{30}$$

with

$$d = \frac{a_4^2}{2a_2} - \frac{a_3^2}{4a_1} \quad \text{and} \quad e = -\frac{3a_4^2}{2a_2}, \quad \text{hence} \quad d + e = -\frac{a_3^2}{4a_1} - \frac{a_4^2}{a_2}. \tag{31}$$

As is known –see for instance [6] or [11], §7.3– the analysis of the equilibria as well as of their stability can be directly done on $\tilde{\Phi}$. Here we follow the pattern presented in [6] or [11], §7.4.6, where a potential of the form (30) without the last term is considered. According to that analysis, for $y_i, i = 1, \dots, 6$ near zero there are certain equilibria of rhombohedral symmetry and others of orthorhombic symmetry, and all these are unstable. Here the last, sixth-order term in $\tilde{\Phi}$ is introduced to re-stabilize the orthorhombics.

First of all let us consider the jacobian

$$\frac{\partial(I_4, I_5, I_6)}{(y_4, y_5, y_6)} = 8(y_4^2 - y_5^2)(y_5^2 - y_6^2)(y_6^2 - y_4^2). \tag{32}$$

There are no equilibria where this jacobian does not vanish. Indeed, there the equilibrium conditions are equivalent to the vanishing of the partial derivatives of $\tilde{\Phi}$ with respect to I_4, I_5 and I_6; which is impossible because $\frac{\partial \tilde{\Phi}}{\partial I_5} = c$ and $\frac{\partial \tilde{\Phi}}{\partial I_6} = e$. The (reduced) invariance of $\tilde{\Phi}$ described above implies that we can restrict the equilibrium and stability analysis to one of the subsets of $\mathbb{R}^3$ where one of the factors in (32) vanishes. Since the choice of the cubic invariant I_7 is related to giving the simplest representation, namely $y_2 = 0$, to the tetragonal configurations whose tetragonal axis is the z axis in $\mathbb{R}^3$, we choose the couple of planes where $y_4^2 - y_5^2 = 0$. We next intoduce the new variables $X = y_6, Y = y_4^2 = y_5^2$ and express $\tilde{\Phi}$ in terms of these:

$$F(X, Y, \theta) = b(X^2 + 2Y) + cXY + d(X^2 + 2Y)^2 + e(X^4 + 2Y^2) + f(X^2 + 2Y)^3. \tag{33}$$

Where the jacobian

$$\frac{\partial(X, Y)}{(y_6, y_4)} = 2y_4 \tag{34}$$

does not vanish the equilibrium equations –which consist in the vanishing of the partial derivatives of F with respect to X and Y– and the invariance of $\tilde{\Phi}$ imply that the solutions have rhombohedral symmetry and are of the form

$$y_4 = y_5 = \pm x, \quad y_6 =: x, \tag{35}$$

with x solving the single relevant equilibrium equation

$$0 = 2b + cx + 4(3d + e)x^2 + 54fx^4. \tag{36}$$

Using this equation to express $2b$ in terms of x and denoting partial derivatives with a comma –for instance $\tilde{\Phi}_{,4} := \frac{\partial \tilde{\Phi}}{\partial y_4}, \tilde{\Phi}_{,45} := \frac{\partial^2 \tilde{\Phi}}{\partial y_4 \partial y_5}$ etc.– we get the following elements in the hessian along the solutions (35) with the plus sign:

$$\tilde{\Phi}_{,44} = \tilde{\Phi}_{,55} = \tilde{\Phi}_{,66} = -cx + 8(d+e)x^2 + 72fx^4, \tag{37}$$

$$\tilde{\Phi}_{,45} = \tilde{\Phi}_{,56} = \tilde{\Phi}_{,64} = cx + 8dx^2 + 72fx^4, \tag{38}$$

while for the solutions with the minus signs only $\tilde{\Phi}_{,56}$ and $\tilde{\Phi}_{,64}$ change sign. In either case the positive definiteness of the hessian is guaranted by the following inequalities, where the above second derivatives are used:

$$\tilde{\Phi}_{,44} > 0, \quad (\tilde{\Phi}_{,44})^2 - (\tilde{\Phi}_{,45})^2 > 0, \quad (\tilde{\Phi}_{,44})^3 - 3(\tilde{\Phi}_{,44})(\tilde{\Phi}_{,45})^2 + 2(\tilde{\Phi}_{,45})^3 > 0. \tag{39}$$

We will return to these equilibria after having examined the orthorhombic equilibria for $y_4 = 0 = y_5$ and having guaranteed their stability. In this case the nontrivial equilibria solve the following single equation for $y := y_6$:

$$0 = b + 2(d+e)y^2 + 3fy^4; \tag{40}$$

this produces the desired subcritical bifurcation because $d + e < 0$ by (31):

$$y^2 = -\frac{d+e}{3f}\left(1 \pm \sqrt{1 - \frac{3fb}{(d+e)^2}}\right) \tag{41}$$

A schematic diagram of this bifurcation is given in figure 2. By analogy, the critical, Maxwell and upmost martensitic existence values for b are denoted by b_c, b_M and $\check{b}$, respectively. Similar notation is used for the corresponding values of the strain parameter y. Having no information on the temperature coefficient α in (28) I prefer to deal with b directly rather than θ. In our case, the index O denoting stable orthorhombic,

$$b_c = 0, \quad y_{cO}^2 = -\frac{2(d+e)}{3f}, \quad \check{b} = \frac{(d+e)^2}{3f}, \quad \check{y}_O^2 = -\frac{d+e}{3f} = \frac{1}{2}y_{cO}^2, \tag{42}$$

$$b_M = \frac{(d+e)^2}{4f} = \frac{3}{4}\check{b}, \quad y_{MO}^2 = -\frac{d+e}{2f} = \frac{3}{2}\check{y}_O^2. \tag{43}$$

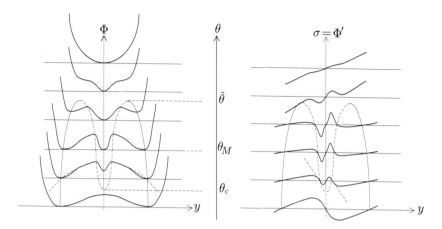

Fig. 2 A schematic 1-dimensional example based on the orthorhombic sextic included in (30). The dotted parts of the M-shaped curves indicate instability. Indicated are also the temperatures θ_c, θ_M and $\check{\theta}$: the critical and the Maxwell temperatures, and the upmost temperature of existence of the orthorhombic phase, respectively. Φ is the free energy, and its derivative, σ, represents stress.

Besides the condition $d + e < 0$, which is generally necessary to get a subcritical orthorhombic bifurcation from a potential of the form (30), and is in this case automatically satisfied by (31)$_3$, the extra conditions for stability of the bifurcated branches, obtained as usual from the hessian of $\tilde{\Phi}$, are

$$y^2 > -\frac{d+e}{3f} = \check{y}_O^2, \quad e < 0, \quad -16\frac{d+e}{3f}\left(\frac{e}{c}\right)^2 > 1. \qquad (44)$$

The first condition characterizes the continuous part of the M-shaped lines in figure 2; the second is automatically satisfied in our case by (31)$_2$; so, the third is the actual condition on the coefficients of $\tilde{\Phi}$ we have to satisfy.

Concerning the rhombohedral solutions, we know that they exist and are unstable for b in some neighborhood of 0 but determining their existence and stability character more in general is complicated by the high order of the algebraic equations and inequalities involved in the analysis.

To reduce the indeterminateness in the above model we consider, as an example, the austenitic and martensitic lattice parameters given in [5] for Ti$_{50}$Ni$_{39}$Pd$_{11}$, assuming further that they represent the phases at the Maxwell temperature:

$$a_0 = 3.0499, \quad a = 2.8304, \quad b = 4.3135, \quad c = 4.6041. \qquad (45)$$

This choice is justified, at least partially, by the fact that for this alloy, among those considered in [5], the best compatibility of a single variant of martensite with austenite is achieved; hence, presumably, the bulk energy landscape is most important in governing the phase equilibrium. In general, also the energy associated to the interfaces should be somehow taken into account. For instance, this is the case of the model for hysteresis in Ti$_{50}$Ni$_{50-x}$Pd$_x$ and similar alloys proposed in [15].

In terms of the strains

$$i = \frac{b+c}{2\sqrt{2}a_0}, \quad j = \frac{-b+c}{2\sqrt{2}a_0}, \quad k = \frac{a}{a_0} \tag{46}$$

the nonzero components of the lagrangian strain tensor $\mathbf{E}$ are

$$E_{11} = E_{22} = \frac{1}{2}(i^1 + j^2 - 1), \quad E_{12} = ij, \quad E_{33} = \frac{1}{2}(k^2 - 1). \tag{47}$$

Using the data (45) and the formulae (21) we obtain the following values for the y_i:

$$y_1 = 2.3 \times 10^{-4}, \quad y_2 = 0, \quad y_3 = 0.085, \quad y_4 = 0 = y_5, \quad y_6 = 0.049. \tag{48}$$

By means of these values, of (29), (31) and (43)$_3$, we can express d, e and f in terms of a_1 and a_2, which remain undetermined but must be positive; also c is undetermined but subject to the inequality (44)$_4$. The choices in (49) below are somewhat arbitrary but, because of three free parameters being involved, I believe they are representative of a nontrivial subset of the possibilities. So, let

$$a_1 = 0.1, \quad a_2 = 1, \quad \text{and} \quad c = -50. \tag{49}$$

Correspondingly

$$d = 626.6, \quad e = -1880, \quad d + e = -1253.3, \quad \text{and} \quad f = 260995, \tag{50}$$

hence the inequality (44)$_4$ is satisfied and the stability of the orthorhombic martensitic branch corresponding to the $+$ sign in (41) guaranteed. Furthermore, the value of $\check{b}$ is 2.006; which, together with the value of the temperature hysteresis of $13\,^\circ K$ reported in [5], if interpreted as $\check{\theta} - \theta_c$, implies a coefficient α in (28) of 0.154. Also, by (43)$_2$, $b_M \approx 1.5$.

We now return to the stability of the possible rhombohedral equilibria. Using the equilibrium condition (36) to express b in terms of the remaining quantities, we plot $\tilde{\Phi}_{\cdot 44}$ and the other quantities in (39) whose positivity is a condition for stability, versus the strain parameter x. The plots, which are not shown here, indicate that the rhombohedral equilibria are never stable when they exist. For the existence we solve numerically the quartic (36) for the strain x as a function of b. It turns out that there are two real solutions but only for $b \lesssim 0.18$. Plots of the orthorhombic and of the rhombohedral solutions, as well as of the corresponding energies are given in figures 3 to 6 below, while figure 6 is an energy landscape polar contour plot for the value of b, 0.15, which corresponds to the intersection of the lower rhombohedral equilibrium energy with the b axis –see fig. 6– that is, where the rhombohedral equilibrium of lower energy has the same energy as the still metastable cubic equilibrium at the same temperature.

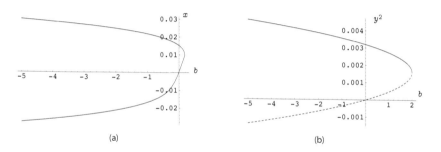

Fig. 3 Equilibrium configurations in the $y_4^2 = y_5^2$ subset. (a) The rhombohedral ones, solving (36); they are all unstable. (b) The orthorhombic ones, solving (40). The dotted line represents the unstable orthorhombics.

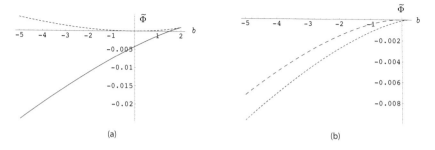

Fig. 4 The free energy along the equilibria in Fig. 3. (a) Orthorhombic energy. The dotted line corresponds to the unstable orthorhombics. (b) Rhombohedral energy.

Fig. 5 Plot of all the free energy branches along the equilibria. The intersection of the continuous line of stable orthorhombics with the b axis corresponds to the Maxwell value $b_M \approx 1.5$ according to (43)$_2$.

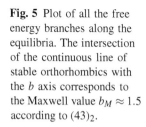
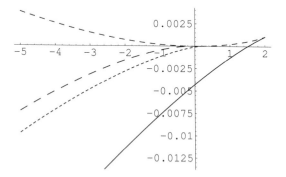

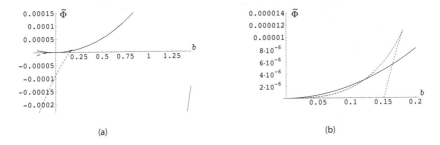

(a)

(b)

Fig. 6 (a) Detail of Fig. 5. On the lower right corner a small part of the stable orthorhombic branch is shown. (b) Further detail of (a). The dashed lines correspond to the rhombohedral equilibria while the continuous curve corresponds to the unstable orthorhombic ones.

5 Conclusions

First of all, the conclusions drawn here from the model proposed above are not 'generic': while they are insensitive to small changes in the parameters, they do not hold for arbitrary choices of them. For instance, for the choices of a_1, a_2 proposed in (49) not all the values of c compatible with the stability of the orthorhombic martensite guarantee the instability of the rhombohedral one. In fact, in the range $|c| \lesssim 300$, which corresponds to orthorhombic stability, rhombohedral instability is only guaranteed in the range $|c| \lesssim 85$. The absolute value is due to the fact that c is not involved in the orthorhombic analysis, and in the rhombohedral one changing its sign produces the negative of an already obtained equilibrium for the same b, with the same stability character –see (37), (38). Only with additional information on the alloy, for instance on the elastic moduli, can we decide if the proposed choice of the free energy coefficients, or of the energy itself, is adequate.

With this proviso we draw the following conclusions. Figures 4-6 show that the unstable rhombohedrals quickly become energetically favored over the unstable orthorhombics. In particular one rhombohedral branch –the one marked with a thicker dashing in figs 5-6– acquires lower energy than the unstable orthorhombic very soon after it begins to exist. Then so does also the other rhombohedral branch nearer the critical point. So, it seems that, if a rank-1 connection is dynamically interpreted as a continuous deformation path and a mountain-pass argument rules the preferred path from one orthorhombic well to another, passing through the unstable rhombohedrals is preferable below the critical point and also above it but not too far. We suggest that in this range of temperatures compound twins are less likely than Type-1 or Type-2 twins. For this purpose consider also the polar contour plots in figure 7. There, the stable orthorhombics for $y_4 = 0 = y_5$ correspond to the two darkest stripes at the top and at the bottom, co-latitudes 0 and π, while the other four orthorhombic wells appear in the middle, co-latitude $\pi/2$. The compound twins correspond, modulo cubic symmetry, to rank-1 connections between co-latitudes 0 and π orthorhombics, while any rank-1 connection of a 0 or π to a $\pi/2$ orthorhombic well corresponds

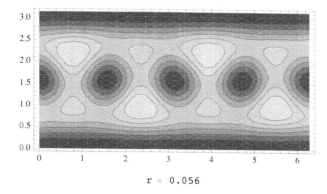

$$r = 0.056$$

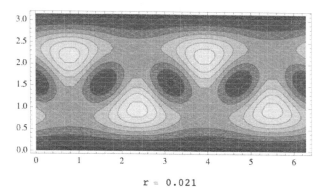

$$r = 0.021$$

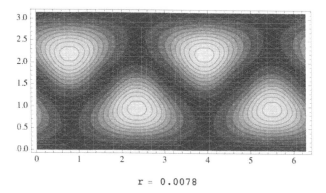

$$r = 0.0078$$

Fig. 7 Energy landscape for $b = 0.15$; polar contour plot in the y_4, y_5, y_6 space, the polar parameter r being $\sqrt{y_4^2 + y_5^2 + y_6^2}$. Within each frame darker means of lower energy. For the given b the value $r = 0.056$ [0.078] is obtained from (41) with the plus sign [the minus sign] and corresponds to the stable [unstable] orthorhombic martensite, while the value 0.021 is obtained from the lower-energy numerical solution of (36) and corresponds to the unstable rhombohedral martensite represented by the thicker dashed line in figs 5-6.

to a pair of Type-1 and Type-2 twins. According to the energetics in figs 5-6 the latter have a lower montain-pass energy that the former from shortly above the critical temperature down below it. Also from the topological point of view compound twins seem to be much less likely. In fact, according to fig 7, a path starting, say, at co-latitude 0 and trying to reach co-latitude π gets very likely trapped into one of the basins of attraction of co-latitude $\pi/2$ stable orthorhombic wells. The fact that such a path encounters the unstable rhombohedrals before the unstable orthorhombics, and is more likely driven into one of the $\pi/2$ stable orthorhombics may justify the preference for twinning modes with a larger shear magnitude, in spite of it.

The proposed model shows that the small shear criterion for twinning selection may be too naive. Indeed in our case the preferred twinning mode corresponds to a larger shear, but correspondingly to a lower energy barrier. This also corrects the possible a priori prejudice that smaller shear may imply a lower energy barrier. This result, and the others, follow from the simplest polynomial potential in a traditional Landau-type approach. It is not obvious that the energy landscape so obtained might have been guessed beforehand if the stiff polynomial class in the Landau approach would have been replaced by more flexible families of functions, like certain splines.

A final remark: the likelihood of the compound twins in this alloy seems to be low well above the critical point by the compatibility with austenite argument, while near and below the critical point it seems to remain low by the pure bulk energy model proposed above. Compound twins may become more likely if, unlike what is done here, one also considers the energy of an applied load, assumed to force the transition from one well to another. The topology of the energy landscape suggests that a properly directed load may indeed drive a co-latitude 0 orthorhombic into a co-latitude π one without falling into any of the co-latitude $\pi/2$ wells. In this line of thinking, it might be interesting to check the behavior of the Pd$_{11}$ alloy under load below the critical temperature. Moreover,[2] since the proposed model works well below the critical point, when the cubic phase is not even metastable, hence not observable, its consequences may be thought of applying to newly formed *mechanical* twins appearing in the alloy when it sits well into the martensitic phase. The analysis of such mechanical twins may provide another experimental test for the model presented here.

Acknowledgements. Part of this work was conceived during the activities of the MULTI-MAT Marie Curie Research Training Network within the 6th Framework Programme of the EU.

References

1. Mathematica, 6.0 edn. Wolfram Research, Inc., Champaign, Illinois (2007)
2. Ball, J.M., James, R.D.: Fine phase mixtures as minimizers of energy. Arch. Rational Mech. Anal. 100, 13–52 (1987)

---

[2] For this remark I am endebted to G. Zanzotto.

3. Bhattacharya, K.: Microstructure of Martensite: Why It Forms and How It Gives Rise to the Shape-Memory Effect. Oxford Univ. Press, Oxford (2003)
4. Delville, R.: TEM investigation of microstructures in $Ti_{50}Ni_{50-x}Pd_x$ alloys with special lattice parameters. In: Multi-scale modelling and characterization of materials, periodic meeting of the network, MULTIMAT European Network, Aula Convegni, CNR Building, Piazzale Aldo Moro 7, Roma, April 2-4 (2008)
5. Delville, R., Schryvers, D., Zhang, Z., James, R.D.: Transmission electron microscopy investigation of microstructures in low-hysteresis alloys with special lattice parameters. Scripta Materialia 60, 293–296 (2009)
6. Ericksen, J.L.: Local bifurcation theory for thermoelastic Bravais lattices. In: Ericksen, J.L., James, R.D., Kinderlehrer, D., Luskin, M. (eds.) Microstructure and phase transition. The IMA Volumes in Mathematics and its Applications, vol. 54. Springer, New York (1993)
7. Hane, K.F., Shield, T.W.: Microstructure in a cubic to orthorhombic transition. J. of Elasticity 59, 267–318 (2000)
8. James, R.D., Hane, K.T.: Martensitic transformations and shape-memory materials. Acta Mater. 48, 197–222 (2000)
9. Klassen-Nekliudova, M.V.: Mechanical twinning of crystals. Consultant Bureau, New York (1964)
10. Pitteri, M.: Some remarks on the fcc-fct phase transformation in InTl alloys. Mathematics and Mechanics of Solids, Online First (to appear, 2009)
11. Pitteri, M., Zanzotto, G.: Continuum models for phase transitions and twinning in crystals. CRC/Chapman & Hall, Boca Raton (2002)
12. Salje, E.K.H.: Phase transitions in ferroelastic and co-elastic crystals. Cambridge University Press, Cambridge (1990)
13. Schwarz, G.W.: Smooth functions invariant under the action of a compact Lie group. Topology 14, 63–68 (1975)
14. Truesdell, C.A.: A first course in rational continuum mechanics, vol. 1. Academic Press, New York (1977)
15. Zhang, Z., James, R.D., Müller, S.: Energy barriers and hysteresis in martensitic phase transformations. Preprint series, n. 87, Max Planck Institut für Mathematik in den Naturwissenschaften Leipzig (2008)

Analysis of Shear Banding with a Hypoplastic Constitutive Model for a Dry and Cohesionless Granular Material

Erich Bauer

Abstract. For stress paths related to plane strain biaxial compression the possibility of a spontaneous development of shear bands in dry and cohesionless granular materials is investigated based on a hypoplastic continuum model and the general bifurcation theory given by Rudnicki and Rice. The volume of the solid grains is assumed to remain constant so that state changes of the granular body with empty void space can be modelled based on a single component continuum. The basics of hypoplasticity are outlined and a specific model by Gudehus and Bauer for granular materials like sand is presented. The micro-structure of the material is taken into account in a simplified manner with the current void ratio which is related to the pressure dependent limit void ratios and the critical void ratio. For the analysis of shear band bifurcation particular attention is paid to the influence of the initial density and the stress level on the onset and orientation of possible shear bands.

1 Introduction

The possibility of a shear band bifurcation results from an instability in the constitutive description of the inelastic material behaviour. In this context instability means that the constitutive relation permits the homogeneous deformation of an initially uniform material up to a bifurcation point at which non-uniform deformation in a local domain can start under conditions of continuing equilibrium and continuing homogeneous deformation outside the localized domain (Rudnicki and Rice, 1975; Vardoulakis and Sulem, 1995). Predictions of shear bands strongly depend on the constitutive model assumed and on the quantities of the constitutive constants involved. In the present paper the modelling of the non-linear and inelastic stress-strain behaviour of dry and cohesionless granular materials like sand is considered.

Erich Bauer
Graz University of Technology, 8010 Graz, Austria
e-mail: erich.bauer@tugraz.at

For the description of the multicomponent structure of granular bodies usually the theory of immiscible mixtures (e.g. Bowen, 1976; Truesdell, 1984) and the concept of volume fractions is used (e.g. de Boer, 2000; Ehlers, 1996; Wilmanski, 2008; Albers 2009). In the following the assumption is made that under load changes the volume of the solid mass of the grains remains constant, which means that the volume change of the void space is equal to the volume change of the granular body. The so-called void ratio, which is defined as the ratio of the volume of the voids to the volume of solid particles, is introduced as a state quantity reflecting in a simplified manner the micro-structure of the granular material. With respect to the foregoing assumption that the mass density of the solid grains remains constant, the void ratio is not an independent state variable, i.e. the evolution of the void ratio is determined by the balance equation of mass, and consequently the granular body with empty voids can be modelled based on a single component continuum. It can be noted that grain abrasion and grain crushing is not excluded by this assumption. From experiments it is well known that the void ratio has a significant influence on the incremental stiffness and the volume strain behaviour. Depending on the arrangement of the grains in the skeleton, cohesionless granular materials can show different void ratios under the same pressure. As there is no unique relation between the pressure and the void ratio, it is convenient for constitutive modelling to relate the current void ratio to the pressure dependent limit void ratios. Such relative densities are embedded in the hypoplastic constitutive model by Bauer (1996) and Gudehus (1996) which is considered in the present paper. The model is applicable for loading and unloading paths with the exception of non-symmetric stress-strain cycles (Bauer and Wu, 1993; Niemunis and Herle, 1997). In the hypoplastic model the evolution of the stress is described by an isotropic tensor-valued functions depending on the relative density, the Cauchy stress and the strain rate. Stationary stress states or so-called critical states (Casagrande, 1936, Schofield and Wroth, 1968) are included in the constitutive equation for a simultaneous vanishing of the stress rate and volume strain rate. Since the constitutive equation is incrementally non-linear, irreversible material behaviour can be taken into account using a single constitutive equation. In this context the concept of hypoplasticity differs fundamentally from the concept of elastoplasticity as no decomposition of the strain rate into reversible and irreversible parts is needed (Kolymbas, 1991). Furthermore, the flow rule and the stress limit condition for stationary flow are not described by separate functions but they are included in the evolution equation for the state quantities. The advantage of the hypoplastic concept not only lies in the formulation of the constitutive equations and the implementation in a finite element code but also in an easy adaptation of the constitutive constants to experiments. Based on the general bifurcation theory by Hill et al. (1975), Rudnicki et al. (1975) and Rice et al. (1980) the possibility of a shear band initiation was investigated for different hypoplastic constitutive models by several authors (e.g. Kolymbas, 1981; Chambon and Desrues, 1985; Wu and Sikora, 1991; Bauer, 1999; Bauer et al. 2004). As shown by Chambon et al. (1990) the application of the general theory to a certain class of incrementally non-linear constitutive models needs no assumptions regarding the possibility of different stiffnesses inside and outside the shear band. Thus, the question as to whether the material outside the shear band is in a loading/unloading state or behaves as a rigid body is irrelevant in hypoplasticity. In contrast to a constitutive model proposed by Chambon et al. (1993) no additional parameter has been introduced in the present constitutive equation to adjust the onset of shear band localization to the experiments. Thus, the results of the present shear band analysis, i.e. the stress state for which a shear band bifurcation may occur and the inclination of the shear band, are a pure prediction of the hypoplastic constitutive model. In the present paper the analytical investigations are restricted the onset of shear strain localization within a material element under plane strain compression. In order to study the evolution of shear bands enhanced constitutive models are necessary as shown for instance within the framework of micro-polar hypoplasticity by Tejchman and Gudehus (2001), Huang et al. (2002), Huang and Bauer (2003).

The paper is organized as follows: In Section (2) the concept of hypoplasticity is briefly outlined. Section (3) describes the properties of the hypoplastic constitutive model by Gudehus (1996) and Bauer (1996). In Section (4) an analytical criterion for shear band bifurcation is derived and analysed for stress paths related to plane strain compressions starting from various initial densities and pressures. The results are discussed and compared with experiments.

Throughout the paper compression stresses and strains are defined as negative. Bold lower case, bold upper case and calligraphic letters denote vectors, tensors of second order and of fourth order, respectively. In particular, the identity tensor of second order is denoted by $\mathbf{I}$ and the identity tensor of fourth order is denoted by $\mathscr{I}$. For vector and tensor components indices notation with respect to a rectangular Cartesian basis $\mathbf{e}_i$ $(i = 1, 2, 3)$ is used. Operations and symbols are defined as: $\mathbf{a}\mathbf{b} = a_i b_i$, $\mathbf{A}\mathbf{b} = A_{ij} b_j \mathbf{e}_i$, $\mathbf{a} \otimes \mathbf{b} = a_i b_j \mathbf{e}_i \otimes \mathbf{e}_j$, $\mathbf{I} = \delta_{ij} \mathbf{e}_i \otimes \mathbf{e}_j$, $\mathscr{I} = \mathbf{I} \odot \mathbf{I} = \delta_{ik} \delta_{jl} \mathbf{e}_i \otimes \mathbf{e}_j \otimes \mathbf{e}_k \otimes \mathbf{e}_l$, $\mathbf{A}\mathbf{B} = A_{ik} B_{kj} \mathbf{e}_i \otimes \mathbf{e}_j$, $\mathscr{A} : \mathbf{B} = A_{ijkl} B_{kl} \mathbf{e}_i \otimes \mathbf{e}_j$ and $\operatorname{tr}\mathbf{A} = \mathbf{I} : \mathbf{A} = A_{ii}$. Herein δ_{ik} denotes the Kronecker delta and the summation convention over repeated indices is employed. A superimposed dot indicates a time derivative, i.e. $\dot{\mathbf{A}} = d\mathbf{A}/dt$, and the symbol $[\![\mathbf{A}]\!]$ denotes the jump of the field quantity $\mathbf{A}$ immediately on the plus side and on the minus side of a discontinuity, i.e. $[\![\mathbf{A}]\!] = \mathbf{A}^+ - \mathbf{A}^-$.

2   Concept of Hypoplasticity

Within the framework of a classical hypoplastic continuum the objective stress rate tensor $\overset{\circ}{\mathbf{T}}$ is described by an tensor-valued function $\mathbf{H}$ depending in the simplest case on the current Cauchy stress tensor $\mathbf{T}$ and the symmetric part of the velocity gradient or so-called strain rate tensor $\mathbf{D}$, i.e.

$$\overset{\circ}{\mathbf{T}} = \mathbf{H}(\mathbf{T}, \mathbf{D}) . \qquad (1)$$

In order to specify the function $\mathbf{H}(\mathbf{T}, \mathbf{D})$ in Eq. (1) several requirements must be fulfilled which are based on continuum mechanics and on the general mechanical behaviour of granular materials detected in experiments. For instance:

(i)   Frame invariance requires that **H** is an isotropic tensor-valued function of its arguments (e.g. Rivlin and Ericksen, 1955).

(ii)   To describe inelastic behaviour function **H** must be a non-linear function of **D** so that

$$\mathbf{H}(\mathbf{D}) \neq -\mathbf{H}(-\mathbf{D}).$$

(iii)   For a rate-independent material behaviour tensor function **H** must be positively homogeneous of the first order in **D**, i.e.

$$\mathbf{H}(\lambda \mathbf{D}) = \lambda \mathbf{H}(\mathbf{D}) \text{ for any scalar } \lambda > 0.$$

(iv)   For a constant **D** the corresponding stress path asymptotically tends towards an orientation which is independent of the initial stress state if the constitutive equation is homogeneous in **T**, i.e.

$$\mathbf{H}(\lambda \mathbf{T}) = \lambda^m \mathbf{H}(\mathbf{T}),$$

where λ is an arbitrary scalar and m denotes the order of homogeneity.

(v)   In order to model so-called critical states (Casagrande 1936, Schofield and Wroth, 1968), i.e. states in which a granular body can be deformed continuously at a constant stress and a constant volume, there must be a certain **D** for which the conditions

$$\overset{\circ}{\mathbf{T}} = \mathbf{0} \quad \text{and} \quad \operatorname{tr}\mathbf{D} = 0$$

are satisfied simultaneously .

2.1 Modelling Inelastic Material Properties

Requirement (i), (ii) and (iii) are fulfilled by a decomposition of the isotropic tensor valued function $\mathbf{H}(\mathbf{T},\mathbf{D})$ in Eq. (1) into the sum of the following two parts (Wu and Kolymbas, 1990; Kolymbas, 1991):

$$\mathbf{H}(\mathbf{T},\mathbf{D}) = \mathscr{A}(\mathbf{T}) : \mathbf{D} + \mathbf{B}(\mathbf{T})\,||\mathbf{D}||\,. \tag{2}$$

Herein the function $\mathscr{A}(\mathbf{T}) : \mathbf{D}$ is linear in **D** and the function $\mathbf{B}(\mathbf{T})\,||\mathbf{D}||$ is non-linear in **D** with respect to the Euclidean norm of **D**, i.e. $||\mathbf{D}|| = \sqrt{\mathbf{D}:\mathbf{D}}$. The constitutive equation (2) is homogeneous of the first order in **D** and therefore it describes a rate independent material behaviour. It is easy to prove that the basic concept of hypoplasticity in the form of the constitutive equation (2) is incrementally non-linear, e.g. for two particular strain rates $\mathbf{D}_a$ and $\mathbf{D}_b$ with the same norm, i.e. $||\mathbf{D}_a|| = ||\mathbf{D}_b||$, but opposite signs, i.e. $\mathbf{D}_b = -\mathbf{D}_a$, the corresponding stress rates are:

$$\mathbf{H}(\mathbf{T},\mathbf{D}_a) \neq -\mathbf{H}(\mathbf{T},\mathbf{D}_b) \quad \text{or equivalently} \quad \mathbf{H}(\mathbf{T},\mathbf{D}_a) \neq -\mathbf{H}(\mathbf{T},-\mathbf{D}_a).$$

Therefore an inherent inelastic material behaviour is described with a single constitutive equation and there is no need to decompose the deformation into elastic and plastic parts.

2.2 Modelling SOM Properties

One of the fundamental properties of granular materials is reflected by the experimental finding that homogeneous and proportional deformations, i.e. paths with a constant $\mathbf{D}$, asymptotically lead to almost proportional stress paths (e.g. Goldscheider and Gudehus, 1973, Goldscheider, 1982). While the stress path at the beginning may be influenced by both the initial density and the initial stress, the memory of the material of the initial state declines with continuous proportional deformations. Isochoric drained deformations with a constant $\mathbf{D}$ asymptotically reach a stationary stress ratio which depends on the direction of the stress deviator but is independent of the initial stress state and initial density. Such so-called SOM-states are characterized by a sweeping out of memory (Gudehus et al., 1977) and play an important role for the constitutive modelling and the calibration. In order to model a transition from an arbitrary initial state to SOM-stress-states the tensorial function $\mathscr{A}(\mathbf{T})$ of rank four and the tensorial function $\mathbf{B}(\mathbf{T})$ of rank two of the constitutive equation (2) must be homogeneous of the same order in $\mathbf{T}$ because of requirement (iv). Therefore the terms of Eq. (2) can for instance be factorized using the dimensionless tensor $\hat{\mathbf{T}} = \mathbf{T}/\mathrm{tr}\mathbf{T}$. Then Eq. (2) can be represented as:

$$\mathbf{H}(\mathbf{T},\mathbf{D}) = (\mathrm{tr}\mathbf{T})^m \left[\mathscr{L}(\hat{\mathbf{T}}) : \mathbf{D} + \mathbf{N}(\hat{\mathbf{T}}) \|\mathbf{D}\| \right] . \tag{3}$$

Herein m denotes the order of homogeneity in $\mathbf{T}$. Relations (2) and (3) represent the basic concept of hypoplasticity. The embedding of this concept into the framework of thermodynamics is discussed for instance by Schneider and Hutter (2009). Based on Eq. (3) various specific representations of the functions $\mathscr{L}$ and $\mathbf{N}$ have been proposed in the literature. Limitations and extensions of the basic concept are discussed by Bauer and Wu (1993, 1995), Gudehus (1996), Niemunis and Herle (1997). A comprehensive historical review can be found for instance in Wu and Kolymbas (2000) as well as in Bauer and Herle (2000). In this context it can be noted that limit stress states or so-called stationary stress states are generally included in a constitutive equation of the type of Eq. (3). For modelling critical states, however, additional requirements must be fulfilled for the specific representation of $\mathscr{L}$ and $\mathbf{N}$ as outlined in the next section.

2.3 Modelling Limit Stress States and Critical States

In order to model critical states the specific representation for $\mathscr{L}(\hat{\mathbf{T}}) : \mathbf{D}$ and $\mathbf{N}(\hat{\mathbf{T}}) \|\mathbf{D}\|$ must fulfil two conditions (Bauer, 1995; von Wolffersdorff 1996; Bauer

2000), which can be derived by applying definition (v) for critical states to the general form of the constitutive relation (3). In particular the requirement of a constant stress, i.e. $\mathring{\mathbf{T}} = \mathbf{0}$, leads to

$$\mathscr{L}(\hat{\mathbf{T}}) : \mathbf{D} + \mathbf{N}(\hat{\mathbf{T}})||\mathbf{D}|| = \mathbf{0} \,, \tag{4}$$

and consequently to

$$\hat{\mathbf{D}} = \frac{\mathbf{D}}{||\mathbf{D}||} = -\mathscr{L}^{-1}(\hat{\mathbf{T}}) : \mathbf{N}(\hat{\mathbf{T}}), \tag{5}$$

where $\hat{\mathbf{D}}$ is a non-vanishing dimensionless tensor. Inserting Eq. (5) into the identity $\hat{\mathbf{D}} : \hat{\mathbf{D}} = 1$ leads to a scalar equation depending only on the quantity $\hat{\mathbf{T}}$, i.e.

$$\left[\mathscr{L}^{-1}(\hat{\mathbf{T}}) : \mathbf{N}(\hat{\mathbf{T}})\right] : \left[\mathscr{L}^{-1}(\hat{\mathbf{T}}) : \mathbf{N}(\hat{\mathbf{T}})\right] - 1 = 0 \,. \tag{6}$$

The set of all stresses which fulfil this condition can be represented by a surface in the stress space which is called limit stress surface (Wu and Kolymbas, 1990, Wu and Niemunis, 1996). It is worth noting that Eq. (6) first fulfils only the requirement for a vanishing stress rate and it is only related to critical stress states, i.e. to the so-called critical stress surface (Bauer, 1996), if the second requirement for a vanishing volume strain rate is also fulfilled, i.e.

$$\mathrm{tr}\hat{\mathbf{D}} = \mathrm{tr}\left[-\mathscr{L}^{-1}(\hat{\mathbf{T}}) : \mathbf{N}(\hat{\mathbf{T}})\right] = 0. \tag{7}$$

For modelling critical states requirements (6) and (7) are necessary conditions for the tensorial functions $\mathscr{L}(\hat{\mathbf{T}}) : \mathbf{D}$ and $\mathbf{N}(\hat{\mathbf{T}})||\mathbf{D}||$, which are fulfilled for the following specific relations (Bauer, 1995, 1996):

$$\mathscr{L}(\hat{\mathbf{T}}) = \hat{a}^2 \mathscr{I} + \hat{\mathbf{T}} \otimes \hat{\mathbf{T}} \tag{8}$$

and

$$\mathbf{N}(\hat{\mathbf{T}}) = \hat{a}(\hat{\mathbf{T}} + \hat{\mathbf{T}}^*) \,. \tag{9}$$

Then the tensor function $\mathbf{H}(\mathbf{T},\mathbf{D})$ in Eq. (3) can be represented as:

$$\mathbf{H}(\mathbf{T},\mathbf{D}) = (\mathrm{tr}\mathbf{T})^m \left[\hat{a}^2 \mathbf{D} + \mathrm{tr}(\hat{\mathbf{T}}\mathbf{D})\hat{\mathbf{T}} + \hat{a}(\hat{\mathbf{T}} + \hat{\mathbf{T}}^*)||\mathbf{D}||\right] \,. \tag{10}$$

Herein the normalized tensor $\hat{\mathbf{T}}^* = \hat{\mathbf{T}} - \mathbf{I}/3$ is the deviatoric part of tensor $\hat{\mathbf{T}}$, $\mathbf{I}$ and $\mathscr{I}$ respectively denote identity tensors of rank two and four. A similar representation for $\mathscr{L}(\hat{\mathbf{T}}) : \mathbf{D}$ and $\mathbf{N}(\hat{\mathbf{T}})||\mathbf{D}||$ was proposed by von Wolffersdorff (1996), which is included as a special case of the more general representation in Eq. (8) and Eq. (9) as discussed by Bauer and Herle (2000). In Eq. (10) factor $\hat{a}$ is related to critical stress states which can be easily shown by applying the definition for critical states to the hypoplastic constitutive equation. In particular for a vanishing stress rate, i.e. $\mathring{\mathbf{T}} = 0$, the tensor function $\mathbf{H}(\mathbf{T},\mathbf{D})$ reduces to:

$$\hat{a}^2\,\mathbf{D} + \hat{\mathbf{T}}\,\mathrm{tr}(\hat{\mathbf{T}}\cdot\mathbf{D}) + \hat{a}\,(\hat{\mathbf{T}} + \hat{\mathbf{T}}^*)\,||\mathbf{D}|| = \mathbf{0}\,. \tag{11}$$

Division of relation (11) with $||\mathbf{D}||$ and using the dimensionless tensor $\hat{\mathbf{D}} = \mathbf{D}/||\mathbf{D}||$ leads to

$$\hat{\mathbf{D}} = -\frac{1}{\hat{a}^2}\,\hat{\mathbf{T}}\,\mathrm{tr}(\hat{\mathbf{T}}\cdot\hat{\mathbf{D}}) - \frac{1}{\hat{a}}\,[\hat{\mathbf{T}} + \hat{\mathbf{T}}^*]\,. \tag{12}$$

For stationary states the volume strain rate must also vanish simultaneously with the stress rate, i.e.

$$\mathrm{tr}\hat{\mathbf{D}} = -\frac{1}{\hat{a}^2}\,\mathrm{tr}\hat{\mathbf{T}}\,\mathrm{tr}(\hat{\mathbf{T}}\cdot\hat{\mathbf{D}}) - \frac{1}{\hat{a}}\,[\mathrm{tr}\hat{\mathbf{T}} + \mathrm{tr}\hat{\mathbf{T}}^*] = 0\,,$$

and with respect to $\mathrm{tr}\hat{\mathbf{T}} = \mathrm{tr}(\hat{\mathbf{T}}/\mathrm{tr}\hat{\mathbf{T}}) = 1$ and $\mathrm{tr}\hat{\mathbf{T}}^* = \mathrm{tr}(\mathbf{T}/\mathrm{tr}\mathbf{T} - 1/3) = 1 - 1 = 0$ the equation reduces to

$$\mathrm{tr}(\hat{\mathbf{T}}\cdot\hat{\mathbf{D}}) + \hat{a} = 0\,. \tag{13}$$

Substituting (13) into Eq. (12) yields

$$\hat{\mathbf{D}}_c = -\frac{1}{\hat{a}}\,\hat{\mathbf{T}}^*\,, \tag{14}$$

which reflects coaxiality between $\hat{\mathbf{D}}$ and $\hat{\mathbf{T}}^*$ for stationary states. Since $\mathrm{tr}\hat{\mathbf{T}}^* = 0$ holds independent of $\hat{\mathbf{T}}^*$, relation (14) also fulfils the requirement for a vanishing volume strain rate, i.e.

$$\mathrm{tr}\hat{\mathbf{D}}_c = -\frac{1}{\hat{a}}\,\mathrm{tr}\hat{\mathbf{T}}^* = 0\,.$$

Inserting (14) into the identity $||\hat{\mathbf{D}}||^2 = \mathrm{tr}(\hat{\mathbf{D}}^2) = 1$ leads to an equation for the critical stress surface (Bauer, 1996), i.e.

$$\hat{a} = ||\hat{\mathbf{T}}^*||\,. \tag{15}$$

With respect to the relation $\hat{\mathbf{T}}^* = \mathbf{T}/\mathrm{tr}(\mathbf{T}) - (1/3)\mathbf{1}$ it is obvious that the critical stress surface is a cone with its apex at the origin of the principal stress component space. Herein the dimensionless factor $\hat{a}$ determines the shape of the critical stress surface and can be expressed as a function of the Lode-angle θ, i.e. $\hat{a}_c = \hat{a}_c(\theta)$ as illustrated in Fig. (1a). It was shown by Bauer (2000) that factor $\hat{a}$ in Eq. (15) can be related to various stress limit conditions without loss of the general form of the constitutive equation. In the present paper $\hat{a}$ is adapted to the stress limit condition given by Matsuoka and Nakai (1977), i.e.

$$\hat{a} = \frac{\sin\varphi_c}{3 + \sin\varphi_c}\left[\sqrt{\frac{(8/3) - 3\,||\hat{\mathbf{T}}^*|| + \sqrt{3/2}\,||\hat{\mathbf{T}}^*||^3\,\cos(3\,\theta)}{1 + \sqrt{3/2}\,||\hat{\mathbf{T}}^*||\,\cos(3\,\theta)}} + ||\hat{\mathbf{T}}^*||\right], \tag{16}$$

with the Lode-angle θ, which is defined as:

$$\cos(3\,\theta) = -\sqrt{6}\,\frac{\mathbf{I}:\hat{\mathbf{T}}^{*3}}{[\mathbf{I}:\hat{\mathbf{T}}^{*2}]^{3/2}}\,.$$

In Eq. (16) the only constitutive constant is the so-called critical friction angle φ_c which is defined for the triaxial compression test in the critical state (Bauer, 2000).

3 Specific Hypoplastic Model for Sand by Gudehus and Bauer

In the following a specific hypoplastic model by Gudehus (1996) and Bauer (1996) is presented which was developed for modelling the essential mechanical properties of cohesionless granular materials like sand. In order to take into account the influence of the packing density of the grains in the skeleton on the incremental stiffness the constitutive equation (1) is extended by the current void ratio e, i.e.

$$\mathring{\mathbf{T}} = \mathbf{H}(\mathbf{T},\mathbf{D},e)\,. \tag{17}$$

Herein the void ratio e is defined as the ratio of the volume of the voids to the volume of the solid particles and can be related to the mass density ρ_s of the solid grains and to the bulk density ρ of the granular material according to:

$$e = \frac{\rho_s}{\rho} - 1\,. \tag{18}$$

Assuming that the volume change of the solid grains under load changes can be neglected, the volume change of the granular material is determined by the volume change of the void space. It can be noted that this assumption does not exclude the possibility of grain plastification and grain breakage. With Eq. (18) and the assumption that $\rho_s = $ constant one obtains from the balance relation of mass the following relation for the rate of the void ratio:

$$\dot{e} = (1+e)\,\mathrm{tr}\mathbf{D}\,. \tag{19}$$

It follows from relation (19) that the void ratio is not an independent state variable, and that for an empty void space with $\rho_s = $ constant the granular body can be treated as a single component continuum.

In the hypoplastic constitutive model proposed by Gudehus (1996) and Bauer (1996) the influence of the current void ratio e on the mechanical behaviour is incorporated with the so-called stiffness factor f_s and the density factor f_d in the following form:

$$\mathring{\mathbf{T}} = f_s\left[\hat{a}^2\mathbf{D} + \mathrm{tr}(\hat{\mathbf{T}}\mathbf{D})\hat{\mathbf{T}} + f_d\,\hat{a}\,(\hat{\mathbf{T}} + \hat{\mathbf{T}}^*)\,||\mathbf{D}||\right]\,. \tag{20}$$

In particular the stiffness factor f_s is defined as:

$$f_s = \left(\frac{e_i}{e}\right)^{\beta} f_b ,\qquad(21)$$

and the density factor f_d reads:

$$f_d = \left(\frac{e - e_d}{e_c - e_d}\right)^{\alpha} ,\qquad(22)$$

with the constants $\alpha < 0.5$ and $\beta > 1$. Factors f_s and f_d represent relative densities where the current void ratio e is related to the pressure dependent maximum void ratio e_i, minimum void ratio e_d and critical void ratio e_c. Herein the maximum void ratio e_i is defined for an isotropic compression starting from the loosest possible skeleton with grain contacts and approximated by the following compression relation (Bauer, 1995):

$$e_i = e_{i0} \exp\left[-(3p/h_s)^n\right] .\qquad(23)$$

Herein $p = -\mathrm{tr}\mathbf{T}/3$ denotes the mean pressure and e_{i0} denotes the value of e_i for $p = 0$. The solid hardness h_s has the dimension of stress, and the exponent n is a dimensionless constant. For high pressures the void ratio in Eq. (23) tends to zero, which can be explained by grain plastification and crushing. The same pressure dependence for e_i on p is postulated by Gudehus (1996) for the void ratios e_c, and e_d, i.e.

$$\frac{e_c}{e_i} = \frac{e_{c0}}{e_{i0}} , \quad \frac{e_d}{e_i} = \frac{e_{d0}}{e_{i0}} ,\qquad(24)$$

wherein e_{c0} and e_{i0} and e_{d0} are the appropriate void ratios for $p = 0$ as shown in Fig. (1b) in the so-called phase diagram of grain skeletons (Gudehus, 1997).

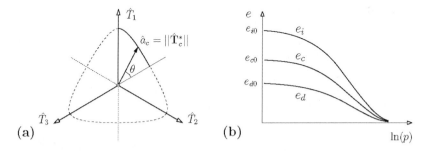

Fig. 1 (a) Stress limit condition in the deviator-plane; (b) Decrease of the maximum void ratio e_i, the critical void ratio e_c and the minimum void ratio e_d with increasing mean pressure p.

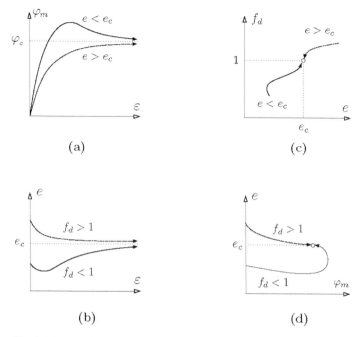

Fig. 2 Monotonic shearing under constant mean pressure for different initial void ratios: (a) Mobilized friction angle φ_m versus shear strain ε; (b) Void ratio e versus shear strain ε; (c) Density factor f_d versus void ratio e; (d) Void ratio e versus mobilized friction angle φ_m.

With the density factor f_d strain softening and the dependence of the peak friction angle on the void ratio is included in the constitutive equation (20). However, the assumption of strain softening as a material behaviour must not be confused with the geometrical softening of a discrete mechanical structure caused by strain localization. In the present model strain softening is not accounted for by an additional constitutive parameter but it follows from the asymptotical property of the constitutive equation. In other words, for unlimited monotonic shearing a critical state is reached asymptotically, independently of the initial void ratio as sketched in Fig.(2). For $e > e_c$ Eq. (22) yields $f_d > 1$ while for $e < e_c$ a value of $f_d < 1$ is obtained (Fig.2c). The peak friction angle is higher for a void ratio $e < e_c$ (Fig.2a). With advanced monotonic shearing the density factor tends to $f_d = 1$ as a consequence of dilatancy. For $e > e_c$ the material shows contractancy up to the critical state (Fig.2b) and for $e = e_c$ the value of the density factor becomes $f_d = 1$ both for an initially dense and for an initially loose state (Fig.2c) so that for critical states the constitutive equation (20) reduces again to Eq. (11).

A lower void ratio e means a denser state of the material and a higher stiffness factor f_s as described by Eq. (21). Consequently, at the beginning of shearing the incremental stiffness obtained for a lower void ratio is higher (Fig.2a). Moreover the incremental stiffness also increases with an increase of the pressure. The pressure dependence of the stiffness is taken into account by factor f_b in Eq. (21), which can be determined by a consistency condition. According to Gudehus (1996) the consistency condition links the compression relation (23) to the constitutive equation (20). In particular for the same void ratio e_i, there must be a coincidence between the rate of the mean pressure under isotropic proportional compression calculated from (23), i.e.

$$3\,\dot{p} = \frac{\dot{e}_i}{e_i}\frac{h_s}{n}\left(\frac{3p}{h_s}\right)^{1-n} ,$$

and from (20), i.e.

$$3\,\dot{p} = f_b\,h_i\frac{\dot{e}_i}{(1+e_i)} .$$

Thus, factor f_b can be obtained as

$$f_b = \frac{h_s}{n\,h_i}\frac{1+e_i}{e_i}\left(\frac{3p}{h_s}\right)^{1-n} , \tag{25}$$

with the constant:

$$h_i = \frac{1}{3}\left[\frac{8\sin^2\varphi}{(3+\sin\varphi)^2} + 1 - \frac{2\sqrt{2}\sin\varphi}{3+\sin\varphi}\left(\frac{e_{i0}-e_{d0}}{e_{c0}-e_{d0}}\right)^{\alpha}\right] .$$

Since the current void ratio e is related to the limit void ratios by the pressure dependent functions f_s and f_d, the constitutive constants are not restricted to a certain initial density or stress state. As long as the mechanical behaviour does not change substantially due to grain abrasion, the parameters of the hypoplastic model remain constant for one granular material (Herle and Gudehus, 1999). Thus, the mechanical behaviour of initially dense or initially loose sand can be described using one set of constitutive constants. More details about the performance of the present hypoplastic model and the calibration of the constitutive constants involved are outlined for instance by Gudehus (1996, 1997) and Bauer (1996, 1999).

For the numerical investigations in the next section the calibration of the constants are based on data from compression tests and triaxial tests for medium quartz sand. The following values are used: $\varphi_c = 30°$, $n = 0.4$, $h_s = 190$ MPa, $\alpha = 0.11$, $\beta = 1.05$, $e_{i0} = 1.2$, $e_{d0} = 0.51$, $e_{c0} = 0.82$.

4   Shear Band Bifurcation Analysis

The spontaneous formation of a shear band or so-called discontinuity plane is characterized by a different velocity gradient $\nabla \mathbf{v}$ on either side of this plane (Rudnicki and Rice 1975). The jump of the velocity gradient, i.e.

$$[\![\nabla \mathbf{v}]\!] = \mathbf{g}\otimes\mathbf{n}\neq\mathbf{0} , \tag{26}$$

can be represented by the dyadic product of the unit normal $\mathbf{n}$ of the discontinuity plane and the vector $\mathbf{g}$ defining the discontinuity mode of the velocity gradient. Continuous equilibrium across the discontinuity requires

$$[[\dot{\mathbf{T}}]]\mathbf{n} = \mathbf{0}. \tag{27}$$

Herein the jump of the stress rate can be related to the jump of the Jaumann stress rate, i.e.

$$[[\dot{\mathbf{T}}]] = [[\overset{\circ}{\mathbf{T}}]] + [[\mathbf{W}]]\mathbf{T} - \mathbf{T}[[\mathbf{W}]],$$

where $\overset{\circ}{\mathbf{T}}$ is the response of the hypoplastic model (20) and $\mathbf{W}$ denotes the antisymmetric part of the velocity gradient. Inserting the Jaumann stress rate into Eq. (27) leads to the following relation:

$$f_s(\mathscr{L} : [[\mathbf{D}]])\mathbf{n} + [[||\mathbf{D}||]]f_s f_d \mathbf{N}\mathbf{n} + [[\mathbf{W}]]\mathbf{T}\mathbf{n} - \mathbf{T}[[\mathbf{W}]]\mathbf{n} = 0, \tag{28}$$

The jump of the stretching and the spin tensor are related to the jump of the velocity gradient (26), i.e.

$$[[\mathbf{D}]] = [\mathbf{g}\otimes\mathbf{n}+\mathbf{n}\otimes\mathbf{g}]\frac{1}{2} \quad \text{and} \quad [[\mathbf{W}]] = [\mathbf{g}\otimes\mathbf{n}-\mathbf{n}\otimes\mathbf{g}]\frac{1}{2}.$$

At the onset of shear banding the stress $\mathbf{T}$ and the void ratio e are the same on either side of the discontinuity plane and therefore the quantities f_s, f_d, $\mathscr{L}$ and $\mathbf{N}$ are also the same. It is a peculiarity in hypoplasticity that the possibility of different incremental stiffnesses due to a different velocity gradient on either side of the discontinuity is taken into account by a single relation (28) and that there is no need to distinguish whether the material outside the shear band undergoes loading or unloading (e.g. Chambon and Desrues 1985). Relation (28) can be rewritten as (Bauer and Huang 1997; Bauer, 1999):

$$\mathbf{K}\mathbf{g} - \lambda\mathbf{r} = \mathbf{0},$$

with:

$$\mathbf{K} = f_s\left[\hat{a}^2(\mathbf{I}+\mathbf{n}\otimes\mathbf{n})\frac{1}{2} + (\hat{\mathbf{T}}(\mathbf{n}\otimes\mathbf{n}))\hat{\mathbf{T}}\right] +$$
$$+\frac{1}{2}\left[(\mathbf{n}(\mathbf{T}\mathbf{n}))\mathbf{I} - (\mathbf{n}\otimes\mathbf{n})\mathbf{T} - \mathbf{T} + \mathbf{T}(\mathbf{n}\otimes\mathbf{n})\right],$$

$$\mathbf{r} = -f_s f_d \hat{a}(\hat{\mathbf{T}} + \hat{\mathbf{T}}^*)\mathbf{n} \quad \text{and} \quad \lambda = [[||\mathbf{D}||]].$$

Inserting $\mathbf{g} = \lambda\mathbf{K}^{-1}\mathbf{r}$ into

$$\gamma = ||[[\mathbf{D}]]|| = \sqrt{[[\mathbf{D}]] : [[\mathbf{D}]]} = \sqrt{[\mathbf{g}\mathbf{g}+(\mathbf{g}\mathbf{n})^2]/2}$$

leads to the bifurcation condition:

$$\sqrt{\frac{(\mathbf{K}^{-1}\mathbf{r})\,(\mathbf{K}^{-1}\mathbf{r})+((\mathbf{K}^{-1}\mathbf{r})\,\mathbf{n})^2}{2}} - \frac{\gamma}{|\lambda|} = 0. \tag{29}$$

The inequality $|[\![\,||\mathbf{D}||\,]\!]| \le ||[\![\mathbf{D}]\!]||$, i.e. $|\lambda| \le \gamma$, is valid regardless of the amount of the strain rate on either side of the discontinuity. Eq. (28) is positively homogeneous of the first order with respect to $\mathbf{g}$ so that γ is arbitrary and can be set to $1\ s^{-1}$. It can be shown that the lowest bifurcation stress ratio is obtained for $|\lambda| = \gamma = 1$ (Bauer and Huang, 1997). With respect to a fixed co-ordinate system the components of vector $\mathbf{n}$ can be expressed by the unknown shear band inclination angle ϑ. Thus, relation (29) represents an equation for ϑ, in which only real solutions to (29) indicate the possibility of a spontaneous shear band formation. Real solutions for ϑ appear in pairs which are symmetric with respect to the principal stress directions.

Fig. (3a) shows the results obtained from numerical examinations of the bifurcation condition (29) for stress paths which are related to homogeneous plane strain compressions, where T_1 is the constant lateral stress and T_2 denotes the compression stress starting from an isotropic stress state. The lowest bifurcation stress ratio at which bifurcation is possible is higher for an initially lower void ratio and it is reached for a smaller vertical strain rather than for a higher void ratio. States beyond the first bifurcation point again fulfil condition (29), in which additional real solutions are possible for $|\lambda| < \gamma$. The influence of the minimum principal stress T_1 and the void ratio on the shear band inclination ϑ is shown in Fig. (3b). For the same stress T_1 the angle ϑ increases with a decrease of the void ratio but for an increase of T_1 the shear band inclination decreases. The predictions are in accordance with the experiments performed by Yoshida et al. (1993).

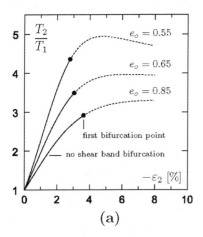
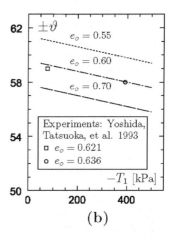

(a)   (b)

Fig. 3 Onset of shear band formation in uniform biaxial compressions under constant lateral pressures T_1: (a) stress ratio T_2/T_1 versus vertical strain ε_2 for various initial void ratios and $T_1 = -100$ kPa, (b) Shear band orientation ϑ versus minimum stress T_1.

5 Conclusion

For the analytical investigation of shear banding in cohesionless and rate-independent granular materials a continuum model based on the concept of hypoplastic has been considered. With a unified description of the interaction between pressure level and the limit void ratios the present hypoplastic model enables the analysis of a shear band localization for a wider range of pressures and densities using a single set of constitutive constants. The constitutive constants involved have a clear physical meaning and they are not restricted to a certain initial density or stress state. Since the constitutive equation is incrementally non-linear, the question as to whether the velocity gradient on either side of the discontinuity is related to loading, unloading or to a rigid body motion is irrelevant for the analytical bifurcation analysis. From the bifurcation analysis it can be concluded that the lowest possible bifurcation stress ratio occurs before the peak. The analysis also shows the influence of the initial density and pressure level on the possibility of a shear band bifurcation. In particular for a lower void ratio the bifurcation stress ratio is higher than for a higher void ratio and it is reached for a smaller vertical strain. At the lowest bifurcation stress two symmetric shear bands are obtained. The inclination of shear bands with regard to the direction of the minor principal stress decreases with an increase of the initial void ratio and the pressure.

References

1. Albers, B.: Modeling and Numerical Analysis of Wave Propagation in Saturated and Partially Saturated Porous Media. Shaker Press, Aachen (2009)
2. Bauer, E., Wu, W.: A hypoplastic model for granular soils under cyclic loading. In: Proc. of the Int. Workshop on Modern Approaches to Plasticity, pp. 247–258. Elsevier, Amsterdam (1993)
3. Bauer, E.: Constitutive Modelling of Critical States in Hypoplasticity. In: Proceedings of the Fifth International Symposium on Numerical Models in Geomechanics, Davos, Switzerland, Balkema, pp. 15–20 (1995)
4. Bauer, E., Wu, W.: A hypoplastic constitutive model for cohesive powders. Powder Technology 85, 1–9 (1995)
5. Bauer, E.: Calibration of a comprehensive hypoplastic model for granular materials. Soils and Foundations 36(1), 13–26 (1996)
6. Bauer, E., Huang, W.: The dependence of shear banding on pressure and density in hypoplasticity. In: Adachi, O., Yashima (eds.) Proc. of the 4th Int. Workshop on Localisation and Bifurcation Theory for Soils and Rocks, pp. 81–90. Balkema Press (1997)
7. Bauer, E.: Analysis of shear band bifurcation with a hypoplastic model for a pressure and density sensitive granular material. Mechanics of Materials 31, 597–609 (1999)
8. Bauer, E.: Conditions for embedding Casagrande's critical states into hypoplasticity. Mechanics of Cohesive-Frictional Materials 5, 125–148 (2000)
9. Bauer, E., Herle, I.: Stationary states in hypoplasticity. In: Kolymbas (ed.) Constitutive Modelling of Granular Materials, pp. 167–192. Springer, Heidelberg (2000)
10. Bauer, E., Huang, W., Wu, W.: Investigation of shear banding in an anisotropic hypoplastic material. Solids and Structures 41, 5903–5919 (2004)
11. de Boer, R.: Theory of Porous Media. Springer, Berlin (2000)

12. Bowen, R.M.: Theory of mixtures. In: Continuum Physics, vol. 3, pp. 2–127. Academic Press, New York (1976)
13. Casagrande, A.: Characteristics of cohesionless soils affecting the stability of earth fills (1936); J. Boston Soc. Civil Engrs. Contribution to Soil Mech., 1925–1945 (2004)
14. Chambon, R., Desrues, J.: Bifurcation par localisation et non linéarité incrémentale: un exemple heuristique d'analyse compléte. In: Presses de l'ENPC (ed.) Plastic Instability, Paris, pp. 101–113 (1985)
15. Chambon, R., Charlier, R., Desrues, J., Hammad, W.: A rate type constitutive law including explicit localisation: development and implementation in a finite element code. In: 2nd Eur. Spec. Conf. on Num. Meth. in Geotechn. Engn., Santander (1990)
16. Chambon, R., Desrues, J., Tillard, D.: Shear moduli identification versus experimental localization data. In: Proc. of the 3th Int. Workshop on Localization and Bifurcation Theory of Soils and Rock, France, pp. 101–111. Balkema (1993)
17. Ehlers, W.: Grundlegende Konzepte in der Theorie Poröser Medien. Technische Mechanik, 63–76 (1996)
18. Goldscheider, M., Gudehus, G.: Rectilinear extension of dry sand: Testing apparatus and experimental results. In: Proc. 8th Int. Conf. on Soil Mech. and Found. Eng., Moscow, vol. 1/21, pp. 143–149 (1973)
19. Goldscheider, M.: True triaxial tests on dense sand. In: Workshop on Constitutive Relations for Soils, Grenoble, pp. 11–54 (1982)
20. Gudehus, G., Goldscheider, M., Winter, H.: Mechanical Properties of Sand and Clay and Numerical Integration Methods: Some Sources of Errors and Bounds of Accuracy. In: Gudehus (ed.) Finite Elements of Geomechanics, pp. 121–150. John Wiley, New York (1977)
21. Gudehus, G.: A comprehensive constitutive equation for granular materials. Soils and Foundations 36(1), 1–12 (1996)
22. Gudehus, G.: Attractors, percolation thresholds and phase limits of granular soils. In: Behringer, Jenkins (eds.) Proc. Powder and Grains, pp. 169–183. Balkema (1997)
23. Herle, I., Gudehus, G.: Determination of parameters of a hypoplastic constitutive model from grain properties. Mechanics of Cohesive-Frictional Materials 4, 461–486 (1999)
24. Hill, R.J., Hutchinson, J.W.: Bifurcation phenomena in the plane tension test. J. Mech. Phys. Solids 23, 239–264 (1975); Engineering Methods, Tucson, Arizona
25. Huang, W., Nübel, K., Bauer, E.: A polar extension of hypoplastic model for granular material with shear localization. Mechanics of Materials 34, 563–576 (2002)
26. Huang, W., Bauer, E.: Numerical investigations of shear localization in a micro-polar hypoplastic material. Int. J. for Numerical and Analytical Methods in Geomechanics 27, 325–352 (2003)
27. Kolymbas, D.: Bifurcation analysis for sand samples with non-linear constitutive equation. Ingenieur-Archive 50, 131–140 (1981)
28. Kolymbas, D.: An outline of hypoplasticity. Archive of Applied Mechanics 61, 143–151 (1991)
29. Matsuoka, H., Nakai, T.: Stress-strain relationship of soil based on the 'SMP'. In: Proc. of Speciality Session 9, IX Int. Conf. Soil Mech. Found. Eng., Tokyo, pp. 153–162 (1977)
30. Niemunis, A., Herle, I.: Hypoplastic model for cohesionless soils with elastic strain range. Mechanics of Cohesive-Frictional Materials 2(4), 279–299 (1997)
31. Rice, J., Rudnicki, J.W.: A note on some features on the theory of localization of deformation. Int. J. Solids Structures 16, 597–605 (1980)
32. Rivlin, R., Ericksen, J.: Stress-deformation relations for isotropic materials. J. Rat. Mech. Anal. 4, 323–425 (1955)

33. Rudnicki, J., Rice, J.: Conditions for the localization of deformation in pressure sensitive dilatant materials. J. Mech. Phys. Solids 23, 371–394 (1975)
34. Schneider, L., Hutter, K.: Solid-Fluid Mixtures of Frictional Materials in Geophysical and Geotechnical Context. Springer, Heidelberg (2009)
35. Schofield, A.N., Wroth, C.P.: Critical State Soil Mechanics. McGraw-Hill, London (1968)
36. Tejchman, J., Gudehus, G.: Shearing of a narrow granular strip with polar quantities. J. Num. and Anal. Methods in Geomechanics 25, 1–18 (2001)
37. Truesdell, C.: Thermodynamics of diffusion. In: Rational Thermodynamics, 2nd edn., pp. 219–236. Springer, New York (1984)
38. Vardoulakis, I., Sulem, J.: Bifurcation analysis in geomechanics. Chapman & Hall, Blackie Academic & Professional, Glasgow (1995)
39. Wilmanski, K.: Continuum Thermodynamics. Part I: Foundations. World Scientific, Singapore (2008)
40. von Wolffersdorff, P.A.: A hypoplastic relation for granular materials with a predefined limit state surface. Mechanics of Cohesive-Frictional Materials 1, 251–271 (1996)
41. Wu, W., Sikora, Z.: Localized bifurcation in hypoplasticity. Int. J. Engng. Sci. 29(2), 195–201 (1991)
42. Wu, W., Kolymbas, D.: Numerical testing of the stability criterion for hypoplastic constitutive equations. Mechanics of Materials 9, 245–253 (1990)
43. Wu, W., Niemunis, A.: Failure criterion, flow rule and dissipation function derived from hypoplasticity. Mechanics of Cohesive-Frictional Materials 1, 145–163 (1996)
44. Wu, W., Kolymbas, D.: Hypoplasticity then and now. In: Kolymbas, D. (ed.) Constitutive Modelling of Granular Materials, pp. 57–105. Springer, Heidelberg (2000)
45. Yoshida, T., Tatsuoka, F., Siddiquee, M.S.A., Kamegai, Y., Park, C.S.: Shear banding in sands observed in plane strain compression. In: Proc. of the 3th Int. Workshop on Localisation and Bifurcation Theory for Soils and Rocks, Grenoble, France, p. 170. Balkema (1993)

Principal Axes and Values of the Dispersion Coefficient in the 2D Axially Symmetric Porous Medium

Leonid G. Fel and Jacob Bear

Abstract. We perform the diagonalization of the 2nd rank dispersion tensor D_{ij} in the 2-dim anisotropic porous medium with two mirror lines, which are perpendicular to each other. We derive the eigenvalues of the matrix and calculate an angular measure of anisotropy.

1 Introduction

The coefficients of dispersion and dispersivity appear in the expression for the flux, **J**, of a solute in saturated flow through porous media. At every point within the fluid, a solute concentration, γ, and a fluid velocity, V, can be identified. A mathematical model describing the solute's transport within the void space requires information on the fluid-solid interface that bounds the fluid occupied domain. However, in natural materials, such as soil and fractured rock, this information is not available. Therefore, the problem of solute transport is usually described in terms of averaged velocity, $\overline{V}$, and averaged solute concentration, $\overline{\gamma}$. These averaged values, taken over some representative elementary volume, are assigned to *every* point in the porous medium domain.

With the total (advective) solute flux at a point *inside* the void-space defined by γV, and at a point in the porous medium domain, regarded as a continuum, defined by $\overline{\gamma V}$, we have [2],

$$\overline{\gamma V} = \overline{\gamma}\,\overline{V} + \overline{\gamma' V'}, \quad \text{where} \quad \gamma' = \gamma - \overline{\gamma}, \quad V' = V - \overline{V},$$

are deviations from the average values, with $\overline{\gamma'} = \overline{V'} = 0$. The above equality states that the total flux of the solute at a point in the porous medium domain, is made up

Leonid G. Fel · Jacob Bear
Department of Civil Engineering and Environmental Engineering,
Technion - Israel Institute of Technology, Haifa 32000, Israel
e-mail: {lfel,cvrbear}@techunix.technion.ac.il

of an advective flux (= product of average concentration and average velocity), and another flux, called dispersive flux, equal to the average of the product of the two deviations, $\mathbf{J} = \overline{\gamma' \mathbf{V}'}$.

The above discussion can be extended to any extensive quantity, with γ representing its intensive quantity. Thus, the phenomenon of dispersion occurs whenever an extensive quantity (mass of a solute, in the example considered above) is transported in a fluid phase that flows in the void space of a porous medium. It is a consequence of using an average velocity and average concentration to describe the advective flux of the solute in the void space. The total flux is the sum of the advective flux and the dispersive one. In this article, we use the mass of dissolved chemical species as the considered extensive quantity. For short, we denote the average values: $\overline{\gamma} = c$, $\overline{V} = V$.

The coefficient of dispersion, D_{ij}, is a symmetric tensor of 2nd rank, which appears in the Fickian-type expression for the dispersive flux of a solute (mass of solute per unit area of void space in a cross-section of the porous medium) [2],

$$J_i = - \sum_{j=1}^{3} D_{ij} \nabla_j c, \quad D_{ij} = D_{ji}, \quad i, j = 1, 2, 3, \tag{1}$$

where J_i denotes the ith component of the solute's 3-dim flux vector, $\mathbf{J}$, and c is the solute's average concentration. Equation (1) is valid for the general case of any anisotropic porous medium, with isotropic media as a special case.

In contrast to similar linear flux laws, e.g., Ohm's and Fourier's laws, with corresponding tensorial coefficients of electroconductivity and of thermoconductivity, respectively, which depend only on properties of the considered medium, here, in the case of the dispersive flux in saturated flow through a porous medium, the dispersion tensor D_{ij} depends also on the fluid's average velocity $\mathbf{V}$ in the void space [1], [4],

$$D_{ij} = \frac{1}{U} \sum_{k,l=1}^{3} a_{ijkl} V_k V_l, \quad \mathbf{V}^2 = \sum_{k=1}^{3} V_k^2, \quad \mathbf{V}^2 = V^2, \tag{2}$$

where a_{ijkl} is a property of the porous medium only, called *dispersivity*, and V_i is the ith component of the fluid's average velocity vector $\mathbf{V}$.

2 Dispersion in Two-Dimensional Anisotropic Porous Media

We study the dispersion tensor in a 2-dim porous medium domain, which is obtained by intersecting a 3-dim uniaxial porous medium with a plane that is parallel to the uniaxial axis. In the resulting 2-dim porous medium domain, there are two mirror lines, which are perpendicular to each other. By these symmetry elements, it is easy to recognize the dihedral point symmetry group $\mathscr{D}_2$. In a recent paper [3], we have derived the 3D dispersion tensor D_{ij} in the form

$$VD_{ij} = a_1 \mathbf{V}^2 \delta_{ij} + a_2 V_i V_j + a_3 \mathbf{V}^2 e_i e_j + a_4 \langle \mathbf{e}, \mathbf{V} \rangle^2 \delta_{ij}$$
$$+ a_5 \langle \mathbf{e}, \mathbf{V} \rangle (V_i e_j + e_i V_j) + a_6 \langle \mathbf{e}, \mathbf{V} \rangle^2 e_i e_j, \tag{3}$$

where the a_i's, $1 \leq i \leq 6$, stand for the dispersivity moduli, which satisfy six inequalities:

$$a_1 + a_2 > 0, \quad a_1 + a_3 > 0, \quad a_1 + a_4 > 0,$$
$$a_1 + a_2 + a_3 + a_4 + 2a_5 + a_6 > 0, \quad a_1 > 0, \tag{4}$$
$$a_1^2 + a_1(3a_2 + a_3 + a_4 + 2a_5 + a_6) + 2a_2(a_3 + a_4 + a_6) > 2a_5^2.$$

The elements of the 2D tensor (3), D_{ij}, $i, j = 1, 2$, read

$$VD_{11} = a_1 \mathbf{V}^2 + a_2 V_1^2 + a_3 \mathbf{V}^2 e_1^2 + a_4 \langle \mathbf{e}, \mathbf{V} \rangle^2 + 2a_5 \langle \mathbf{e}, \mathbf{V} \rangle e_1 V_1 + a_6 \langle \mathbf{e}, \mathbf{V} \rangle^2 e_1^2,$$
$$VD_{22} = a_1 \mathbf{V}^2 + a_2 V_2^2 + a_3 \mathbf{V}^2 e_2^2 + a_4 \langle \mathbf{e}, \mathbf{V} \rangle^2 + 2a_5 \langle \mathbf{e}, \mathbf{V} \rangle e_2 V_2 + a_6 \langle \mathbf{e}, \mathbf{V} \rangle^2 e_2^2,$$
$$VD_{12} = a_2 V_1 V_2 + a_3 \mathbf{V}^2 e_1 e_2 + a_5 \langle \mathbf{e}, \mathbf{V} \rangle (e_1 V_2 + e_2 V_1) + a_6 \langle \mathbf{e}, \mathbf{V} \rangle^2 e_1 e_2. \tag{5}$$

Diagonalizing this 2×2 matrix, $D_{i,j}$, with the two invariants,

$$\Delta_1 = V(D_{11} + D_{22}) = (2a_1 + a_2 + a_3) \mathbf{V}^2 + (2a_4 + 2a_5 + a_6) \langle \mathbf{e}, \mathbf{V} \rangle^2,$$
$$\Delta_2 = V^2(D_{11}D_{22} - D_{12}^2) = a_1^2 \mathbf{V}^4 + a_1 \mathbf{V}^2 \left[(2a_4 + 2a_5 + a_6) \langle \mathbf{e}, \mathbf{V} \rangle^2 + (a_2 + a_3) \mathbf{V}^2 \right]$$
$$+ \langle \mathbf{e}, \mathbf{V} \rangle^2 \left[a_4(a_4 + 2a_5 + a_6) \langle \mathbf{e}, \mathbf{V} \rangle^2 + a_3 a_4 \mathbf{V}^2 - a_5^2 [\mathbf{e} \times \mathbf{V}]^2 \right]$$
$$+ a_2 \left[a_4 \langle \mathbf{e}, \mathbf{V} \rangle^2 \mathbf{V}^2 + a_3 \mathbf{V}^2 [\mathbf{e} \times \mathbf{V}]^2 + a_6 [\mathbf{e} \times \mathbf{V}]^2 \langle \mathbf{e}, \mathbf{V} \rangle^2 \right], \tag{6}$$

we find its eigenvalues, $\lambda_{1,2}$, by solving the equation

$$\det \{ D_{ij} - \lambda \delta_{ij} \} = \det \begin{Bmatrix} D_{11} - \lambda & D_{12} \\ D_{12} & D_{22} - \lambda \end{Bmatrix} = 0, \tag{7}$$

where $[\mathbf{e} \times \mathbf{V}]^2 = (V_1 e_2 - V_2 e_1)^2$ and $\langle \mathbf{e}, \mathbf{V} \rangle^2 = (V_1 e_1 + V_2 e_2)^2$. These eigenvalues are

$$\lambda_{1,2} = \frac{1}{2} \left(\Delta_1 \pm \sqrt{\Delta_3} \right), \quad \Delta_3 = \Delta_1^2 - 4\Delta_2 = V^2 \left[(D_{11} - D_{22})^2 + 4D_{12}^2 \right] \geq 0, \tag{8}$$

where

$$\Delta_3 = (a_2 + a_3)^2 \mathbf{V}^4 + 2(a_2 + a_3)(2a_5 + a_6) \mathbf{V}^2 \langle \mathbf{e}, \mathbf{V} \rangle^2 + (2a_5 + a_6)^2 \langle \mathbf{e}, \mathbf{V} \rangle^4$$
$$+ 4 \left[(a_5^2 - a_2 a_6) \langle \mathbf{e}, \mathbf{V} \rangle^2 - a_2 a_3 \mathbf{V}^2 \right] [\mathbf{e} \times \mathbf{V}]^2. \tag{9}$$

From (8) it follows that the values of both λ_1 and λ_2 are positive.

Another important entity is the diagonalization angle, ϕ, with

$$\tan 2\phi = \frac{2D_{12}}{D_{22} - D_{11}}. \tag{10}$$

where

$$V(D_{22} - D_{11}) = a_2 \left(V_2^2 - V_1^2\right) + \left(a_3 \mathbf{V}^2 + a_6 \langle \mathbf{e}, \mathbf{V} \rangle^2\right) \left(e_2^2 - e_1^2\right)$$
$$+ 2a_5 \langle \mathbf{e}, \mathbf{V} \rangle \left(e_2 V_2 - e_1 V_1\right),$$

(11)

which gives the directions of the principal axes of the dispersion matrix.

Consider two different cases: $\langle \mathbf{e}, \mathbf{V} \rangle = 0$ and $[\mathbf{e} \times \mathbf{V}] = 0$, in which the vectors $\mathbf{e}$ and $\mathbf{V}$ are either perpendicular or parallel to each other, respectively. Hence, in the 1st case we obtain

$$\lambda_{1,2} = \frac{V}{2} \left(2a_1 + a_2 + a_3 \pm (a_2 - a_3)\right) = \begin{cases} V(a_1 + a_2), \\ V(a_1 + a_3) \end{cases}.$$

(12)

$$\tan 2\phi = 2 \frac{a_2 V_1 V_2 + a_3 \mathbf{V}^2 e_1 e_2}{a_2 \left(V_2^2 - V_1^2\right) + a_3 \mathbf{V}^2 \left(e_2^2 - e_1^2\right)},$$

(13)

with $\tan 2\phi = 0$ when $V_1 = e_2 = 0$.

In the 2nd case, $[\mathbf{e} \times \mathbf{V}] = 0$, we obtain

$$\lambda_{1,2} = \frac{V}{2} \left(2a_1 + a_2 + a_3 + 2a_4 + 2a_5 + a_6 \pm (a_2 + a_3 + 2a_5 + a_6)\right)$$
$$= \begin{cases} V(a_1 + a_2 + a_3 + a_4 + 2a_5 + a_6), \\ V(a_1 + a_4) \end{cases}.$$

(14)

$$\tan 2\phi = 2 \frac{a_2 V_1 V_2 + (a_3 + a_6)\mathbf{V}^2 e_1 e_2 + a_5 V (e_1 V_2 + e_2 V_1)}{a_2 \left(V_2^2 - V_1^2\right) + (a_3 + a_6)\mathbf{V}^2 \left(e_2^2 - e_1^2\right) + 2a_5 V (e_2 V_2 - e_1 V_1)},$$

(15)

and $\tan 2\phi = 0$ when $V_1 = e_1 = 0$. Note that all eigenvalues in (12) and (14) are positive in accordance with (4).

In an isotropic case, the relationships (12), (13) and (14), (15) give

$$\lambda_1 = V(a_1 + a_2), \quad \lambda_2 = Va_1, \quad \tan 2\phi = \frac{2V_1 V_2}{V_2^2 - V_1^2},$$

(16)

i.e., two principal axes of the matrix D_{ij} coincide with the tangent and normal vectors to the streamlines. However, in the general case (10), applied to an anisotropic porous medium, this is not true.

We finish the paper with a discussion on one more question: for a given direction of the $\mathbf{e}$-axis, are there nontrivial (not parallel to $\mathbf{e}$-axis) direction and nonzero magnitude of a uniform velocity, $\mathbf{V}$, such that $\tan 2\phi = 0$?

By (10), this occurs when $D_{12} = 0$. In such a case, in accordance with (5), we have

$$a_2 V_1 V_2 + \left(a_3 \mathbf{V}^2 + a_6 \langle \mathbf{e}, \mathbf{V} \rangle^2\right) e_1 e_2 + a_5 \langle \mathbf{e}, \mathbf{V} \rangle (e_1 V_2 + e_2 V_1) = 0.$$

(17)

In the $(e_1 - e_2)$-plane, (17) describes an algebraic curve of the 4th degree. However, in the $(V_1 - V_2)$-plane, it is only of 2nd degree,

$$V_1^2 \left(a_3 + a_5 + a_6 e_1^2\right) e_1 e_2 + V_2^2 \left(a_3 + a_5 + a_6 e_2^2\right) e_1 e_2$$
$$+ V_1 V_2 \left(a_2 + a_5 + 2a_6 e_1^2 e_2^2\right) = 0 . \tag{18}$$

It has two real roots when

$$\left(a_2 + a_5 + 2a_6 e_1^2 e_2^2\right)^2 \geq 4 \left(a_3 + a_5 + a_6 e_1^2\right) \left(a_3 + a_5 + a_6 e_2^2\right) e_1^2 e_2^2 ,$$

or

$$4 \left[(a_3 + a_5)^2 + a_6 (a_3 - a_2)\right] e_1^2 e_2^2 \leq (a_2 + a_5)^2 . \tag{19}$$

Otherwise equation (17) holds if and only if $V_1 = V_2 = 0$.

Keeping in mind the parameterization $e_1 = \cos\alpha$, $e_2 = \sin\alpha$, we obtain

$$\sin^2 2\alpha \leq \frac{(a_2 + a_5)^2}{(a_3 + a_5)^2 + a_6 (a_3 - a_2)} , \tag{20}$$

which can be satisfied for all α when the r.h.s. of (20) exceeds one, i.e.,

$$(a_2 + a_3 + 2a_5 + a_6)(a_2 - a_3) \geq 0 . \tag{21}$$

As a special case, the requirements (20) and (21) are obtained when $e_1 = e_2 = 1/\sqrt{2}$. In an isotropic case, (21) is always satisfied, since $a_2^2 \geq 0$.

References

1. Bear, J.: On the Tensor Form of Dispersion. J. Geophys. Res. 66, 1185–1197 (1961)
2. Bear, J.: Dynamics of Fluids in Porous Media, p. 764. American Elsevier (1972)
3. Fel, L.G., Bear, J.: Dispersion and Dispersivity Tensors in Saturated Porous Media with Uniaxial Symmetry. Submitted to TIPM (preprint, 2009), http://arXiv.org/abs/physics.flu-dyn/0904.3447
4. Scheidegger, A.E.: General Theory of Dispersion in Porous Media. J. Geophys. Res. 66, 3273–3278 (1961)

The Importance of Sand in Earth Sciences

Dimitrios Kolymbas

Abstract. Being a solid, sand can sustain shear stresses at rest but it can also undergo large plastic deformations without considerable change of its properties, behaving thus like a fluid. As a product of erosion, sand cannot be broken into parts because it is already a broken ('clastic') material. The pronounced deformability of sand gave rise not only to a large diversity of experimental investigations in Soil Mechanics but rendered also sand a model material for physical simulations of deformation processes of the earth crust: Sand box models serve to simulate not only folding and faulting processes of the earth crust but also processes of deformation of the earth mantle. There are also similarities between magma volcanism and the so-called sand boils or sand volcanoes that appear subsequent to liquefaction of water-saturated loose sand. The complex behaviour of sand is a permanent object of study not only by Soil Mechanics but —in recent time— also by Physics. In this paper, the ability of sand to model the behaviour of other geomaterials is elucidated and a new theoretical frame is presented to describe mathematically the behaviour of sand based on its asymptotic properties.

1 The Perception of Sand Outside the Earth Sciences

Specialists in Soil Mechanics having spent a large part of their lives trying to decipher the mysteries of sand do quite well understand Antoine de Saint-Exupéry, who wrote in his book "The Wisdom of the Sands (Citadelle)"

> Nous nous sommes nourris de la magie des sables, d'autres peut-être y creuseront leurs puits de pétrole et s'enrichiront de leurs marchandises. Mais ils seront venus trop tard. Car les palmeraies interdites ou la poudre vierge des coquillages nous ont livré leur part la plus précieuse : elles n'offraient qu'une vaine ferveur, et c'est nous qui l'avons vécue.

Dimitrios Kolymbas
University of Innsbruck, Institute of Infrastructure, Division of Geotechnical and
Tunnel Engineering, Technikerstr. 13, A-6020 Innsbruck, Austria
e-mail: dimitrios.kolymbas@uibk.ac.at

In fact, DAVID MUIR WOOD named his remarkable Bjerrum lecture of 2005 "The magic of sand" [21]. In 2008 the German journal DER SPIEGEL published an article entitled "The magic of intelligent sand" [7]. There, a special sand is envisioned whose grains are equipped with mini-computers such that, say, an initial cube of sand can attain upon demand the form of a trumpet, see Fig. 1. Judging from the difficulties to decipher the mechanical behaviour of sand we may state that sand is indeed an intelligent material, but his intelligence resides not upon the individual grains but upon the interaction and the resulting mean behaviour of large assemblies of grains.

The concept of matter consisting of innumerable small particles has been introduced by DEMOCRIT and obtained an important role in modern physics, which succeeded in explaining the behaviour of e.g. gases as resulting from the motion of molecules. The related branch of physics, statistical physics, grew to an admirable refinement but it left aside sand, whose behaviour is completely different from ideal gas. Sand became the realm of Soil Mechanics, a branch of Civil Engineering, and followed a completely different trajectory from ideal and real gases. Although its foundations were laid down by COULOMB, the father of electrostatics, sand was disregarded by physicists until recently, when they detected that its behaviour is fascinating. Prof. M. Liu (Theoretical Physics, University of Tübingen) stated that sand is more difficult than fluid helium or liquid crystals. Another statement is that *"Many properties of sand and other granular materials are as puzzling to scientists*

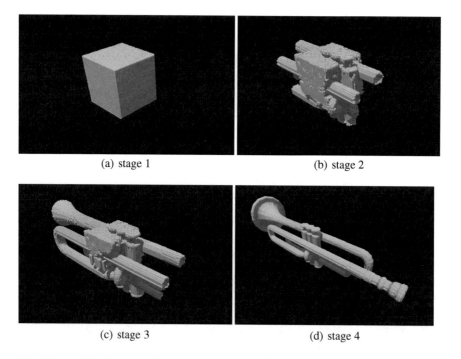

(a) stage 1   (b) stage 2

(c) stage 3   (d) stage 4

Fig. 1 Several fictitious transformation stages of a cube made of 'intelligent sand' [7]

as the big bang is" [20]. In the meanwhile, Soil Mechanics has carried out an amazingly large collection of laboratory experiments and obtained a large collection of theoretical results.

2 Similarities to Other Geomaterials

2.1 Similarity to Rock

Sand covers large parts of the earth and is, thus, in itself an important geomaterial. Another important soil type, clay, consists also of particles and its behaviour is in many aspects identical to sand. More precisely, dense sand is similar to so-called overconsolidated clay, and loose sand is similar to normally consolidated clay. The majority of rocks consist also of grains (e.g. granite), as shown in the thin cut of sandstone shown in Fig. 2. In difference to soils, the grains of rock are densely packed and are also 'glued' (cemented) to each other. However, density and cementation do not hinder large deformation of rock associated with grain re-arrangement, provided that the deformation is sufficiently slow.

0,5 mm

Fig. 2 Thin cut of sandstone. Courtesy of Prof. Chr. Lempp, Halle.

Another similarity of sand and rock can be found in the fact that both are frictional materials, i.e. their strength increases with normal stress. BYERLEE [5] observed that the coefficient of friction of almost every type of crustal rock falls within the surprisingly narrow range of 0.6 - 1.0, a range that is commonly referred to as Byerlee's Law;

> "The experimental results show that at the low stresses encountered in most civil engineering problems the friction of rock can vary between very wide limits and the variation is mainly because at these low stresses friction is strongly dependent on surface roughness. At intermediate pressure such as encountered in mining engineering problems and at high stresses involved during sliding on faults in the deep crust the initial surface roughness has little or no effect on friction. At normal stresses up to 2 kbar ($= 2 \cdot 10^5$ kN/m^2) the shear stress required to cause sliding is given approximately by the equation $\tau = 0.85\sigma_n$[1]. At normal stresses above 2 kbar the friction is given approximately by $\tau = 0.5 + 0.6\sigma_n$. These equations are valid for initially finely ground surfaces, initially totally interlocked surfaces or on irregular faults produced in initially intact rocks. Rock types have little or no effect on friction"[5].

The capability of sand to undergo large plastic deformations renders it an ideal material to scale down (in geometry and time) tectonic deformation processes. This fact has been exploited in the so-called sandbox models (Figs. 3, 4 and 5). Also in numerical simulations sand proves to be an appropriate material to model tectonic processes (Figs. 6 and 7).

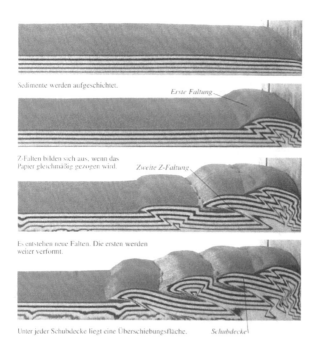

Sedimente werden aufgeschichtet.

Erste Faltung

Z-Falten bilden sich aus, wenn das Papier gleichmäßig gezogen wird.

Zweite Z-Faltung

Es entstehen neue Falten. Die ersten werden weiter verformt.

Unter jeder Schubdecke liegt eine Überschiebungsfläche.

Schubdecke

Fig. 3 Sandbox model for the simulation of tectonic processes [1]

---

[1] i.e. $\tau = 1.7 \cdot 10^5$ kN/m^2. Compare with the cohesion of granite of the order of magnitude of $2 \cdot 10^3$ kN/m^2.

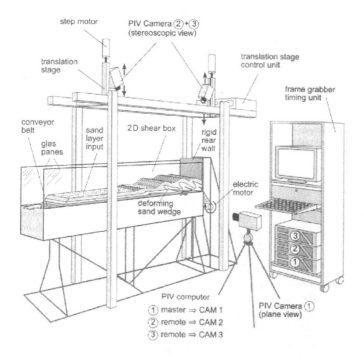

Fig. 4 Sandbox apparatus [2]

Fig. 5 Modelling of extension of lithosphere [27]

Also clay is a good material to model deformations of rock strata. The first to conduct simulations of tectonic processes with clay cakes was the geologist HANS CLOOS, see Fig. 8.

The simulation of tectonic processes in sandbox models shows that sand is, in fact, a good model for rock strata but does not explain why this is so. BYERLEE's law explains that in large geometric scale the strength of rock is controlled by friction, which is a genuine property of sand, whereas cohesion is of minor importance. Nevertheless, it is still difficult to grasp how rock can undergo large deformation without exhibiting brittle failure, as one would expect based on the experience with

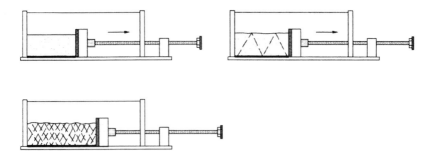

Fig. 6 Stretching the base of a granular layer [22]. Compare the obtained pattern with the numerical simulation shown in Fig. 7.

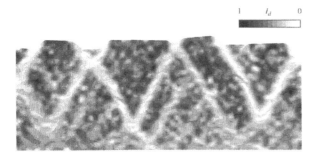

Fig. 7 Calculated pressure-adjusted density index I_d after base deformation $\Delta l / l = 20\%$ [22]

small geometry and time scales. Even if we recognise both materials, sand and rock (e.g. sandstone), as granular ones we have to admit that in rock the grains are (i) in a much denser state, and (ii) cemented to each other, so that grain-rearrangement is not possible without breakage of the bonds. It turns however, that ductile grain re-arrangement is not possible *in short time scales* but still possible for very slow deformation. Non-brittle, i.e. ductile deformation, of rock is considered, mainly in Structural Geology [30]. In macroscopic terms, ductile deformation is often manifested as creep. For stationary creep, two cases are distinguished:

1. Power-law creep (for low and moderate stress): $|\mathbf{D}| \propto \left(\sqrt{\mathrm{tr} \mathbf{T}^{\star 2}} \right)^n \cdot \exp \frac{-E}{RT}$
2. Exponential creep law (for high stress): $|\mathbf{D}| \propto \exp(\beta \sqrt{\mathrm{tr} \mathbf{T}^{\star}}) \cdot \exp \frac{-E}{RT}$.

Herein, $\mathbf{D}$ indicates the so-called stretching tensor, i.e. the symmetric part of the velocity gradient grad $\mathbf{v} = v_{i,j}$, and $\mathbf{T}^{\star}$ is the deviatoric part of the CAUCHY stress $\mathbf{T}$. The term $\exp \frac{-E}{RT}$ indicates that creep is a thermally activated process. In microscopic terms, several types of creep have been identified, such as diffusion and solution creep, including the so-called Nabarro-Herring creep, coble creep and others. It must be remarked, however, that these types of creep are not (yet) integrated

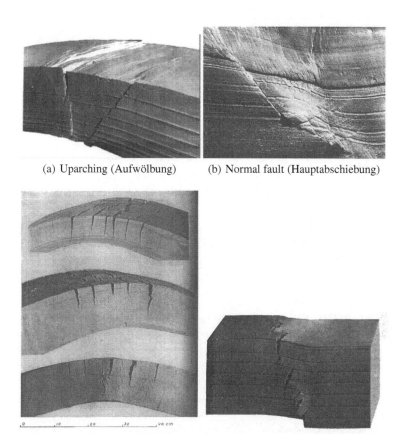

(a) Uparching (Aufwölbung)   (b) Normal fault (Hauptabschiebung)

(c) Gap volcanoes (Spaltenvulkane)   (d) Cracking (Zerspaltung)

Fig. 8 Experiments of CLOOS: Simulations of tectonic patterns with clay [6]

into a global model, so that their consideration appears rather confusing from a macroscopic point of view.

Note that Structural Geology uses some terms different from Soil Mechanics:

Granular flow: Deformation based on grain re-arrangement. This is the main type of deformation considered in Soil Mechanics. In large depths, dilatancy and grain re-arrangement may be suppressed by the large prevailing pressure. However, large pore fluid pressures may reduce the effective stresses so that granular flow can freely develop.

Cataclastic flow: Granular flow with grain crushing. Mainly assumed to occur within faults.

Large ductile deformation can be explained by the fact that rock is capable to heal manifestations of brittle failure when deformed slowly enough. This can be demonstrated by the cobble shown in Fig. 9.

Fig. 9 Cobble collected by the author at the Greek island of Kythera. The slip surface does not affect the integrity of the cobble.

2.2   Similarity to the Earth Mantle

The mechanical properties of the earth mantle can only be indirectly assessed by interpretation of seismograms. It is generally accepted that the mantle is in perpetual motion, which is the cause of continental drift and other tectonic deformations of the earth crust. What is the reason of this perpetual motion? — Besides thermal convection, tidal deformation has been assumed (e.g. by WEGENER) to sustain a monotonic motion of the mantle. In fact, the daily rotation of the earth in the gravitational fields of the sun and the moon causes a cyclic loading of the earth. If the earth behaved elastically, cyclic loading would cause a (certainly minute) cyclic deformation but no monotonic deformation of the earth. If, however, the earth (mainly, the earth mantle) behaves anelastically, then cyclic loading may result in a perpetual monotonic motion of the mantle. The daily anelastic deformation, how small it may be, accumulates over centuries and attains considerable values. REVUZHENKO has carried out a series of spectacular tests with sand, as a modelling material of the earth mantle [26]. Unable to simulate tidal forces (i.e. mass forces with rotating direction) in the laboratory, he invented a 2D test where a flexible containment was filled with dry sand and subjected to cyclic deformation by means of an imposed rotating mold with elliptical cross section (Figs. 10 a-c). Thus, the mass forces exerted by the attraction of moon and sun were replaced by surface tractions that produced more or less the same deformation. The result was a monotonic rotational deformation of the sand, as shown with embedded coloured sand in Figs. 11 and 12. Similar results were obtained with 3D tests with an analogous deformation of a sphere (Fig. 10 d).

2.3   Similarities of Lava Volcanoes to Sand Volcanoes

There is a striking similarity of magma and sand volcanoes. Soil is a granular material consisting of grains and pores, the later being usually filled with water. The analogy with melting rock has been pointed out by MCKENZIE [19]: Partially molten rock consists of a deformable matrix (corresponding to the grain-skeleton of soils)

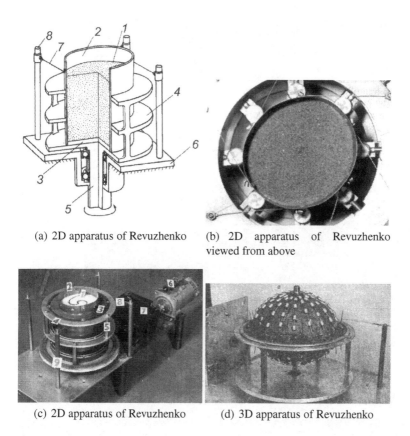

(a) 2D apparatus of Revuzhenko   (b) 2D apparatus of Revuzhenko viewed from above

(c) 2D apparatus of Revuzhenko   (d) 3D apparatus of Revuzhenko

Fig. 10 Apparatuses of REVUZHENKO [26] intended to simulate the tidal deformation of the earth by the application of surface tractions to dry sand bodies.

and pore fluid, which in this case is molten rock. Thus, both materials, i.e. partially molten rock and water-saturated soil, are two-phase materials consisting of a solid matrix and interstitial fluid.

Whereas the detrimental action of magma volcanoes is widespread known, sand volcanoes (also called 'sand boils') are less known. They appear in liquefied soil. Soil liquefaction is a peculiar phenomenon that occurs when loose water-saturated sand undergoes rapid deformation, as this is the case due to e.g. earthquakes or (underground) explosions. Sand volcanoes accompany liquefaction and the resulting craters are (besides the devastation of buildings) the only remnant of liquefaction in green fields. Japanese authors [18] describe sand volcanoes as follows:

"The most violent water spouting during the Earthquake No. 4 was observed at the site ... near the Shonai river where water with sand ejected high above 2 m from wells and sands deposited over the roofs of nearby houses."

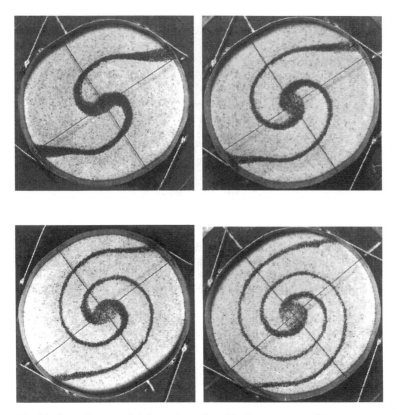

Fig. 11 Several stages of deformation of sand in the apparatus of REVUZHENKO [26]

Fig. 12 Initial and deformed sand bodies in the 2D apparatus of REVUZHENKO [26]

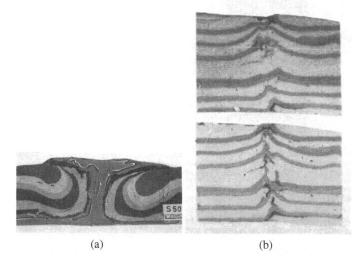

(a)   (b)

Fig. 13 Simulations with clay-oil mixtures by RAMBERG [25]

Fig. 14 Sand boil after liquefaction. Photograph taken by the author. The liquefaction
was released by underground explosion.

Fig. 15 Crater from sand boil

Similar reports are available from India (Bihar earthquake, 1934, cited in [10]):

> "...then water spouts, hundreds of them throwing up water and sand, were to be ob-
> served on the whole face of the country, the sand forming miniature volcanoes, while
> the water spouted out of the craters; some of the spouts were were quite six feet high.
> In a few minutes, on both sides of the road as far as the eye could see, was vast expanse
> of sand and water ..."

The places of appearance of sand volcanoes are unpredictable and, therefore, photos
can be rarely taken (Fig. 14). Craters remain after the eruption to witness liquefaction (Fig. 15).

3 Mechanical Behaviour of Sand

3.1 *The Physics of the Grain Skeleton*

It is tempting to conceive a grain skeleton as a framework. In case of elastic grains,
a linear relation between the forces $\Delta\mathbf{F}$ and displacements $\Delta\mathbf{x}$ would then prevail:
$\Delta\mathbf{F} = \mathbf{K}\,\Delta\mathbf{x}$, i.e. the stiffness $\mathbf{K}$ depends on the deformation $\Delta\mathbf{x}$. However, the
individual grains are not fixed to each other (except for the case of cementation)
and, thus, the grain skeleton is not a framework but rather a sort of bad-quality
ball-bearing. The individual grains are displaced and rotated during deformation
and the points of contact permanently change, i.e. deformation is *always accompanied with grain re-arrangement* i.e. change of geometry. This means that the force-
displacement relation is non-linear: $\Delta\mathbf{F} = \mathbf{K}(\Delta\mathbf{x})\Delta\mathbf{x}$. From a microscopic point of
view, this non-linearity is geometric, but from a macroscopic point of view, which
disregards grain re-arrangement, the non-linearity is inherent (material) and 'incremental', i.e. it cannot be linearised 'in the small'.

   If we idealise a granular material as a continuum, then the considered displace-
ments $\mathbf{u}$ are a mean field that is always accompanied with a scatter $\Delta\mathbf{u}$ due to grain
re-arrangement. The mean square of this scatter can be interpreted as 'granular temperature' T_g and this notion can be used to extend thermodynamic concepts to the
mechanics of sand. Since any macroscopic deformation is connected with granular
temperature, it is natural to set $T_g \propto \sqrt{\mathrm{tr}\mathbf{D}^2}$.

   The inner friction of granulates has little or nothing to do with the friction between two adjacent grains. If the contacts were controlled by friction and if there
was no grain re-arrangement, then a proportional loading of a skeleton consisting of
stiff grains would not cause any considerable deformation. In reality, however, every
loading (including proportional ones) causes grain re-arrangement and macroscopic
deformation.

   Internal friction (and, hence, shear strength) of sand is not a manifestation of
intergranular friction but rather a result of force diffusion within the irregular grain
skeleton.

4   Strain Localisation and Pattern Formation

At first glance, sand being an amorphous material, seems incapable to exhibit any pronounced structure. All the same, sand was the first medium to store information (remember of Archimedes drawing his sketches on sand), thanks to its irreversible or hysteretic mechanical behaviour. Besides of imposed patterns, sand is capable to spontaneously produce fascinating patterns. These are only visible if the sand has been marked before (say by colour) or if the pattern appears at the surface, as is the case in Figs. 16 and 17.

Fig. 16 Pattern formation on the horizontal surface of sheared sand (REVUZHENKO [26])

The majority of the patterns are related to strain localisation into thin shear bands. Their spontaneous formation, especially in the course of tests with initially homogeneous deformation of the samples, has been extensively studied. The sample deformation is imposed to the sample via prescribed displacements of the sample boundaries, say within a HAMBLY-type biaxial apparatus (Fig. 18).

It is remarkable that the imposed information propagates from the boundaries to the interior of the sample and manages that the deformation is equally distributed all over the sample. This propagation occurs with wave velocity, even if the deformation is quasistatic and no waves are perceivable. However, if a wave velocity happens to vanish, homogeneous deformation is no more possible. It then happens something peculiar: the sample gets subdivided into several blocks, each of which obeys the displacement imposed upon its own external boundary. The several blocks then glide relatively to each other along shear bands.

Fig. 17 Localisation pattern in nature (Photograph: Intern. Soc. Rock. Mech.). The pattern formation in this photograph is similar to the one in Fig. 16 that was obtained in the laboratory.

Fig. 18 Biaxial apparatus of the HAMBLY type [17, 28]

4.1 Experimental Observations—Proportional Loading

From true triaxial tests carried out with sand GOLDSCHEIDER inferred two rules.[2] They refer to proportional paths (PP)[3] [9]:

---

[2] GOLDSCHEIDER, impressed by the complexity of the stress and strain paths obtained with a true triaxial apparatus, expressed (in the seventies of the past century) to the author the opinion that it is, presumably, impossible to find a constitutive relation of sand. It is, though, remarkable to see that the constitutive relation for sand is more or less included in the aforementioned rules, as will be shown in this paper.

[3] Proportional stress and strain paths are characterised by constant ratios of the principal values $\sigma_1 : \sigma_2 : \sigma_3$ and $\varepsilon_1 : \varepsilon_2 : \varepsilon_3$, respectively.

Rule 1: Proportional strain paths (PεP, also called 'consolidations', as they are connected with densification) starting from the stress[4] $\mathbf{T} = \mathbf{0}$ are associated to proportional stress paths (PσP).

Rule 2: Proportional strain paths starting from $\mathbf{T} \neq \mathbf{0}$ lead asymptotically to the corresponding proportional stress paths obtained when starting at $\mathbf{T} = \mathbf{0}$ (which act, so to say, as attractors).

Rule 1 also applies to much simpler materials, e.g. linear elastic ones, for which straight stress paths are obtained with PεP starting from *any* stress state. The speciality of cohesionless granular materials arises from the fact that the feasible stress states and paths are limited within a fan with apex at $\mathbf{T} = \mathbf{0}$. Therefore, PεP starting from $\mathbf{T} \neq \mathbf{0}$ correspond to stress paths which, being limited within this fan, either approach asymptotically PσP's or end at $\mathbf{T} = \mathbf{0}$ (Fig. 19). The latter ones are characterised by volumetric dilation.

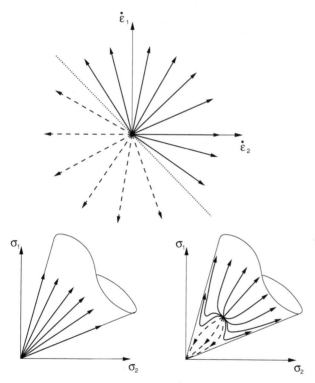

Fig. 19 Proportional strain paths (above) and corresponding stress paths starting from $\mathbf{T} = \mathbf{0}$ (below left) and $\mathbf{T} \neq \mathbf{0}$ (below right)

The second rule is, more or less, a consequence of the limitation of feasible stress states within the aforementioned fan. In this paper it is shown how a constitutive equation for granular materials can be inferred from these two rules. A constitutive

---

[4] In compliance with [29], the CAUCHY stress is denoted with $\mathbf{T}$ instead of σ.

relation that holds for all possible PεP's starting from *arbitrary* stress states can be expected to describe the entire mechanical behaviour of granular materials.

4.2 Other Experimental Evidence

In Figures 20 to 23 are shown test results that demonstrate some principal features of the mechanical behaviour of sand. Figure 20 shows the behaviour of sand at oedometric compression, i.e. at compression with no lateral strain. There are shown the stress path and the stress-strain curve for two load cycles of loading and unloading.

Figure 21 shows the behaviour of sand at drained triaxial compression, i.e. compression in axial direction with lateral expansion. Note that a constant-stress condition prevails at the mantle of the cylindrical sand sample. The upper curve shows a series of four loadings and unloadings whereas the lower curve shows the corresponding volumetric strain ε_{vol} ($= \Delta V / V$) vs. the axial strain ε_1.

Figure 22 shows the behaviour of sand at undrained (i.e. constant volume) triaxial deformation of sand at various stress levels. Figure 23 shows experimental results that reveal the rate sensitivity of sand, and Fig. 24 shows sand at cyclic undrained triaxial deformation.

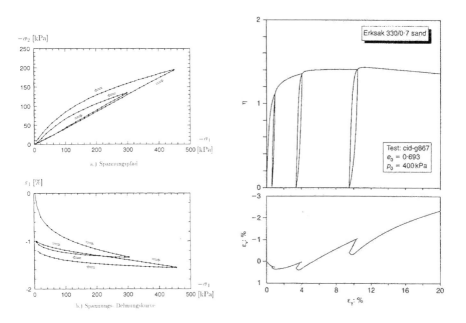

Fig. 20 Oedometric test with dense Karlsruher Sand (initial void ratio $e_0 = 0.55$). BAUER, 1992 [4]

Fig. 21 Triaxial test with Erksak Sand, JEFFERIES, 1997 [11]. $\eta := \dfrac{\sigma_1 - \sigma_2}{(\sigma_1 + \sigma_2 + \sigma_3)/3}$

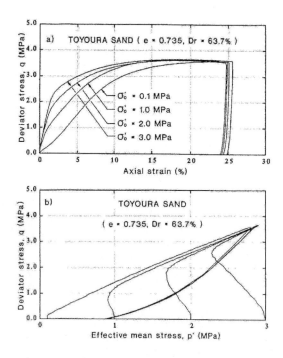

Fig. 22 Undrained triaxial tests with $e = 0.833$, VERDUGO & ISHIHARA, 1996 [31]

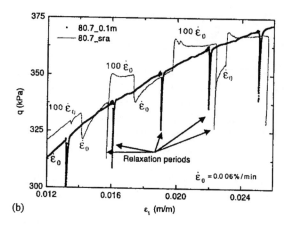

Fig. 23 CD triaxial tests on dry Hostun sand with jumps of deformation rate, PHAM VAN BANG et al., 2007 [3]

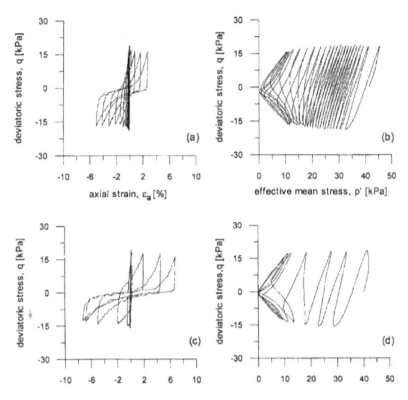

Fig. 24 Gravel and sand, reconstituted (water sedimentation (top row), air pluviation (bottom row)), GHIONNA & PORCINO, 2006 [8]

5 Mathematical Models

There is a large number of mathematical models to describe the mechanical behaviour of sand. Most of them are formulated in the frame of the theory of elasto-plasticity, i.e. elastic behaviour is assumed to prevail as long as the stress is within the so-called yield surface. Plastic strains occur only when the stress lies upon the yield surface. The pertinent definitions of loading, unloading and reloading are built upon considerations of the yield surface and its transformations in the stress space. Contrary to elastoplastic theories, hypoplasticity [13, 14, 15, 16] concentrates on the representation of the constitutive relation as a single evolution equation for stress, built along the principles of Rational Mechanics [29]. In this paper is introduced 'barodesy' , a new branch of hypoplasticity .

5.1 Barodesy

5.1.1 Mathematical Model of Goldscheider's First Rule

As known, the stretching tensor **D** is defined as the symmetric part of the velocity gradient grad**v**. For rectilinear extensions, **D** can be conceived as the time rate of the logarithmic strain, i.e. $\mathbf{D} = \dot{\varepsilon}_{ij}$.[5] A proportional strain path (PεP) is obtained by the application of a constant stretching $\mathbf{D}$. The equations of barodesy are here derived on the basis of rectilinear extensions, motions with rotation of the principal axes are, however, automatically included in the obtained equations, which are written for fully occupied 3×3 tensors. All proportional stress paths (PεP), i.e. all consolidations, form a fan $\mathscr{F}_\varepsilon$ in the strain space, and the corresponding PσP's form a fan $\mathscr{F}_\sigma$ in the stress space.

GOLDSCHEIDER's first rule relates $\mathbf{D}$ with a tensor $\mathbf{R}$ which points in the direction of the associated stress path. This relation can be expressed as:

$$\mathbf{R}(\mathbf{D}) := (\mathrm{tr}\mathbf{D}^0)\, \mathbf{1} + \alpha \exp\left(\beta\, \mathbf{D}^0\right) \quad , \tag{1}$$

where α and β are material constants. Equation 1 has been obtained by trial and error. Herein, the exponent 0 denotes normalisation of a tensor $\mathbf{X}$, i.e. $\mathbf{X}^0 := \mathbf{X}/|\mathbf{X}|$, with $|\mathbf{X}| := \sqrt{\mathrm{tr}\mathbf{X}^2}$. Obviously, $\mathbf{R}(\mathbf{D})$ is homogeneous of the zeroth degree with respect to $\mathbf{D}$.

5.1.2 Adaptation to the Second Rule

Using equ. 1, a constitutive law of the rate type that is valid only for PP starting from $\mathbf{T} = \mathbf{0}$ can be formulated:

$$\dot{\mathbf{T}} = \mathbf{R} \cdot |\mathbf{T}| \cdot |\mathbf{D}| . \tag{2}$$

Obviously, the constitutive relation 2 is homogeneous of the first degree with respect to $\mathbf{T}$ and $\mathbf{D}$.[6] Now, adding to the right part of equ. 2 a term which is proportional to stress $\mathbf{T}$ will incorporate also rule 2 without breaking rule 1. Rule 2 states that if we start from a stress state $\mathbf{T} \neq \mathbf{0}$ and apply a PεP, i.e. a constant stretching $\mathbf{D}$, the stress will asymptotically approach the line $\mathbf{T} = \mu\, \mathbf{R}$. In other words, the obtained stress rate $\dot{\mathbf{T}}$ will fulfil the equation $\mathbf{T} + \dot{\mathbf{T}}\, \Delta t = \mu \mathbf{R}$.[7] This equation can also be written as

$$\dot{\mathbf{T}} = a_1 \mathbf{R} + a\, \mathbf{T} . \tag{3}$$

To preserve homogeneity of the first degree with respect to $\mathbf{D}$, the scalar factor a must be homogeneous of the first degree with respect to $\mathbf{D}$. It proves that $a = a_2\, |\mathbf{D}|$ is the best choice:

$$\dot{\mathbf{T}} = |\mathbf{T}| \cdot (a_1 \mathbf{R} + a_2 \mathbf{T}^0) \cdot |\mathbf{D}| . \tag{4}$$

5.1.3 Homogeneity with Respect to Stress

The subsequent modification of equation 4 is to change the degree of homogeneity in $\mathbf{T}$ from 1 to γ:

---

[5] In the general case, $\mathbf{D}$ is *not* the time derivative of any strain measure.

[6] As known, homogeneity of the degree 1 with respect to $\mathbf{D}$ implies rate-independence.

[7] Note that for rectilinear extensions, as considered in this section, $\overset{\circ}{\mathbf{T}} \equiv \dot{\mathbf{T}}$ holds, with $\overset{\circ}{\mathbf{T}}$ being a co-rotational stress rate.

$$\dot{\mathbf{T}} = |\mathbf{T}|^{\gamma} \cdot (a_1 \mathbf{R} + a_2 \mathbf{T}^0) \cdot |\mathbf{D}| \ . \tag{5}$$

In this way, equation 5 complies with OHDE's relation $d\sigma/d\varepsilon \propto \sigma^{\gamma}$ [12, 24]. For *proportional paths* (i.e. $\mathbf{D}$ =const), integration of equ. 5 or 2 can be obtained as follows: Setting $\sigma := |\mathbf{T}|$, $\dot{\varepsilon} := |\mathbf{D}|$ or $\varepsilon := |\int_0^t \mathbf{D}dt'|$, $\zeta_{\mathbf{D}} := |a_1 \mathbf{R} + a_2 \mathbf{T}^0| \cdot |\mathbf{D}|/\mathrm{tr}\mathbf{D}$ and $\eta_{\mathbf{D}} := |a_1 \mathbf{R} + a_2 \mathbf{T}^0|$ and using the known relation[8] $\dot{e} = (1+e)\,\mathrm{tr}\mathbf{D}$ leads to the differential equations

$$\dot{\sigma} = \sigma^{\gamma} \zeta_{\mathbf{D}} \frac{\dot{e}}{1+e} \tag{6}$$

and

$$\dot{\sigma} = \sigma^{\gamma} \eta_{\mathbf{D}} \dot{\varepsilon} \tag{7}$$

with the general solutions

$$\sigma^{1-\gamma} = \sigma_0^{1-\gamma} + (1-\gamma)\zeta_{\mathbf{D}} \log \frac{1+e}{1+e_0} \tag{8}$$

and

$$\sigma^{1-\gamma} = \sigma_0^{1-\gamma} + (1-\gamma)\eta_{\mathbf{D}}(\varepsilon - \varepsilon_0) \ . \tag{9}$$

Taking into account that for proportional stress paths the initial stress and strain vanish, $\sigma_0 = 0, \varepsilon_0 = 0$, yields finally:

$$\sigma^{1-\gamma} = (1-\gamma)\zeta_{\mathbf{D}} \log \frac{1+e}{1+e_0} \tag{10}$$

and

$$\sigma^{1-\gamma} = (1-\gamma)\eta_{\mathbf{D}}\varepsilon \ . \tag{11}$$

e_0 is the maximum void ratio of the considered sand.

5.1.4 Virgin Strain

The 'virgin' strain $\hat{\mathbf{E}} = \hat{\varepsilon}_{ij} = \hat{\varepsilon}_{ij}(\mathbf{T})$ can be defined as the strain that *uniquely* corresponds to a particular stress $\mathbf{T}$ obtained with a PP starting at $\mathbf{T} = \mathbf{0}$ and $e = e_0$.[9] To determine $\hat{\varepsilon}_{ij}(\mathbf{T})$ we need to know the $\mathbf{D}$-tensor pertinent to $\mathbf{T}$ according to equ. 1. This tensor we call $\hat{\mathbf{D}}$. It can be obtained by numerical inversion of equ. 1.[10]

With $\hat{\mathbf{D}}$ we can easily obtain $\hat{\mathbf{E}}$:

$$\hat{\mathbf{E}} = \varepsilon\hat{\mathbf{D}} \tag{12}$$

with

$$\varepsilon = \frac{\sigma^{1-\gamma}}{(1-\gamma)\eta_{\hat{\mathbf{D}}}} \ . \tag{13}$$

---

[8] This relation holds for incompressible grains.

[9] Such a path can be considered as 'virgin consolidation'.

[10] According to the theorem of CAYLEY-HAMILTON, the quasi-elastic relation $\hat{\mathbf{E}}(\mathbf{T})$ can be represented as $\hat{\mathbf{E}} = \xi_1(\mathbf{T})\mathbf{1} + \xi_2(\mathbf{T})\mathbf{T} + \xi_3(\mathbf{T})\mathbf{T}^2$. However, the analytical expression of this relation on the basis of equ.s 1 and 4 is extremely difficult, if not impossible.

In a similar way we can obtain the virgin void ratio $\hat{e}$ from equation 10:

$$\hat{e} = \exp\left(\frac{\sigma^{1-\gamma}}{(1-\gamma)\zeta_{\hat{\mathbf{D}}}}\right)(1+e_0) - 1 \tag{14}$$

with $e_0 = e_{max}$.[11] Equations 13 and 14 are only seemingly explicit in ε and $\hat{e}$, because $\eta_{\mathbf{D}}$ and $\zeta_{\mathbf{D}}$ depend on a_2 and hence on c, which (via equation 17) depends on $\hat{e}$. Thus, the determination of $\hat{e}$ on the basis of equations 13 has to be achieved numerically.

5.1.5 Barotropy and Pyknotropy

The terms barotropy and pyknotropy denote the dependence of the mechanical behaviour (stiffness) of soil on stress level and on density, respectively.[12] This dependence marks the more intricate aspects of soil mechanics. A crucial notion in this context is the so-called critical void ratio e_c, which denotes the void ratio at a state where the stress remains constant under isochoric (undrained) deformation.

Let us consider the fan $\mathscr{F}_\sigma$. Its boundary consists of PσP's that correspond to isochoric (or undrained) consolidations[13]. However, 'undrained consolidation' is a contradiction in itself, because in standard soil mechanics 'consolidation' is associated with volume decrease. How to resolve this contradiction? – In reality, undrained consolidations can be traced if the actual void ratio is smaller than the critical one. To model this fact, it is necessary that the void ratio e appears in the constitutive equation.

We re-write equ. 5 in the form

$$\dot{\mathbf{T}} = a\,|\mathbf{T}|^\gamma \cdot (b\,\mathbf{R} + c\,\mathbf{T}^0) \cdot |\mathbf{D}| \tag{15}$$

with $ab = a_1$ and $ac = a_2$, and require that $c = 1$ for critical states, i.e. for $e = e_c$ and isochoric motions (i.e. $\mathrm{tr}\mathbf{D} = 0$). To obtain vanishing stiffness at virgin states with isochoric deformation we then only need to set

$$b = -\frac{1}{|\mathbf{R}(\mathbf{D}^{0\star})|} \quad, \tag{16}$$

with

$$\mathbf{D}^{0\star} := \mathbf{D}^0 - \frac{1}{3}\mathrm{tr}\mathbf{D}^0\,\mathbf{1} \quad.$$

As mentioned, the scalar quantity c must depend on void ratio e and on stress $\mathbf{T}$ in such a way that $c = 1$ for virgin states. No explicit relation for the critical void ratio e_c will be used, as it proves that the virgin void ratio is a more general concept. To meet the requirements for c we set:

---

[11] The experimental determination of e_{max} is based on convention, thus the so obtained values of e_{max} should be used with caution.

[12] It is remarkable that barotropy and pyknotropy are interrelated and, therefore, cannot be considered separately.

[13] The term 'consolidation' refers to the fact that along such PσP's the mean stress $p := -(T_1 + T_2 + T_3)/3$ increases.

$$c = \exp[c_1(\hat{e} - e)] \quad . \tag{17}$$

With the consitutive relation of barodesy (i.e. with equ.s 17, 18 and 19) the simulations of element tests shown in Figs. 25, 26 and 27 are obtained.

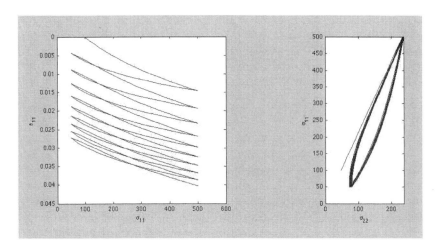

Fig. 25 Stress-strain curve and stress path for cyclic oedometric test on loose sand.

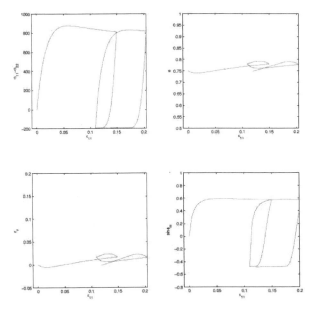

Fig. 26 Simulation of a conventional triaxial test.

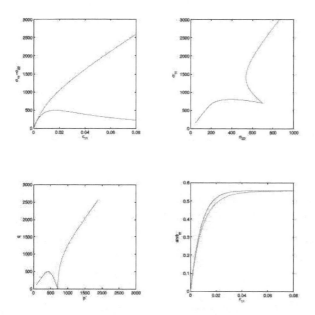

Fig. 27 Undrained triaxial compressions of dense and loose sand.

5.2   Plasticity Theory without Yield Surfaces

In 1973 PALMER and PEARCE published a paper titled "Plasticity theory without yield surfaces" [23]. Some sentences of this paper deserve being quoted here:

> It was quite natural that the idea of a yield surface should assume such importance in a theory built on experience with metals , since in most metals yield occurs at a fairly well-defined stress level. ...

> In soil mechanics the status of the yield surface concept is quite different, both in theory and experiment...

> ...strain measurements in clay depend on direct observation of boundary displacements, so that only quite large strain increments are reliably measurable, creep and pore-pressure diffusion confuse results...

> ...yield surface motions during strain-hardening are often too complex for the results to be helpful in constructing usable stress-strain relations.

> Might it be possible to resolve this (dilemma) by constructing a different kind of plasticity model, in which the yield surface concept had been dropped or relegated to a minor role?

> ...it might be useful to idealise clay as a material in which the yield surface has shrunk to a point, so that all deformations are plastic and *any* changes of stress from the current state will produce plastic strain increments.

PALMER and PEARCE present in their paper a concept for a plasticity theory without yield surfaces. This concept is based on two postulates by ILYUSHIN which are, in a sense, precursors of GOLDSCHEIDER's theorems:

Isotropy postulate: If the strain path is rotated in strain space, then the corresponding stress path is rotated by the same amount. This postulate has actually nothing to do with isotropy, since it considers rotations in the strain and stress spaces, not in the natural space. It is controversial and certainly not valid in the full stress and strain spaces. It is only approximately valid in the deviatoric subspaces: This postulate implies that the deviatoric directions of proportional strain and stress paths coincide. This is, however, not true, according to experimental results by GOLDSCHEIDER [9].

Delay postulate: The stress at some instant in a loading history does not depend on the whole previous history, but only on the last part of it. This is a postulate of fading memory and is similar to GOLDSCHEIDER's second theorem.

Based on ILYUSHIN's postulates, PALMER and PEARCE present the following concept:

> The deviatoric stress has two components. The magnitude of the first component is a function of the octahedral shear strain, and its direction coincides with the principal strain vector (referring strain to an isotropically-consolidated initial state). The magnitude of the second component is constant, and its direction coincides with the current strain rate ...
>
> Reversal of the strain path would reverse the second component but not the first ...

The very last sentence appears prophetic, as it strongly resembles to a basic concept of hypoplasticity, to which certainly the authors would have concluded, had they used rate equations instead of finite ones.

References

1. Die visuelle Geschichte der Erde und des Lebens, Gerstenbergs visuelle Enzyklopädie. Gerstenberg Verlag, Hildesheim (1999)
2. Adam, J., Urai, J., Wieneke, B., Oncken, O., Pfeiffer, K., Kukowski, N., Lohrmann, J., Hoth, S., van der Zee, W., Schmatz, J.: Shear localisation and strain distribution during tectonic faulting–new insights from granular-flow experiments and high-resolution optical image correlation techniques. Journal of Structural Geology 27(2), 283–301 (2005)
3. Bang, D.P.V., Benedetto, H.D., Duttine, A., Ezaoui, A.: Viscous behaviour of dry sand. International Journal for Numerical and Analytical Methods in Geomechanics 31(15), 1631–1658 (2007)
4. Bauer, E.: Zum mechanischen Verhalten granularer Stoffe unter vorwiegend ödometrischer Beanspruchung. No. 130 in Veröffentlichungen des Institutes für Bodenmechanik und Felsmechanik der Universität Karlsruhe (1992)
5. Byerlee, J.: Friction of rocks. Pure and Applied Geophysics 116, 615–626 (1978)
6. Cloos, H.: Hebung – Spaltung – Vulkanismus. Geologische Rundschau - XXX - Zwischenheft 4A (1939)
7. Dworschak, M.: Magie des schlauen Sandes. Der SPIEGEL 6, 126 (2009)

8. Ghionna, V.N., Porcino, D.: Liquefaction resistance of undisturbed and reconstituted samples of a natural coarse sand from undrained cyclic triaxial tests. Journal of Geotechnical and Geoenvironmental Engineering 132(2), 194–202 (2006)
9. Goldscheider, M.: Grenzbedingung und Fließregel von Sand. Mech. Res. Comm. 3, 463–468 (1976)
10. Housner, G.: The mechanisms of sandblows. Bulletin of the Seismological Society of America 48, 155–161 (1958)
11. Jefferies, M.: Plastic work and isotropic softening in unloading. Géotechnique 47, 1037–1042 (1997)
12. Jefferies, M., Been, K.: Implications for critical state theory from isotropic compression of sand. Géotechnique 50(4), 419–429 (2000)
13. Kolymbas, D.: A rate-dependent constitutive equation for soils. Mech. Res. Comm. 4, 367–372 (1977)
14. Kolymbas, D.: Computer-aided design of constitutive laws. Int. J. Numer. Anal. Methods Geomech. 15, 593–604 (1991)
15. Kolymbas, D.: An outline of hypoplasticity. Archive of Applied Mechanics 61, 143–151 (1991)
16. Kolymbas, D.: Introduction to hypoplasticity. In: Advances in Geotechnical Engineering and Tunnelling, vol. 1. Balkema, Rotterdam (2000)
17. Kuntsche, K.: Materialverhalten von wassergesättigten Tonen bei ebenen und zyklischen Verformungen. No. 91 in Veröffentlichungen des Institutes für Bodenmechanik und Felsmechanik der Universität Karlsruhe (1982)
18. Kuribayashi, E., Tatsuoka, F.: History of earthquake-induced soil liquefaction in Japan. No. 38 in Bulletin of Public Works Research Institute (1977)
19. McKenzie, D.: The generation and compaction of partially molten rock. Journal of Petrology 25(3), 713–765 (1984)
20. Morsch, O.: Die Physik der Körner. Neue Züricher Zeitung (13.02.2008)
21. Muir Wood, D.: The magic of sands — the 20th Bjerrum Lecture presented in Oslo, 25 November 2005. Can. Geotech. J. 44, 1329–1350 (2007)
22. Nübel, K.: Experimental and Numerical Investigation of Shear Localisation in Granular Material. No. 159 in Veröffentlichungen des Institutes für Bodenmechanik und Felsmechanik der Universität Karlsruhe (2002)
23. Palmer, A., Pearce, J.: Plasticity theory without yield surface. In: Palmer, A. (ed.) Symposium on the Role of Plasticity in Soil Mechanics, Cambridge (1973)
24. Pestana, J., Whittle, A.: Compression model for cohesionless soils. Géotechnique 45(4), 611–631 (1995)
25. Ramberg, H.: Gravity, deformation and the earth's crust. Academic Press, London (1967)
26. Revuzhenko, A.: Mechanics of Granular Media. Springer, Heidelberg (2006)
27. Sokoutis, D., Corti, G., Bonini, M., Brun, J., Cloetingh, S., Mauduit, T., Manetti, P.: Modelling the extension of heterogeneous hot lithosphere. Tectonophysics 444, 63–79 (2007)
28. Topolnicki, M.: Observed stress-strain behaviour of remoulded saturated clay and examination of two constitutive models. No. 107 in Veröffentlichungen des Institutes für Bodenmechanik und Felsmechanik der Universität Karlsruhe (1987)
29. Truesdell, C., Noll, W.: The Non-Linear Field Theories of Mechanics, 2nd edn. Springer, Heidelberg (1992)
30. Twiss, R., Moores, E.: Structural Geology. W.H. Freeman and Company, New York (1992)
31. Verdugo, R., Ishihara, K.: The steady state of sandy soils. Soils and Foundations 2, 81–91 (1996)

Author Index

Index